Methods in Molecular Biology

Series Editor
John M. Walker
School of Life and Medical Sciences
University of Hertfordshire
Hatfield, Hertfordshire, UK

For further volumes:
http://www.springer.com/series/7651

For over 35 years, biological scientists have come to rely on the research protocols and methodologies in the critically acclaimed *Methods in Molecular Biology* series. The series was the first to introduce the step-by-step protocols approach that has become the standard in all biomedical protocol publishing. Each protocol is provided in readily-reproducible step-by-step fashion, opening with an introductory overview, a list of the materials and reagents needed to complete the experiment, and followed by a detailed procedure that is supported with a helpful notes section offering tips and tricks of the trade as well as troubleshooting advice. These hallmark features were introduced by series editor Dr. John Walker and constitute the key ingredient in each and every volume of the *Methods in Molecular Biology* series. Tested and trusted, comprehensive and reliable, all protocols from the series are indexed in PubMed.

Lentiviral Vectors

Applications for Basic and Clinical Neuroscience

Edited by

Kazuto Kobayashi

Department of Molecular Genetics, Fukushima Medical University School of Medicine, Fukushima, Japan

Hirokazu Hirai

Department of Neurophysiology, Gunma University, Maebashi, Japan

Hiroyuki Hioki

Department of Neuroanatomy, Juntendo University, Tokyo, Japan

Humana Press

Editors
Kazuto Kobayashi
Department of Molecular Genetics
Fukushima Medical University School of Medicine
Fukushima, Japan

Hirokazu Hirai
Department of Neurophysiology
Gunma University
Maebashi, Japan

Hiroyuki Hioki
Department of Neuroanatomy
Juntendo University
Tokyo, Japan

ISSN 1064-3745 ISSN 1940-6029 (electronic)
Methods in Molecular Biology
ISBN 978-1-0716-4806-3 ISBN 978-1-0716-4807-0 (eBook)
https://doi.org/10.1007/978-1-0716-4807-0

This Humana imprint is published by the registered company Springer Science+Business Media, LLC, part of Springer Nature.
The registered company address is: 1 New York Plaza, New York, NY 10004, U.S.A.

Preface

Lentiviral vectors exhibit tropism for the nervous system and enable stable and long-term expression of transgenes in host cells. Pseudotyping of these vectors with various viral envelope glycoproteins alters the transduction preferences among neuronal, glial, and neural stem cells and can also influence the entry sites of vectors in neuronal cells, including their cell bodies and axon terminals. In particular, engineered envelope glycoproteins have been developed to achieve highly efficient retrograde gene transfer via uptake at the nerve terminals. These features offer a variety of powerful strategies to study basic and clinical neuroscience. This volume provides an overview of lentiviral vector applications and includes detailed laboratory protocols relevant to neuroscience research. In some cases, lentiviral vectors are also used in combination with other types of viral vector systems.

The chapters in this volume are composed of five sections. Part I provides an overview of lentiviral vectors, including practical methods for vector preparation and delivery into the brains of rodents and non-human primates, as well as delivery into non-human primate embryos for transgenic animal production. Parts II to IV focus on the applications of lentiviral vectors for basic neuroscience: anatomical neuroscience (Part II), molecular and developmental neuroscience (Part III), and behavioral and physiological neuroscience (Part IV). Part II introduces both ubiquitous and cell type-specific gene promoters that function in viral vector systems, along with protocols for labeling cellular compartments and selective pathways constituting the neural circuit. Part III provides protocols for transcription factor assay in the brain, lineage tracing of neurons and neural stem cells, and direct reprogramming into neurons. Part IV describes pathway-specific manipulation techniques with lentivirus vectors, including neuronal ablation, synaptic silencing, as well as optogenetic and chemogenetic regulations of neuronal activity. Finally, Section V covers the applications of lentivectors for clinical neuroscience, offering protocols related to gene therapy of motor neuron and ocular diseases. This section also includes an overview of lentiviral vector-mediated transfer of chimeric antigen receptor gene for cell therapy of central nervous system tumors.

All protocol chapters include detailed step-by-step methods and notes addressing technical challenges and troubleshooting, which will be helpful for the efficient execution of experiments and resolution of problems in the laboratory.

Fukushima, Japan — *Kazuto Kobayashi*
Maebashi, Japan — *Hirokazu Hirai*
Tokyo, Japan — *Hiroyuki Hioki*

Contents

Contributors

KENTARO ABE • *Lab of Brain Development, Graduate School of Life Sciences, Tohoku University, Sendai, Miyagi, Japan; Division for the Establishment of Frontier Sciences, Tohoku University, Sendai, Miyagi, Japan*

SATOMI CHIKEN • *Division of System Neurophysiology, National Institute of Physiological Sciences, Okazaki, Aichi, Japan; Section of Multilayer Physiology, National Institute of Physiological Sciences, Okazaki, Aichi, Japan*

MASATSUGU EMA • *Department of Stem Cells and Human Disease Models, Research Center for Animal Life Science, Shiga University of Medical Science, Seta, Tsukinowa-cho, Otsu, Shiga, Japan; Institute for the Advanced Study of Human Biology (ASHBi), Kyoto University, Yoshida-Konoe-cho, Sakyo-ku, Kyoto, Japan*

MITSUGU FUJITA • *Center for Medical Education and Clinical Training, Kindai University Faculty of Medicine, Osaka-Sayama, Japan*

TOMONORI D. FUKUDA • *Laboratory of Brain Development and Regeneration, Graduate School of Biostudies, Kyoto University, Kyoto, Japan; Laboratory of Deconstruction of Stem Cells, Institute for Life and Medical Sciences, Kyoto University, Kyoto, Japan*

TAKAHIRO FURUTA • *Department of Oral Anatomy and Neurobiology, Osaka University Graduate School of Dentistry, Osaka, Japan*

ERIKO HATAI • *Department of Pediatrics, Graduate School of Medical Sciences, Kyushu University, Fukuoka, Japan*

HIDEKI HIDA • *Department of Neurophysiology and Brain Science, Nagoya City University Graduate School of Medical Sciences, Nagoya, Japan*

HIROYUKI HIOKI • *Department of Neuroanatomy, Juntendo University Graduate School of Medicine, Tokyo, Japan*

HIROKAZU HIRAI • *Department of Neurophysiology & Neural Repair, Gunma University Graduate School of Medicine, Maebashi, Gunma, Japan; Viral Vector Core, Gunma University, Initiative for Advanced Research, Maebashi, Gunma, Japan*

TATSUMI HIRATA • *Brain Function Laboratory, National Institute of Genetics, SOKENDAI, Mishima, Shizuoka, Japan*

TAKAHIRO HISAI • *Department of Ophthalmology, Graduate School of Medical Sciences, Kyushu University, Fukuoka, Japan*

YASUHIRO IKEDA • *Department of Ophthalmology, Faculty of Medicine, University of Miyazaki, Miyazaki, Japan*

KEIKO IMAMURA • *Center for iPS Cell Research and Application (CiRA), Kyoto University, Sakyo-ku, Kyoto, Japan; iPSC-based Drug Discovery and Development Team, RIKEN BioResource Research Center (BRC), Kyoto, Japan; Medical-risk Avoidance based on iPS Cells Team, RIKEN Center for Advanced Intelligence Project (AIP), Kyoto, Japan*

ITARU IMAYOSHI • *Center for Living Systems Information Science (CeLiSIS), Graduate School of Biostudies, Kyoto University, Kyoto, Japan; Laboratory of Brain Development and Regeneration, Graduate School of Biostudies, Kyoto University, Kyoto, Japan; Laboratory of Deconstruction of Stem Cells, Institute for Life and Medical Sciences, Kyoto University, Kyoto, Japan*

HARUHISA INOUE • *Center for iPS Cell Research and Application (CiRA), Kyoto University, Sakyo-ku, Kyoto, Japan; iPSC-based Drug Discovery and Development Team, RIKEN BioResource Research Center (BRC), Kyoto, Japan; Medical-risk Avoidance based on iPS Cells Team, RIKEN Center for Advanced Intelligence Project (AIP), Kyoto, Japan*
KEN-ICHI INOUE • *Systems Neuroscience Section, Center for the Evolutionary Origins of Human Behavior, Kyoto University, Inuyama, Japan*
TAKASHI IRIE • *Department of Neurology, Neurological Institute, Graduate School of Medical Sciences, Kyushu University, Fukuoka, Japan*
KAORU ISA • *Department of Neuroscience, Kyoto University Graduate School of Medicine, Kyoto, Japan*
TADASHI ISA • *Department of Neuroscience, Kyoto University Graduate School of Medicine, Kyoto, Japan*
BAT-ERDENE JARGALSAIKHAN • *Department of Stem Cells and Human Disease Models, Research Center for Animal Life Science, Shiga University of Medical Science, Seta, Tsukinowa-cho, Otsu, Shiga, Japan*
HIROSHI KAMEDA • *Department of Cell Biology and Neuroscience, Juntendo University Graduate School of Medicine, Tokyo, Japan*
TAKU KASHIYAMA • *Department of Cellular and Molecular Pharmacology, Juntendo University Graduate School of Medicine, Tokyo, Japan*
SHIGEKI KATO • *Department of Molecular Genetics, Institute of Biomedical Sciences, Fukushima Medical University, Fukushima, Japan*
YUKO KOBAYAKAWA • *Center for Clinical and Translational Research, Kyushu University Hospital, Fukuoka, Japan*
KAZUTO KOBAYASHI • *Department of Molecular Genetics, Fukushima Medical University School of Medicine, Fukushima, Japan*
KENTA KOBAYASHI • *Section of Viral Vector Development, National Institute for Physiological Sciences, Okazaki, Japan; SOKENDAI (The Graduate University for Advanced Studies), Hayama, Japan*
MASATO KOIKE • *Department of Cell Biology and Neuroscience, Juntendo University Graduate School of Medicine, Tokyo, Japan*
AYUMU KONNO • *Department of Neurophysiology & Neural Repair, Gunma University Graduate School of Medicine, Maebashi, Gunma, Japan; Viral Vector Core, Gunma University, Initiative for Advanced Research, Maebashi, Gunma, Japan*
KOHJI KUSANO • *Tsukuba Factory, ID Pharma Co., Ltd., Tsukuba, Japan*
TAITO MATSUDA • *Department of Stem Cell Biology and Medicine, Graduate School of Medical Sciences, Kyushu University, Fukuoka, Japan*
TAKASHI MATSUDA • *Center for One Medicine Innovative Translational Research (COMIT), Institute for Advanced Study, Gifu University, Gifu, Japan; Homeostatic Mechanism Research Unit, Institute of Integrated Research, Institute of Science Tokyo, Tokyo, Kanagawa, Japan*
KANAE MATSUDA-ITO • *Department of Stem Cell Biology and Medicine, Graduate School of Medical Sciences, Kyushu University, Fukuoka, Japan*
NATSUKI MATSUSHITA • *Division of Laboratory Animal Research, Aichi Medical University School of Medicine, Nagakute, Japan*
YASUNORI MATSUZAKI • *Department of Neurophysiology & Neural Repair, Gunma University Graduate School of Medicine, Maebashi, Gunma, Japan; Viral Vector Core, Gunma University, Initiative for Advanced Research, Maebashi, Gunma, Japan*

TIAGO MOREIRA-GOMES • *ABC-Ri, Algarve Biomedical Center Research Institute, Faro, Portugal; Faculdade de Medicina e Ciências Biomédicas, Universidade do Algarve, Faro, Portugal; Doctoral Program in Biomedical Sciences, Faculdade de Medicina e Ciências Biomédicas, Universidade do Algarve, Faro, Portugal*
YUSUKE MURAKAMI • *Department of Ophthalmology, Graduate School of Medical Sciences, Kyushu University, Fukuoka, Japan*
MASANAGA MUTO • *Department of Stem Cells and Human Disease Models, Research Center for Animal Life Science, Shiga University of Medical Science, Seta, Tsukinowa-cho, Otsu, Shiga, Japan*
IKUMI NAGANO • *Laboratory of Brain Development and Regeneration, Graduate School of Biostudies, Kyoto University, Kyoto, Japan; Laboratory of Deconstruction of Stem Cells, Institute for Life and Medical Sciences, Kyoto University, Kyoto, Japan*
SHINJI C. NAGASAKI • *Center for Living Systems Information Science (CeLiSIS), Graduate School of Biostudies, Kyoto University, Kyoto, Japan; Laboratory of Brain Development and Regeneration, Graduate School of Biostudies, Kyoto University, Kyoto, Japan; Laboratory of Deconstruction of Stem Cells, Institute for Life and Medical Sciences, Kyoto University, Kyoto, Japan*
KINICHI NAKASHIMA • *Department of Stem Cell Biology and Medicine, Graduate School of Medical Sciences, Kyushu University, Fukuoka, Japan*
ATSUSHI NAMBU • *Division of System Neurophysiology, National Institute of Physiological Sciences, Okazaki, Aichi, Japan*
CLÉVIO NÓBREGA • *ABC-Ri, Algarve Biomedical Center Research Institute, Faro, Portugal; Faculdade de Medicina e Ciências Biomédicas, Universidade do Algarve, Faro, Portugal*
MASASUKE OHNO • *Department of Neurosurgery, Aichi Cancer Center, Nagoya, Japan*
SHINICHIRO OKAMOTO • *Department of Neuroanatomy, Juntendo University Graduate School of Medicine, Tokyo, Japan*
TARO OKUNOMIYA • *Center for iPS Cell Research and Application (CiRA), Kyoto University, Sakyo-ku, Kyoto, Japan; iPSC-based Drug Discovery and Development Team, RIKEN BioResource Research Center (BRC), Kyoto, Japan*
TAKASHI SAKURAI • *Department of Cellular and Molecular Pharmacology, Juntendo University Graduate School of Medicine, Tokyo, Japan*
HIROMI SANO • *Division of Molecular and Systems Neuropathophysiology, International Center for Brain Science, Fujita Health University, Toyoake, Aichi, Japan; Division of System Neurophysiology, National Institute of Physiological Sciences, Okazaki, Aichi, Japan*
MASAHIKO TAKADA • *Systems Neuroscience Section, Center for the Evolutionary Origins of Human Behavior, Kyoto University, Inuyama, Japan*
TOMOHISA TODA • *Nuclear Architecture in Neural Plasticity and Aging, German Center for Neurodegenerative Diseases (DZNE), Dresden, Germany; Laboratory of Neural Epigenomics, Institute of Medical Physics and Micro-tissue Engineering, Faculty of Medicine, Friedrich-Alexander-Universität Erlangen-Nürnberg, Erlangen, Germany; Max-Planck-Zentrum für Physik und Medizin, Erlangen, Germany*
SHINYA UENO • *Department of Neurophysiology and Brain Science, Nagoya City University Graduate School of Medical Sciences, Nagoya, Japan*
WIM VANDUFFEL • *Laboratory for Neuro-and Psychophysiology, Department of Neurosciences, KU Leuven Medical School, Leuven, Belgium*
AKIYA WATAKABE • *Laboratory for Molecular Mechanism of Brain Development, RIKEN Center for Brain Science, Wako, Japan*

MAYUMI YAMADA • *Center for Living Systems Information Science (CeLiSIS), Graduate School of Biostudies, Kyoto University, Kyoto, Japan; Laboratory of Brain Development and Regeneration, Graduate School of Biostudies, Kyoto University, Kyoto, Japan; Laboratory of Deconstruction of Stem Cells, Institute for Life and Medical Sciences, Kyoto University, Kyoto, Japan; Laboratory of Cell Biology, Institute for Life and Medical Sciences, Kyoto University, Kyoto, Japan*

Chapter 1

Properties and Application of Pseudotyped Lentiviral Vectors

Natsuki Matsushita, Masahiko Takada, and Kazuto Kobayashi

Abstract

Lentiviral (LV) vectors derived from human immunodeficiency virus type 1 (HIV-1) mediate efficient gene transfer into various cell types, including neuronal, glial, and neural stem cells in the central nervous system. The HIV-1-based LV vectors commonly used are self-inactivating and pseudotyped with a variety of viral envelope glycoproteins. Pseudotyping with vesicular stomatitis virus glycoprotein (VSV-G) confers broad tropism to wide variety of cells and enhances vector stability of the vectors. Pseudotyping with other glycoproteins alters the tropism. In particular, the use of rabies virus glycoprotein (RV-G) or fusion glycoproteins (FuGs) consisting of VSV-G and RV-G segments increases the efficiency of retrograde gene delivery into neuronal cells. The LV vector strategy provides useful approaches to genetic manipulation of specific neural pathways for elucidating the structure and function of neural circuits, and also to genome-wide screening of key molecules and gene regulatory elements in development and survival of the nervous system. Such a strategy also provides a powerful tool for clinical applications aimed at gene therapy for neurological and neurodegenerative diseases as well as central nervous system tumors.

Key words Lentiviral vector, Human immunodeficiency virus type 1, Self-inactivating vector, Pseudotyping, Tropism, Envelope glycoprotein, Vesicular stomatitis virus glycoprotein, Rabies virus glycoprotein, Fusion glycoprotein, Retrograde transport, Integrase-deficient lentivirus

1 Introduction

The gene transduction strategy using lentiviral (LV) vectors provides powerful approaches to basic neuroscience research and clinical trials to develop treatments for various neurological and neurodegenerative diseases, as well as tumors of the central nervous system. LV vectors are derived from various species of viruses, such as simian immunodeficiency virus, feline immunodeficiency virus, and equine infectious anemia virus [1], and the most frequently used vectors are derived from human immunodeficiency virus type 1 (HIV-1) [2, 3]. HIV-1 based vectors have a dimeric, plus-stranded linear RNA genome in a viral particle [4]. When the vectors are incorporated into host cells, their RNA is reverse-

Kazuto Kobayashi et al. (eds.), *Lentiviral Vectors: Applications for Basic and Clinical Neuroscience*, Methods in Molecular Biology, vol. 2974, https://doi.org/10.1007/978-1-0716-4807-0_1,

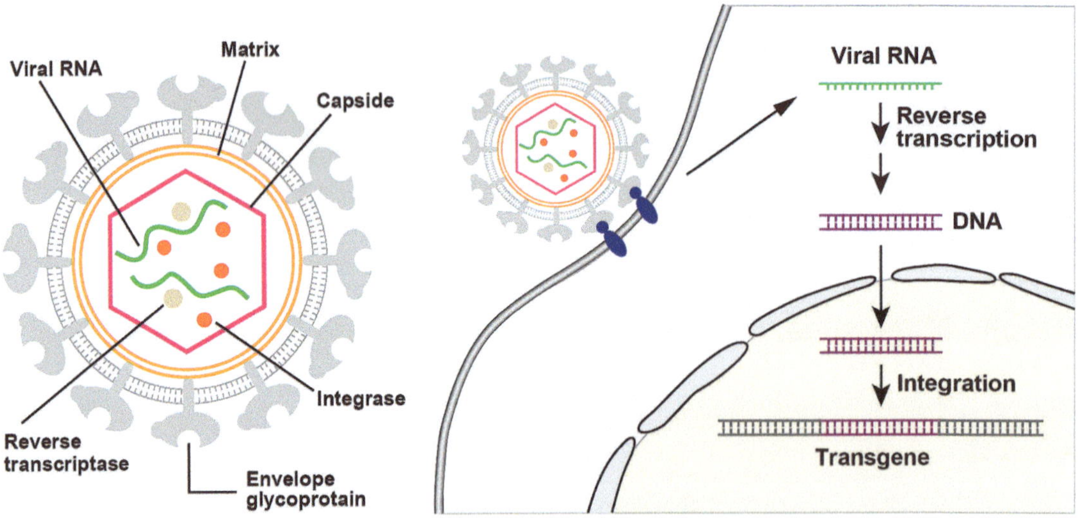

Fig. 1 Gene transfer by lentiviral (LV) vectors. The vector is a viral particle containing a dimeric, plus-stranded RNA genome. The vectors bind to cell surface receptors and are incorporated into host cells. The viral RNA is reverse-transcribed into DNA, which is then integrated into the host genome, resulting in the expression of the transgene

transcribed into DNA and the synthesized DNA is integrated into the host genome (see Fig. 1). LV vectors are capable of efficient gene transfer into both dividing and nondividing cells [3, 5–10], unlike gamma retroviral vectors derived from onco-retroviruses transducing selectively dividing cells [11, 12]. The vectors also enable stable and long-term gene expression because of DNA integration into the genome [9, 10].

The LV vector strategy has been improved to resolve safety issues, particularly to suppress the generation of replication-competent viruses during the production of vectors. The packaging and envelope genes of HIV-1 are placed in separate plasmids, independent of each other [6, 13]. The packaging plasmid has no packaging signal (ψ); thus, the RNA transcribed from this plasmid is not incorporated into viral particles. The use of a four-part plasmid system, consisting of *gag/pol*, *rev*, envelope, and transgene constructs, minimizes the possibility of recombination that generates replication-competent viruses [14]. In addition, part of the 5′-long terminal repeat (LTR) in viral genomes is replaced by a cytomegalovirus or a Rous sarcoma virus promoter, resulting in loss of *tat* gene dependency, and deletion of the promoter/enhancer regions in the 3′-LTR leads to development of the current self-inactivating vectors [15]. All accessory genes (*vpr*, *vif*, *vpu*, and *nef*) of HIV-1 are further removed from the LV packaging system [16]. The removal of these genes does not affect the efficiency of gene transfer into the host cells [17].

HIV-1 envelope glycoprotein binds to CD4, requiring co-receptors for the subsequent viral infection processes, thus exhibiting target-specificity for lymphocytes expressing CD4 [18, 19]. In contrast, LV vectors are generally pseudotyped with different types of viral envelope glycoproteins. When the vectors are pseudotyped with vesicular stomatitis virus glycoprotein (VSV-G), they can transduce a broad range of cell types [6–8], such as neuronal, glial, and neural stem cells in the central nervous system [3–6, 10]. The VSV-G-pseudotyped vectors have resistance to freeze-thaw cycles and ultracentrifugation, and these features are beneficial for vector tools [20, 21]. In addition, pseudotyping with other viral envelope glycoproteins, such as lymphocytic choriomeningitis virus glycoprotein (LCMV-G), murine leukemia virus glycoprotein (MLV-G), and rabies virus glycoprotein (RV-G), alters the tropism of the vectors to different cell types [22]. The use of RV-G and genetically engineered envelope glycoproteins termed fusion glycoproteins (FuGs) consisting of segments from VSV-G and RV-G promotes the entry of vectors into nerve terminal regions of neuronal cells, thereby achieving retrograde transport through axons to their cell bodies [23] (see Fig. 2a).

This chapter describes the properties and tropism of LV vectors pseudotyped with various viral envelope glycoproteins and summarizes the application of these vectors in basic and clinical neuroscience.

2 Pseudotyping with VSV-G

Pseudotyping HIV-1-based vectors with VSV-G is the most typical strategy, leading to high physical stability of the vectors during preparation and enhances their infectious ability, allowing efficient infection of both dividing and nondividing cells [6–8]. This broad tropism is attributed to the interaction of VSV-G with ubiquitous cellular receptors on the plasma membrane, which have been reported to include phospholipids and/or a low-density lipoprotein receptor [24–26]. This explains why the VSV-G-pseudotyped vectors are widely used for many basic and clinical experiments that aim at transducing a wide range of cell types, including proliferating neural stem cells as well as differentiated neuronal and glial cells in the central nervous system [27–30].

However, the broad tropism of VSV-G pseudotyped vectors may pose challenges depending on the research objective [29]. To address this issue, tissue- and cell type-specific promoters enabling spatial control of transgene expression are widely used. Representative promoters are derived from genes encoding calcium/calmodulin-dependent protein kinase II alpha, glutamate decarboxylase 67, or parvalbumin [31–33]. Selective transgene expression can also be induced by the Cre/loxP site-specific recombination system

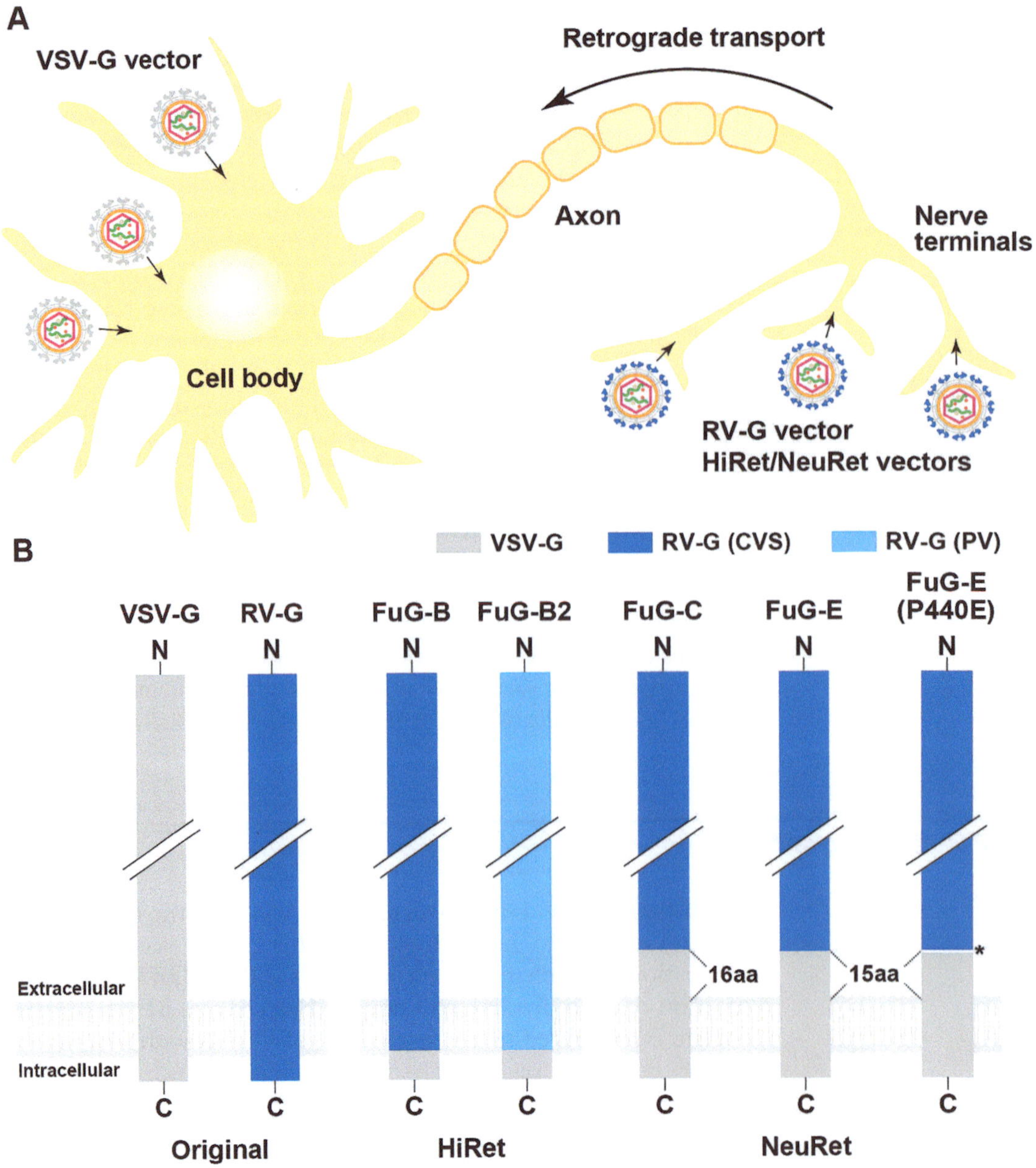

Fig. 2 Gene delivery into neuronal cells by pseudotyped LV vectors. (**a**) Entry site preference of LV vectors. VSV-G-pseudotyped vectors enter the cell body areas of neurons, whereas RV-G- and FuG-pseudotyped vectors enter mainly nerve terminals and are transported retrogradely through the axons to the cell body. (**b**) Structure of viral envelope glycoproteins. The extracellular, transmembrane, and cytoplasmic domains of VSV-G and RV-G are indicated. FuG-B/B2, FuG-C, and FuG-E/E (P440E) contain different sets of VSV-G and RV-G segments. The sequences of RV-G are derived from those of the CVS strain, except for FuG-B2, in which the sequences are derived from the Pasteur virus strain. In FuG-E (P440E), the amino acid residue at position 440 in FuG-E is substituted from proline to glutamic acid

[34–36]. The Cre/loxP system combined with tissue- and cell type-specific promoters is more effective in achieving spatial control of transgene expression. In addition, temporal control of transgene expression can be achieved using the Tet-On/Off system, which utilizes tetracycline-inducible promoters [37–39]. In this system, treatment with tetracycline positively or negatively regulates the activity of transcription factors, resulting in the temporal control of transgene expression. Combining the Tet-On/Off system with the Cre/loxP system will further enhance the spatiotemporal resolution of transgene expression.

3 Pseudotyping with Alternative Viral Envelope Glycoproteins

Other viral envelope glycoproteins can be selected for the pseudotyping of LV vectors, resulting in changes to target cell-type specificity, transduction efficiency, and safety. LCMV is a non-cytopathic arenavirus, which infects rodents and occasionally humans. LV vectors pseudotyped with LCMV-G demonstrate high gene transfer efficiency across various cell types including fibroblasts, epithelial cells, hematopoietic cells, neuroblastoma, and glioma cells [40]. Alpha-dystroglycan, a cellular receptor for LCMV-G, is widely expressed in most cell types [41, 42], enabling the LCMV-G-pseudotyped vectors to introduce the transgenes into a wide range of cell types. However, several studies have reported biases in the cell type specificity of these vectors, such as the selective transduction of neural stem cells in the subventricular zone [43] and solid glioma cells implanted in the brain [44]. LCMV-G-pseudotyped vectors also exhibit a lower toxicity in mouse organs compared with VSV-G-pseudotyped vectors [45, 46].

MLV belongs to the gamma retroviruses, and ecotropic MLV selectively infects mouse and rat cells and causes leukemia. The LV vectors pseudotyped with MLV-G efficiently introduce transgenes into various cells in mice and rats, but not in other mammalian species, including humans [8, 47, 48]. Due to the narrow range of target species, MLV-G-pseudotyped vectors may be preferred for basic experiments for safety reasons. This ecotropic feature can be attributed to the specific interactions of MLV-G with particular receptors, mainly cationic amino acid transporter-1 in rodents [49, 50]; however, MLV-G does not appear to interact with specific molecules expressed in human cells.

RV is an enveloped negative-strand RNA virus that causes a disease in the central nervous system. LV vectors pseudotyped with RV-G have a high affinity for neuronal cells, showing properties of retrograde gene transfer [51, 52]. RV-G has been reported to interact with potential receptors, such as nicotinic acetylcholine receptors, neural cell adhesion molecules, and low-affinity neurotrophic factor receptor p75NTR, although the substantial receptors for RV-G have not yet been identified [53]. Injection of LV vectors, derived from equine infectious anemia virus or HIV-1 and pseudotyped with RV-G, into the striatum induces gene transfer into neurons and retrograde spread of transgenes to the cerebral cortex, thalamus, and ventral midbrain along the neural pathways projecting to the injection site [54–56]. RV-G has been reported to promote retrograde vector transport; however, its gene transfer efficiency remains insufficient [54, 56].

4 Pseudotyping with Fusion Envelope Glycoproteins

As mentioned above, the pseudotyping of LV vectors with RV-G increases the efficacy of retrograde gene transfer of the vectors, while it results in the low efficiency of gene transfer. Some FuGs that combine different segments of RV-G and VSV-G are developed to enhance the efficiency of retrograde gene transfer [23, 57–62] (see Fig. 2b). Pseudotyped vectors with FuGs are classified into two major groups based on their gene transduction properties [23, 61, 62]. The first group is characterized by high efficiency of retrograde gene transfer into neuronal cells, along with transduction into dividing cells, including glial and neural stem cells. This group is referred to as the highly efficient retrograde gene transfer (HiRet) vectors [57, 58]. In HiRet vectors, FuG type B (FuG-B) is used for pseudotyping. FuG-B comprises the extracellular and transmembrane domains of RV-G derived from the challenge virus standard strain and the cytoplasmic domain of VSV-G fused to it. Injection of these vectors into the striatum results in a greater extent of retrograde gene transfer into the brain regions that innervate the striatum [57, 58]. These vectors also introduce the transgene into glial cells around the injection sites and neural stem cells in the subventricular zone [57, 58].

The second group, known as the neuron-specific retrograde gene transfer (NeuRet) vectors, is characterized by highly efficient retrograde gene transfer into neuronal cells with almost no transduction into dividing cells [59]. In NeuRet vectors, FuG type C (FuG-C) is used for pseudotyping, which consists of the N-terminal region (439 amino acids) of the extracellular domain of the RV-G derived from CVS, connected to the short membrane-proximal region (16 amino acids), and the transmembrane/cytoplasmic domains of VSV-G [59]. Injection of the NeuRet vectors with FuG-C into the striatum demonstrates a high level of retrograde gene transfer into the brain regions innervating the striatum, similar to the HiRet vector, whereas the gene transduction of the NeuRet vectors into glial and neural stem cells is much lower compared with that of the HiRet vectors [59]. The retrograde transfer efficiency of NeuRet vectors has been further enhanced by the creation of FuG type-E (FuG-E) with the optimized RV-G and VSV-G linkage sites [63, 64]. FuG-E is composed of the N-terminal region (440 amino acids) of the extracellular domain of RV-G connected to the short membrane-proximal region (15 amino acids) and the transmembrane/cytoplasmic domains of VSV-G, yielding a higher efficiency of retrograde gene delivery in the brain compared with that of FuG-C. Another advantage of the NeuRet vectors over the HiRet vectors is the absence of inflammatory reactions following their injection into the brain, as demonstrated in a nonhuman primate study [65].

5 Integrase-Defective LV Vectors

LV vectors generally provide a powerful tool for gene transfer into various cell and tissue types. While their transgene cassettes are integrated and stably maintained in the host cell genome [8, 13], an important issue arises as to their clinical application to gene therapy both in vivo and ex vivo. DNA integration tends to be biased toward certain genomic regions due to chromatin dynamics [66, 67]. Therefore, the semi-random integration of foreign genes into the host genome may induce oncogenesis through insertional mutations. To address this issue and facilitate clinical application in gene therapy protocols, integrase-defective non-integrating lentiviral (IDLV) vectors have been developed [68, 69]. IDLV vectors generally contain mutations in the HIV-1 *pol* gene that encodes integrase, with various point mutations leading to the substitution of amino acid residues in the integrase [70]. A typical mutation used for IDLV vectors is the substitution of aspartic acid with valine at residue 64 of the integrase [71]. Thus, IDLV vectors can indeed avoid the risks associated with the semi-random integration into the recipient genome and the resulting insertional mutations [72–74]. In IDLV-mediated gene transfer, reverse-transcribed viral genomes become episomal, taking linear or circular double-stranded DNA forms and remaining independent of the host cell genome [75]. These vectors induce a highly effective, transient expression of transgenes in dividing cells [76], and they are also effective for stable, long-term expression in nondividing cells [73, 77].

6 Application of LV Vectors to Neuroscience Research

In the field of neuroscience research, the genetic manipulation strategy of specific neural circuits using LV vectors provides powerful approaches to elucidating the structures and functions of neural circuits [78–81]. Selective targeting of a specific neural pathway is conducted by expressing a recombinant immunotoxin (ITX) receptor via the HiRet vectors and injecting ITX into specific brain regions [58, 82–84]. Silencing of synaptic transmissions in specific neural pathways is induced by expressing a tetracycline-dependent transactivator via the NeuRet vectors and injecting an adeno-associated viral vector encoding tetanus neurotoxin downstream of a tetracycline-inducible promoter into corresponding brain regions [85, 86]. Optogenetic manipulation of specific pathways is selectively performed by expressing channelrhodopsin-2 via HiRet/NeuRet vectors and controlling the activity of targeted neurons in response to light stimulation [80, 87–89]. As an additional application, selective chemogenetic manipulation of certain pathways can be achieved by expressing designer receptors exclusively activated by designer drugs (DREADDs) via HiRet/NeuRet vectors and modulating the activity of targeted neurons in response to an actuator, such as clozapine-N-oxide [82, 90–92].

The LV vector strategy also offers a useful approach to in vivo screening of genome-wide libraries for shRNA and CRISPR-based single guide RNA expression [93, 94]. Both LV libraries contain nearly 100,000 sequences for RNA transcripts with genome-wide coverage. Recent studies have identified some key molecules that play important roles in maintaining neuronal survival in the central nervous system through in vivo screening of these libraries via local injection into mouse brains [93]. Another example of an experimental design is the LV vector library-based massively parallel reporter assay [95], which has been effectively utilized for high-throughput analysis of gene regulatory elements during differentiation of neural progenitor cells [96].

The LV vector system can be used for clinical trials aiming at gene therapy of a variety of neurological and neurodegenerative diseases [97, 98]. For ex vivo gene therapy trials, LV vectors carrying genes for certain transcription factors or neurotrophic factors transduce neural progenitor, embryonic stem, and induced pluripotent stem cells. These cells, or differentiated cells under special culture conditions, are then transplanted into tissues [99–102]. For example, induced pluripotent stem cells transduced with the Lmx1a transcription factor gene efficiently differentiate into dopaminergic neurons, and transplantation of these cells into the brain of a rodent model of Parkinson's disease leads to behavioral improvements [101, 102]. Neural progenitor cells expressing glial cell line-derived neurotrophic factor are transplanted into the motor cortex or spinal cord of a model of amyotrophic lateral sclerosis; where they differentiate into astrocytes contributing to neuronal survival [103, 104]. For in vivo gene therapy trials, LV vectors carrying therapeutic genes are directly injected into the tissues of disease models [97, 105]. Retrograde gene transfer using LV vectors provides an approach for delivering therapeutic genes from presynaptic terminals at the neuromuscular junction to otherwise inaccessible neurons, such as motor neurons in the spinal cord [106]. LV-mediated gene transfer of three transcription factors, Mash1, Nr4a2, and Lmx1a, mediates direct conversion from fibroblasts to dopaminergic neurons in the brain [107, 108]. The LV vector system is also used in chimeric antigen receptor (CAR)-T cell therapy for the treatment of tumors in the central nervous system. This therapy has been shown to be highly effective so far, in that CAR-encoding genes are introduced into T cells via LV vectors, and the resulting CAR-T cells effectively target tumors [30, 109, 110].

A variety of gene transfer strategies using LV vectors have provided and will provide effective tools for research in both basic and clinical neuroscience. However, some challenges remain in the use of the vector system, including the improvement of transduction efficiency, modification of vector properties, and precise delivery into target tissues. Further advances in the pseudotyping and injection strategies of the LV vectors will help overcome some of these limitations and expand their applications in the near future.

Acknowledgments

This work was supported by a Grant-in-Aid for Scientific Research for Transformative Research Areas (A) Adaptive Circuit Census (21H05244) from the Ministry of Education, Science, Sports, and Culture of Japan (N.M. and K.K.); a Research Grant from the Japan Agency for Medical Research and Development (24wm0625103h0001) (K.K.).

Conflict of Interest Statement The authors have no conflicts of interest to declare that are relevant to the content of this chapter.

References

1. Valori CF, Ning K, Wyles M, Azouz M (2008) Development and applications of non-HIV-based lentiviral vectors in neurological disorders. Curr Gene Ther 8:406–418
2. Verma IM, Somia N (1997) Gene therapy – promises, problems and prospects. Nature 389:239–242
3. Naldini L (1998) Lentiviruses as gene transfer agents for delivery to non-dividing cells. Curr Opin Biotechnol 9:457–463
4. Höglund S, Ohagen A, Goncalves J, Panganiban AT, Gabuzda D (1997) Ultrastructure of HIV-1 genomic RNA. Virology 233:271–279
5. Vodicka MA (2001) Determinants for lentiviral infection of non-dividing cells. Somat Cell Mol Genet 26:35–49
6. Naldini L, Blömer U, Gallay P, Ory D, Mulligan R, Gage FH, Verma IM, Trono D (1996) In vivo gene delivery and stable transduction of nondividing cells by a lentiviral vector. Science 272:263–267
7. Akkina RK, Walton RM, Chen ML, Li QX, Planelles V, Chen IS (1996) High-efficiency gene transfer into CD34+ cells with a human immunodeficiency virus type 1-based retroviral vector pseudotyped with vesicular stomatitis virus envelope glycoprotein G. J Virol 70: 2581–2585
8. Reiser J, Harmison G, Kluepfel-Stahl S, Brady RO, Karlsson S, Schubert M (1996) Transduction of nondividing cells using pseudotyped defective high-titer HIV type 1 particles. Proc Natl Acad Sci USA 93: 15266–15271

9. Mochizuki H, Schwartz JP, Tanaka K, Brady RO, Reiser J (1998) High-titer human immunodeficiency virus type 1-based vector systems for gene delivery into nondividing cells. J Virol 72:8873–8883
10. Mitrophanous KA, Yoon S, Rohll JB, Patil D, Wilkes FJ, Kim VN, Kingsman SM, Kingsman AJ, Mazarakis N (1999) Stable gene transfer to the nervous system using a non-primate lentiviral vector. Gene Ther 6:1808–1818
11. Roe T, Reynolds TC, Yu G, Brown PO (1993) Integration of murine leukemia virus DNA depends on mitosis. EMBO J 12:2099–2108
12. Lewis PF, Emerman M (1994) Passage through mitosis is required for oncoretroviruses but not for the human immunodeficiency virus. J Virol 68:510–516
13. Naldini L, Blömer U, Gage FH, Trono D, Verma IM (1996) Efficient transfer, integration and sustained long-term expression of the transgene in adult rat brains injected with a lentiviral vector. Proc Natl Acad Sci USA 93:11382–11388
14. Dull T, Zufferey R, Kelly M, Mandel RJ, Nguyen M, Trono D (1998) A third-generation lentivirus vector with a conditional packaging system. J Virol 72:8463–8471
15. Miyoshi H, Blömer U, Takahashi M, Gage FH, Verma IM (1998) Development of a self-inactivating lentivirus vector. J Virol 72: 8150–8157
16. Kim VN, Mitrophanous KA, Kingsman SM, Kingsman AJ (1998) Minimal requirement for a lentivirus vector based on human immunodeficiency virus type 1. J Virol 72:811–816
17. Zufferey R, Nagy D, Mandel RJ, Naldini L, Trono D (1997) Multiply attenuated lentiviral vector achieves efficient gene delivery in vivo. Nat Biotechnol 15:871–875
18. Dianzani U, Bragardo M, Buonfiglio D, Redoglia V, Funaro A, Portoles P, Rojo J, Malavasi F, Pileri A (1995) Modulation of CD4 lateral interaction with lymphocyte surface molecules induced by HIV-1 gp120. Eur J Immunol 25:1306–1311
19. Bowers K, Pitcher C, Marsh M (1997) CD4: a co-receptor in the immune response and HIV infection. Int J Biochem Cell Biol 29:871–875
20. Burns JC, Friedmann T, Driever W, Burrascano M, Yee JK (1993) Vesicular stomatitis virus G glycoprotein pseudotyped retroviral vectors: concentration to very high titer and efficient gene transfer into mammalian and nonmammalian cells. Proc Natl Acad Sci USA 90:8033–8037
21. Verhoeyen E, Cosset FL (2004) Surface-engineering of lentiviral vectors. J Gene Med 6:S83–S94
22. Cronin J, Zhang XY, Reiser J (2005) Altering the tropism of lentiviral vectors through pseudotyping. Curr Gene Ther 5:387–398
23. Kobayashi K, Kato S, Inoue K, Takada M, Kobayashi K (2016) Altering entry site preference of lentiviral vectors into neuronal cells by pseudotyping with envelope glycoproteins. In: Manfredsson F (ed) Gene therapy of neurological disorders. Methods Mol Biol 1382: 175–186
24. Schlegel R, Tralka TS, Willingham MC, Pastan I (1983) Inhibition of VSV binding and infectivity by phosphatidylserine: is phosphatidylserine a VSV-binding site? Cell 32: 639–646
25. Finkelshtein D, Werman A, Novick D, Barak S, Rubinstein M (2013) LDL receptor and its family members serve as the cellular receptors for vesicular stomatitis virus. Proc Natl Acad Sci USA 110:7306–7311
26. Kim IS, Jenni S, Stanifer ML, Roth E, Whelan SPJ, van Oijen AM, Harrison SC (2017) Mechanism of membrane fusion induced by vesicular stomatitis virus G protein. Proc Natl Acad Sci USA 114:E28–E36
27. Cockrell AS, Kafri T (2007) Gene delivery by lentivius vectors. Mol Biotechnol 36:184–204
28. Hastie E, Cataldi M, Marriott I, Grdzelishvili VZ (2013) Understanding and altering cell tropism of vesicular stomatitis virus. Virus Res 176:16–32
29. Parr-Brownlie LC, Bosch-Bouju C, Schoderboeck L, Sizemore RJ, Abraham WC, Hughes SM (2015) Lentiviral vectors as tools to understand central nervous system biology in mammalian model organisms. Front Mol Neurosci 8:14
30. Bagley SJ, Desai A, Linette GP, June CH, O'Rourke DM (2018) CAR T-cell therapy for glioblastoma: recent clinical advances and future challenges. Neuro-Oncology 20: 1429–1438
31. Dittgen T, Nimmerjahn A, Komai S, Licznerski P, Waters J, Margrie TW, Helmchen F, Denk W, Brecht M, Osten P (2004) Lentivirus-based genetic manipulations of cortical neurons and their optical and electrophysiological monitoring in vivo. Proc Natl Acad Sci USA 101:18206–18211
32. Delzor A, Dufour N, Petit F, Guillermier M, Houitte D, Auregan G, Brouillet E, Hantraye P, Déglon N (2012) Restricted transgene expression in the brain with cell-

type specific neuronal promoters. Hum Gene Ther Methods 23:242–254

33. Sohal VS, Zhang F, Yizhar O, Deisseroth K (2009) Parvalbumin neurons and gamma rhythms enhance cortical circuit performance. Nature 459:698–702
34. Sauer B, Henderson N (1988) Site-specific DNA recombination in mammalian cells by the Cre recombinase of bacteriophage P1. Proc Natl Acad Sci USA 85:5166–5170
35. Schnütgen F, Doerflinger N, Calléja C, Wendling O, Chambon P, Ghyselinck NB (2003) A directional strategy for monitoring Cre-mediated recombination at the cellular level in the mouse. Nat Biotechnol 21:562–565
36. Matsushita N, Nishizawa K, Kato S, Iguchi Y, Fukabori R, Takeuchi K, Miyasaka Y, Mashimo T, Kobayashi K (2022) Catecholaminergic cell type-specific expression of Cre recombinase in knock-in transgenic rats generated by the Combi-CRISPR technology. J Neurosci Methods 381:109707
37. Gossen M, Freundlieb S, Bender G, Müller G, Hillen W, Bujard H (1995) Transcriptional activation by tetracyclines in mammalian cells. Science 268:1766–1769
38. Pluta K, Luce MJ, Bao L, Agha-Mohammadi-S, Reiser J (2005) Tight control of transgene expression by lentivirus vectors containing second-generation tetracycline-responsive promoters. J Gene Med 7:803–817
39. Matsushita N, Matsushita S, Hirakawa S, Higashiyama S (2013) Doxycycline-dependent inducible and reversible RNA interference mediated by a single lentivirus vector. Biosci Biotechnol Biochem 77:776–781
40. Beyer WR, Westphal M, Ostertag W, von Laer D (2002) Oncoretrovirus and lentivirus vectors pseudotyped with lymphocytic choriomeningitis virus glycoprotein: generation, concentration, and broad host range. J Virol 76:1488–1495
41. Cao W, Henry MD, Borrow P, Yamada H, Elder JH, Ravkov EV, Nichol ST, Compans RW, Campbell KP, Oldstone MB (1998) Identification of alpha-dystroglycan as a receptor for lymphocytic choriomeningitis virus and Lassa fever virus. Science 282: 2079–2081
42. Smelt SC, Borrow P, Kunz S, Cao W, Tishon A, Lewicki H, Campbell KP, Oldstone MB (2001) Differences in affinity of binding of lymphocytic choriomeningitis virus strains to the cellular receptor alpha-dystroglycan correlate with viral tropism and disease kinetics. J Virol 75:448–457
43. Stein CS, Martins I, Davidson BL (2005) The lymphocytic choriomeningitis virus envelope glycoprotein targets lentiviral gene transfer vector to neural progenitors in the murine brain. Mol Ther 11:382–389
44. Miletic H, Fischer YH, Neumann H, Hans V, Stenzel W, Giroglou T, Hermann M, Deckert M, Von Laer D (2004) Selective transduction of malignant glioma by lentiviral vectors pseudotyped with lymphocytic choriomeningitis virus glycoproteins. Hum Gene Ther 15:1091–1100
45. Kobinger GP, Deng S, Louboutin JP, Vatamaniuk M, Matschinsky F, Markmann JF, Raper SE, Wilson JM (2004) Transduction of human islets with pseudotyped lentiviral vectors. Hum Gene Ther 15:211–219
46. Park F (2003) Correction of bleeding diathesis without liver toxicity using arenaviral-pseudotyped HIV-1-based vectors in hemophilia A mice. Hum Gene Ther 14:1489–1494
47. Barrette S, Douglas JL, Seidel NE, Bodine DM (2000) Lentivirus-based vectors transduce mouse hematopoietic stem cells with similar efficiency to moloney murine leukemia virus-based vectors. Blood 96:3385–3391
48. Ricks DM, Kutner R, Zhang XY, Welsh DA, Reiser J (2008) Optimized lentiviral transduction of mouse bone marrow-derived mesenchymal stem cells. Stem Cell Dev 17:441–450
49. Kim JW, Closs EI, Albritton LM, Cunningham JM (1991) Transport of cationic amino acids by the mouse ecotropic retrovirus receptor. Nature 352:725–728
50. Wang H, Kavanaugh MP, North RA, Kabat D (1991) Cell surface receptor for ecotropic murine retroviruses is a basic amino acid transporter. Nature 352:729–731
51. Desmaris N, Bosch A, Salaun C, Petit C, Prevost MC, Tordo N, Perrin P, Schwart O, De Rocquigny H, Heard JM (2001) Production and neurotropism of lentivirus vectors pseudotyped with lyssavirus envelope glycoproteins. Mol Ther 4:149–156
52. Watson DJ, Kobinger GP, Passini MA, Wilson JM, Wolfe JH (2002) Targeted transduction patterns in the mouse brain by lentivirus vectors pseudotyped with VSV, Ebola, Mokola, LCMV, or MuLV envelope proteins. Mol Ther 5:528–537
53. Lafon M (2005) Rabies virus receptors. J Neurovirol 11:82–87

54. Mazarakis ND, Azzouz M, Rohll JB, Ellard FM, Wilkes FJ, Olsen AL, Carter EE, Barber RD, Baban DF, Kingsman SM, Kingsman AJ, O'Malley K, Mitrophanous KA (2001) Rabies virus glycoprotein pseudotyping of lentiviral vectors enables retrograde axonal transport and access to the nervous system after peripheral delivery. Hum Mol Genet 10:2109–2121
55. Wong LF, Azzouz M, Walmsley LE, Askham Z, Wilkes FJ, Mitrophanous KA, Kingsman SM, Mazarakis ND (2004) Transduction patterns of pseudotyped lentiviral vectors in the nervous system. Mol Ther 9: 101–111
56. Kato S, Inoue K, Kobayashi K, Yasoshima Y, Miyachi S, Inoue S, Hanawa H, Shimada T, Takada M, Kobayashi K (2007) Efficient gene transfer via retrograde transport in rodent and primate brains using a human immunodeficiency virus type 1-based vector pseudotyped with rabies virus glycoprotein. Hum Gene Ther 18:1141–1151
57. Kato S, Kobayashi K, Inoue K, Kuramochi OT, Yaginuma H, Morimoto K, Shimada T, Takada M, Kobayashi K (2011) A lentiviral strategy for highly efficient retrograde gene transfer by pseudotyping with fusion envelope glycoprotein. Hum Gene Ther 22:197–206
58. Kato S, Kuramochi M, Kobayashi K, Fukabori R, Okada K, Uchigashima M, Watanabe M, Tsutsui Y, Kobayashi K (2011) Selective neural pathway targeting reveals key roles of thalamostriatal projection in the control of visual discrimination. J Neurosci 31: 17169–17179
59. Kato S, Kuramochi M, Takasumi K, Kobayashi K, Inoue K, Takahara D, Hitoshi S, Ikenaka K, Shimada T, Takada M, Kobayashi K (2011) Neuron-specific gene transfer through retrograde transport of lentiviral vector pseudotyped with a novel type of fusion envelope glycoprotein. Hum Gene Ther 22:1511–1523
60. Carpentier DCJ, Vevis K, Trabalza A, Georgiadis C, Ellison SM, Asfahani RI, Mazarakis ND (2012) Enhanced pseudotyping efficiency of HIV-1 lentiviral vectors by a rabies/vesicular stomatitis virus chimeric envelope glycoprotein. Gene Ther 19:761–774
61. Kato S, Kobayashi K (2020) Pseudotyped lentiviral vectors for tract-targeting and application for the functional control of selective neural circuits. J Neurosci Methods 344: 108854
62. Kobayashi K, Inoue K, Tanabe S, Kato S, Takada M, Kobayashi K (2017) Pseudotyped lentiviral vectors for retrograde gene delivery into target brain regions. Front Neuroanat 11:65
63. Kato S, Kobayashi K, Kobayashi K (2014) Improved transduction efficiency of a lentiviral vector for neuron-specific retrograde gene transfer by optimizing the junction of fusion envelope glycoprotein. J Neurosci Methods 227:151–158
64. Kato S, Sugawara M, Kobayashi K, Kimura K, Inoue K, Takada M, Kobayashi K (2019) Enhancement of the transduction efficiency of a lentiviral vector for neuron-specific retrograde gene delivery through the point mutation of fusion glycoprotein type E. J Neurosci Methods 311:147–155
65. Tanabe S, Uezono S, Tsuge H, Fujiwara M, Miwa M, Kato S, Nakamura K, Kobayashi K, Inoue K, Takada M (2019) A note on retrograde gene transfer efficiency and inflammatory response of lentiviral vectors pseudotyped with FuG-E vs. FuG-B2 glycoproteins. Sci Rep 9:3567
66. Papayannakos C, Daniel R (2013) Understanding lentiviral vector chromatin targeting: working to reduce insertional mutagenic potential for gene therapy. Gene Ther 20: 581–588
67. Debyser Z, Christ F, Rijck JD, Gijsbers R (2015) Host factors for retroviral integration site selection. Trends Biochem Sci 40:108–116
68. Lombardo A, Genovese P, Beausejour CM, Colleoni S, Lee YL, Kim KA, Ando D, Urnov FD, Galli C, Gregory PD, Holmes MC, Naldini L (2007) Gene editing in human stem cells using zinc finger nucleases and integrase-defective lentiviral vector delivery. Nat Biotechnol 25:1298–1306
69. Shaw A, Cornetta K (2014) Design and potential of non-integrating lentiviral vectors. Biomedicine 2:14–35
70. Engelman A, Craigie R (1992) Identification of conserved amino acid residues critical for human immunodeficiency virus type 1 integrase function in vitro. J Virol 66:6361–6369
71. Yew CHT, Gurumoorthy N, Nordin F, Tye GJ, Zaman WSWK, Tan JJ, Ng MH (2022) Integrase deficient lentiviral vector: prospects for safe clinical applications. Peer J 10:213704
72. Phillippe S, Sarkis C, Barkats M, Mammeri H, Ladroue C, Petit C, Mallet J, Serguera C (2006) Lentiviral vectors with a defective integrase allow efficient and sustained transgene expression in vitro and in vivo. Proc Natl Acad Sci USA 103:17684–17689
73. Yáñez-Muñoz RJ, Balaggan KS, MacNeil A, Howe SJ, Schmidt M, Smith AJ, Buch P,

MacLaren RE, Anderson PN, Barker SE, Duran Y, Bartholomae C, von Kalle C, Heckenlively JR, Kinnon C, Ali RR, Thrasher AJ (2006) Effective gene therapy with nonintegrating lentiviral vectors. Nat Med 12:348–353
74. Wanisch K, Yáñez-Muñoz RJ (2009) Integration-deficient lentiviral vectors: A slow coming of age. Mol Ther 17:1316–1332
75. Vargas J Jr, Gusella GL, Najfeld V, Klotman ME, Cara A (2004) Novel integrase-defective lentiviral episomal vectors for gene transfer. Hum Gene Ther 15:361–372
76. Nightingale SJ, Hollis RP, Pepper KA, Petersen D, Yu XJ, Yang C, Bahner I, Kohn DB (2006) Transient gene expression by non-integrating lentiviral vectors. Mol Ther 13:1121–1132
77. Saida H, Matsuzaki Y, Takayama K, Iizuka A, Konno A, Yanagi S, Hirai H (2014) One-year follow-up of transgene expression by integrase-defective lentiviral vectors and their therapeutic potential in spinocerebellar ataxia model mice. Gene Ther 21:820–827
78. Sternson SM, Roth BL (2014) Chemogenetic tools to interrogate brain functions. Annu Rev Neurosci 37:387–407
79. Atasoy D, Sternson SM (2018) Chemogenetic tools for causal cellular and neuronal biology. Physiol Rev 98:391–418
80. Fenno L, Yizhar O, Deisseroth K (2011) The development and application of optogenetics. Annu Rev Neurosci 34:389–412
81. Aston-Jones G, Deisseroth K (2013) Recent advances in optogenetics and pharmacogenetics. Brain Res 1511:1–5
82. Kato S, Fukabori R, Nishizawa K, Okada K, Yoshioka N, Sugawara M, Maejima Y, Shimomura K, Okamoto M, Eifuku S, Kobayashi K (2018) Action selection and flexible switching controlled by the intralaminar thalamic neurons. Cell Rep 22:2370–2382
83. Sakayori N, Kato S, Sugawara M, Setogawa S, Fukushima H, Ishikawa R, Kida S, Kobayashi K (2019) Motor skills mediated through cerebellothalamic tracts projecting to the central lateral nucleus. Mol Brain 12:13
84. Inoue K, Koketsu D, Kato S, Kobayashi K, Nambu A, Takada M (2012) Immunotoxin-mediated tract targeting in the primate brain: selective elimination of the cortico-subthalamic “hyperdirect” pathway. PLoS One 7:e39149
85. Kinoshita M, Matsui R, Kato S, Hasegawa T, Kasahara H, Isa K, Watakabe A, Yamamori T, Nishimura Y, Alstermark B, Watanabe D, Kobayashi K, Isa T (2012) Genetic dissection of the circuit for hand dexterity in primates. Nature 487:235–238
86. Kinoshita M, Kato R, Isa K, Kobayashi K, Kobayashi K, Onoe H, Isa T (2019) Dissecting the circuit for blindsight to reveal the critical role of pulvinar and superior colliculus. Nat Commun 10:135
87. Matsuda T, Hiyama TY, Niimura F, Matsusaka T, Fukamizu A, Kobayashi K, Kobayashi K, Noda M (2017) Distinct neural mechanisms for the control of thirst and salt appetite in the subfornical organ. Nat Neurosci 20:230–241
88. Matsuda T, Hiyama TY, Kobayashi K, Kobayashi K, Noda M (2020) Distinct CCK-positive SFO neurons are involved in persistent or transient suppression of water intake. Nat Commun 11:5692
89. Matsuda T, Kobayashi K, Kobayashi K, Noda M (2024) Two parabrachial Cck neurons involved in the feedback control of thirst or salt appetite. Cell Rep 43:113619
90. Roth BL (2016) DREADDs for neuroscientists. Neuron 89:683–694
91. Mitsuhashi M, Yamaguchi R, Kawasaki T, Ueno S, Sun Y, Isa K, Takahashi J, Kobayashi K, Onoe H, Takahashi R, Isa T (2024) Stage-dependent role of interhemispheric pathway for motor recovery in primates. Nat Commun 15:6762
92. Ishida A, Kobayashi K, Ueda Y, Shimizu T, Tajiri N, Isa T, Hida H (2019) Dynamic interaction between cortico-brainstem pathways during training-induced recovery in stroke model rats. J Neurosci 39:7306–7320
93. Wertz MH, Mitchem MR, Pineda SS, Hachigian LJ, Lee H, Lau V, Powers A, Kulicke R, Madan GK, Colic M, Therrien M, Vernon A, Beja-Glasser VF, Hegde M, Gao F, Kellis M, Hart T, Doench JG, Heiman M (2020) Genome-wide in vivo CNS screening identifies genes that modify CNS neuronal survival and mHTT toxicity. Neuron 106:76–89
94. Ruan Z, Takamatsu-Yukawa K, Wang Y, Ushman ML, Labadorf AT, Ericsson M, Ikezu S, Ikezu T (2022) Functional genome-wide short hairpin RNA library screening identifies key molecules for extracellular vesicle secretion from microglia. Cell Rep 39:110791
95. Inoue F, Kreimer A, Ashuach T, Ahituv N, Yosef N (2019) Identification and massively parallel characterization of regulatory elements driving neural induction. Cell Stem Cell 25:713–727
96. Gordon MG, Inoue F, Martin B, Schubach M, Agarwal V, Whalen S, Feng S, Zhao J, Ashuach T, Ziffra R, Kreimer A, Georgakopoulos-Soares I, Yosef N, Ye CJ,

Pollard KS, Shendure J, Kircher M, Ahituv N (2020) lentiMPRA and MPRAflow for high-throughput functional characterization of gene regulatory elements. Nat Protoc 15: 2387–2412

97. Nonou A, Azzouz (2009) Gene therapy for neurodegenerative diseases based on lentiviral vectors. Prog Brain Res 175:187–200

98. Pignataro D, Sucunza D, Rico AJ, Dopeso-Reyes IG, Roda E, Rodríguez-Perez AI, Labandeira-Garcia JL, Broccoli V, Kato S, Kobayashi K, Lanciego JL (2018) Gene therapy approaches in the non-human primate model of Parkinson's disease. J Neural Transm 125:575–589

99. Capowsi EE, Schneider BL, Ebert AD, Seehus CR, Szulc J, Xufferey R, Aebischer P, Svendsen CN (2007) Lentiviral vector-mediated genetic modification of human neural progenitor cells for ex vivo gene therapy. J Neurosci Methods 163:338–349

100. Rust R, Weber RZ, Generali M, Kehl D, Bodenmann C, Uhr D, Wanner D, Zürcher KJ, Saito H, Hoerstrup SP, Nitsch RM, Tackenberg C (2022) Xeno-free induced pluripotent stem cell-derived neural progenitor cells for in vivo applications. J Transl Med 16:421

101. Sánchez-Danés A, Consiglio A, Richaud Y, Rodríguez-Pizà I, Dehay B, Edel M, Bové J, Memo M, Vila M, Raya A, Izpisua Belmonte JC (2012) Efficient generation of A9 midbrain dopaminergic neurons by lentiviral delivery of LMX1A in human embryonic stem cells and induced pluripotent stem cells. Hum Gene Ther 23:56–69

102. de Luzy IR, Niclis JC, Gantner CW, Kauhausen JA, Hunt CPJ, Ermine C, Pouton CW, Thompson LH, Parish CL (2019) Isolation of LMX1a ventral midbrain progenitors improves the safety and predictability of human pluripotent stem cell-derived neural transplants in parkinsonian disease. J Neurosci 39:9521–9531

103. Thomsen GM, Avalos P, Ma AA, Alkaslasi M, Cho N, Wyss L, Vit JP, Godoy M, Suezaki P, Shelest O, Bankiewicz KS, Svendsen CN (2018) Transplantaion of neural progenitor cells expressing glial cell line-derived neurotrophic factor into the motor cortex as a strategy to treat amyotrophic lateral scleorosis. Stem Cells 36:1122–1131

104. Gowing G, Shelley B, Staggenborg K, Hurley A, Avalos P, Victoroff J, Latter J, Garcia L, Svendsen CN (2014) Glial cell line-derived neurotrophic factor-secreting human neural progenitors show long-term survival, maturation into astrocytes, and no tumor formation following transplantation into the spinal cord of immunocompromised rats. Neuroreport 25:367–372

105. Skukan L, Brezak M, Ister R, Klimaschewski L, Vojta A, Zoldoš V, Gajović S (2022) Lentivirus- or AAV-mediated gene therapy interventions in ischemic stroke: A systematic review of preclinical in vivo studies. J Cereb Blood Flow Metab 42:219–236

106. Eleftheriadou I, Trabalza A, Ellison SM, Gharun K, Mazarakis ND (2014) Specific retrograde transduction of spinal motor neurons using lentiviral vectors targeted to presynaptic NMJ receptors. Mol Ther 22:1285–1298

107. Caiazzo M, Dell'Anno MT, Dvoretskova E, Lazarevic D, Taverna S, Leo D, Sotnikova TD, Menegon A, Roncaglia P, Colciago G, Russo G, Carninci P, Pezzoli G, Gainetdinov RR, Gustincich S, Dityatev A, Broccoli V (2011) Direct generation of functional dopaminergic neurons from mouse and human fibroblasts. Nature 476:224–227

108. Aversano S, Palladino R, Caiazzo M (2022) Direct cell conversion of somatic cells into dopamine neurosn: achievements and perspectives. Cell Reprogram 24:259–270

109. Shiina S, Ohno M, Ohka F, Kuramitsu S, Yamamichi A, Kato A, Motomura K, Tanahashi K, Yamamoto T, Watanabe R, Ito I, Senga T, Hamaguchi M, Wakabayashi T, Kaneko MK, Kato Y, Chandramohan V, Bigner DD, Natsume A (2016) CAR-T cells targeting podoplanin reduce orthotopic glioblastomas in mouse brain. Cancer Immunol Res 4:259–268

110. Han J, Chu J, Wing Chan WK, Zhang J, Wang Y, Cohen JB, Victor A, Meisen WH, Kim SH, Grandi P, Wang QE, He X, Nakano I, Chiocca EA, Glorioso JC III, Kaur B, Caligiuri MA, Yu J (2015) CAR-engineered NK cells targeting wild-type EGFR and EGFRvIII enhance killing of glioblastoma and patient-derived glioblastoma stem cells. Sci Rep 5:11483

Chapter 2

Production and Purification of Lentiviral Vectors

Kenta Kobayashi and Natsuki Matsushita

Abstract

Lentiviral vectors are efficient genetic tools for neuroscience research. Among several kinds of lentiviral vectors, human immunodeficiency virus type 1 (HIV-1)-based lentiviral vectors have shown promise. Envelope glycoproteins studding the surface of viral particles play a crucial role in viral infection of host cells. Conventional lentiviral vectors are pseudotyped with vesicular stomatitis virus glycoprotein (VSV-G), and efficiently transfer the gene to different neurons. Lentiviral tropism can be manipulated by genetically engineering glycoproteins. Analyses of specific neural pathways are essential to understand brain functions controlled by complex neural circuits. Viral vectors ferrying transgenes via retrograde axonal transport provide a powerful tool for the analysis of neural pathways. We developed efficient lentiviral vectors for retrograde gene transfer in diverse neural pathways in animal models: highly efficient retrograde gene transfer (HiRet) and neuron-specific retrograde gene transfer (NeuRet) vectors. Both vectors are pseudotyped with fusion glycoproteins composed of rabies virus glycoprotein segments and VSV-G segments. HiRet/NeuRet vectors have been extensively used in neural pathway analysis. In this chapter, we describe the experimental procedures we developed for producing and purifying HIV-1-based VSV-G and HiRet/NeuRet vectors; we also present helpful points for conducting experiments smoothly and effectively.

Key words Lentiviral vector, Envelope glycoprotein, HiRet, NeuRet, Retrograde gene transfer

1 Introduction

Lentiviral toolkits that are used extensively in neuroscience research possess the following properties [1–9]: high-efficiency of transduction into both dividing and nondividing cells, the ability to integrate into host genomes resulting in sustained long-term gene expression, and a high payload of transgenes. Human immunodeficiency virus type 1 (HIV-1)-based lentiviral vectors are the most commonly used vectors in neurological research [7]. Envelope glycoproteins, which are distributed on the periphery of lentivirus virions, are essential for viral entry into cells [3, 6, 7]. Lentiviral tropism can be manipulated by genetically engineering glycoproteins [10–13].

Kazuto Kobayashi et al. (eds.), *Lentiviral Vectors: Applications for Basic and Clinical Neuroscience*, Methods in Molecular Biology, vol. 2974, https://doi.org/10.1007/978-1-0716-4807-0_2,

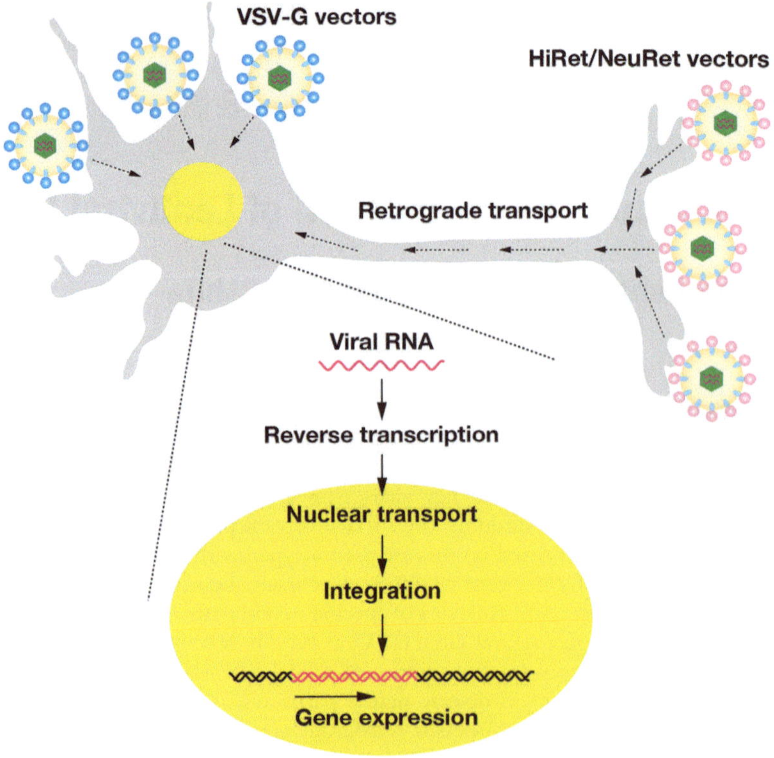

Fig. 1 Lentiviral transduction into neuronal cells. Vesicular stomatitis virus glycoprotein (VSV-G) vector particles infiltrate the cell body. HiRet/NeuRet particles are internalized into nerve terminals and then retrogradely transported along axons to the soma. A viral RNA genome is reverse-transcribed into a double-stranded DNA and integrated into the host cell genome, resulting in sustained expression of the transgene

HIV-1–based conventional lentiviral vectors, which are pseudotyped with vesicular stomatitis virus glycoproteins (VSV-Gs), can efficiently transfer genes into neuronal cells and have been used in a wide range of neuroscience studies (Fig. 1) [14–16]. To understand the mechanisms underlying brain functions regulated by complex neural networks, we need to examine the neural pathways that form the networks. A viral vector system that delivers transgenes in the retrograde direction through axons to cell bodies provides a useful tool for the anatomical and functional analysis of specific neural pathways. Because VSV-G vector-mediated retrograde gene transfer is limited by low potency, we independently developed novel high-performance retrograde lentiviral vectors named highly efficient retrograde gene transfer (HiRet) and neuron-specific retrograde gene transfer (NeuRet) vectors [11–13]. Several powerful retrograde viral vectors have been reported in addition to HiRet/NeuRet vectors, including adeno-associated virus vector [8, 17], rabies virus vector [8, 18], canine adenovirus type-2 vector [8, 19], pseudorabies virus vector [8, 20], and herpes simplex virus type 1 vector [8, 21]. The newly developed HiRet/NeuRet vectors are incorporated at the nerve terminals and undergo retrograde transport to cell bodies; the viral genomes are then integrated into the host genomes via a reverse transcription reaction, thereby providing stable and long-lasting transgene expression (Fig. 1). The HiRet/NeuRet vectors are pseudotyped with fusion glycoproteins (FuGs) composed of rabies virus glycoprotein (RV-G) segments and VSV-G segments [11–13, 22–25]. These vectors show a remarkably higher efficiency of retrograde gene transfer than the parent RV-G-pseudotyped vector [11–13, 22–25]. The HiRet/NeuRet vectors provide highly efficient retrograde gene expression in both rodent and nonhuman primate brains [25–27]. The transduction modality differs significantly between HiRet and NeuRet vectors; HiRet vectors efficiently infect dividing cells such as glial and neural stem/progenitor cells, whereas NeuRet vectors hardly enter these cells [11–13, 22–25]. The inflammation around the injection site is therefore attenuated with the NeuRet vector compared with the HiRet vector. Detailed information about the HiRet/NeuRet vectors is provided in our previous review [13]. HiRet/NeuRet vectors have been applied to the functional analysis of specific neural pathways in diverse animal models [28–35].

Lentiviral vectors remain valuable gene delivery tools for neuroscience research. We therefore provide a detailed description of the production and purification protocols for VSV-G and HiRet/NeuRet vectors for neuroscientists who plan to introduce the lentiviral system into their studies.

2 Materials

1. HEK293T cells.
2. Primaria™ cell culture dish.
3. Phosphate-buffered saline (PBS): Dissolve five tablets (Sigma-Aldrich) in distilled water to a volume of 1000 mL, sterilize in an autoclave, and store at 4 °C.
4. Fetal bovine serum (FBS): Before use, inactivate FBS at 56 °C for 30 min in a water bath, and store at −30 °C.
5. Cell culture medium: Mix 500 mL of DMEM, 56 mL of FBS, and 5 mL of penicillin-streptomycin, and store at 4 °C.
6. HBSP stock solution (10×): Dissolve 40.9 g of NaCl, 29.8 g of HEPES, 1.35 g of Na_2HPO_4-$12H_2O$, 1.9 g of KCl, and 5.4 g of glucose in distilled water to a volume of 500 mL, filtrate the solution with a 0.22-μm filter, and store at 4 °C.

7. HBSP stock solution (2×): Dilute 10 × HBSP stock solution with distilled water, adjust pH to 7.05–7.10 with 1 N NaOH, and store at −30 °C.
8. 2.5 M $CaCl_2$ solution: Dissolve 14.7 g of $CaCl_2$-$2H_2O$ in distilled water to a volume of 40 mL, filter with a 0.22-μm filter, and store at −30 °C.
9. Sucrose solution: Dissolve 20 g or 55 g of sucrose in PBS to a volume of 100 mL, filter the 20% or 55% sucrose solution with a 0.22-μm filter, and store at 4 °C.
10. 1% Bovine serum albumin (BSA) solution: Dissolve 0.5 g of BSA in PBS to a volume of 50 mL, filter with a 0.22-μm filter, and store at 4 °C.
11. NucleoBond Xtra Maxi Plus EF.
12. 0.22-μm filter.
13. 0.45-μm filter.
14. Centrifuge tube.
15. Sepharose Q FF ion-exchange chromatography column.
16. Polypropylene round-bottom tube.
17. Vivaspin 10 K MWCO filter.
18. NucleoSpin RNA virus kit.
19. Lenti-X qRT-PCR titration kit.

3 Methods

3.1 Production of Lentiviral Vectors

Here, we describe the procedures to prepare lentiviral vectors carrying the enhanced green fluorescent protein (EGFP) gene driven by the murine stem cell virus (MSCV) enhancer/promoter. The flowchart of viral preparation is shown in Fig. 2.

1. Day 1: Incubate HEK293T cells (~3.0×10^6 cells/10-cm cell culture dish × 15 dishes) 24 h before transfection (*see* **Note 1**).
2. Day 2: Mix the following four different plasmids as indicated (Fig. 2 and *see* **Notes 2** and **3**): 90 μg of pCAGkGP4.1R, 30 μg of pCAG-RTR2, 30 μg of pCAGGS-glycoprotein, and 150 μg of pCL20c-MSCV-EGFP. Add 750 μL of 2.5 M $CaCl_2$ to the plasmid mixture and add distilled water to a volume of 7.5 mL. Add 7.5 mL of 2 × HBSP while vortexing and let the mixture stand for 3–5 min at room temperature (DNA solution). Add the DNA solution into the culture medium dropwise with even distribution (1 mL of DNA solution/dish) and incubate for 18 h (*see* **Note 4**).

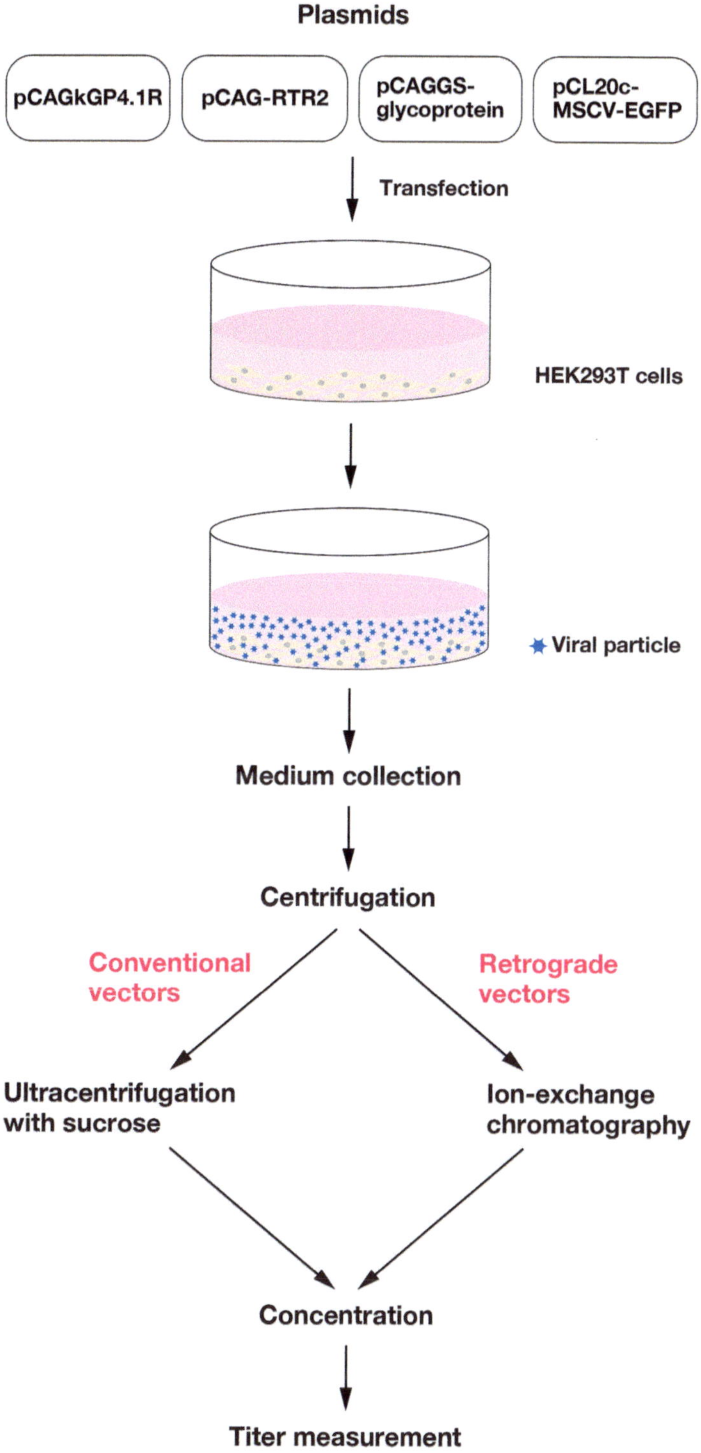

Fig. 2 Flowchart of viral production and purification. Different purification procedures are used for conventional and retrograde lentiviral vectors

3. Day 3: Remove the medium, wash with PBS, add 6 mL of fresh medium, and incubate for 24 h.
4. Day 4: Harvest the medium, centrifuge at 1500×*g* for 5 min at 25 °C, and filter the supernatant with a 0.45-μm filter. Centrifuge the filtered medium at 6000×*g* for 16–20 h at 4 °C.
5. Day 5: Remove the supernatant and gently suspend the pellet including viral particles with 3 mL (for VSV-G vector) or 5 mL (for HiRet/NeuRet vectors) of PBS (*see* **Note 5**). The procedures used for purifying the VSV-G vector or the HiRet/NeuRet vectors are described below.

3.2 Purification of the VSV-G Vector

1. Prepare a layered solution in a centrifuge tube, adding the following solutions in the given order from the bottom: 1 mL of 55% sucrose solution, 2.5 mL of 20% sucrose solution, and 1.5 mL of crude viral solution (Fig. 3a). Prepare two tubes with the layered solution.
2. Centrifuge at 200,000×*g* for 2 h at 4 °C (*see* **Note 6**).

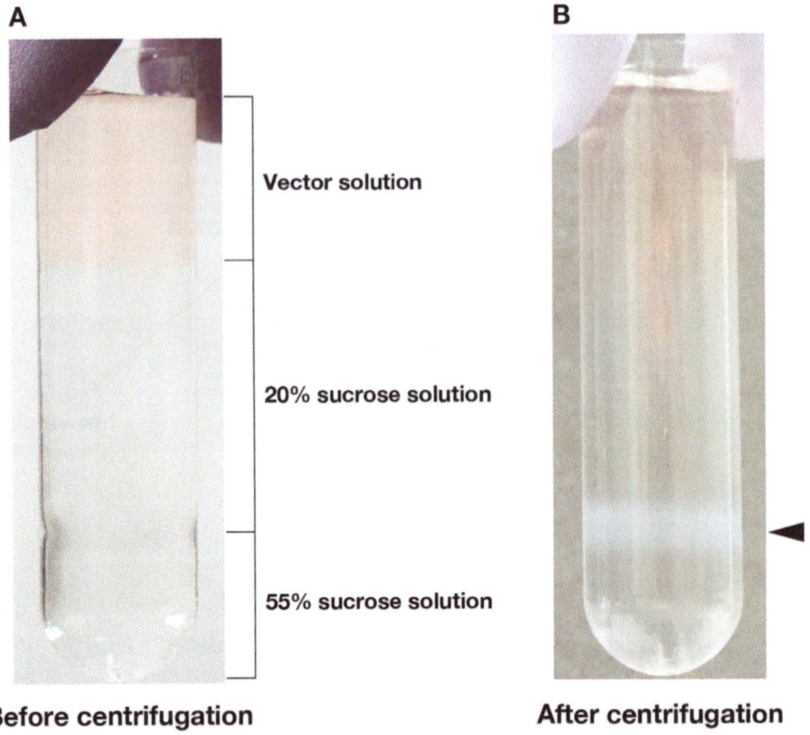

Fig. 3 Purification of VSV-G vectors by ultracentrifugation. (**a**) Before ultracentrifugation, the solution layers are visualized in the centrifuge tube. (**b**) White-colored bands detected after ultracentrifugation contain viral vector particles (arrowhead)

3. The white-colored band contains viral particles (Fig. 3b). Remove an appropriate volume of upper solution (~3 mL) and harvest the white band (750 μL/centrifuge tube).
4. Add 15 mL of 0.01% BSA solution to the purified viral solution (1.5 mL total volume), and concentrate by ultrafiltration using a Vivaspin 10 K MWCO filter* to approximately 100 μL (*see* **Note 7**).
5. Extract the viral RNA genome from 5 μL of concentrated vector solution using the NucleoSpin RNA virus kit following the instructions provided in the manual.
6. Measure the copy number of the RNA genome according to the instruction manual for the Lenti-X qRT-PCR titration kit. The conditions used for quantitative real-time reverse transcription-PCR are as follows: 42 °C for 20 min; 95 °C for 3 min; and 40 cycles each of 95 °C for 15 s and 60 °C for 30 s (*see* **Note 8**).
7. Divide the vector solution into aliquots (10 μL) and store at −80 °C (*see* **Note 9**).

3.3 Purification of HiRet/NeuRet Vectors

1. Apply 5 mL of crude viral solution to a Sepharose Q FF ion-exchange column attached to the AKTA prime plus chromatography system (https://www.cytivalifesciences.com/en/us), and elute with a linear 0.0–1.5 M NaCl gradient. Collect 2 mL of eluate per polypropylene round-bottom tube serially (*see* **Note 10**).
2. From the UV curve of absorbance at 280 nm, harvest the peak fractions containing viral particles (fractions 24–34).
3. Add an equal volume (22 mL) of 0.01% BSA solution to the purified viral vector solution and process by ultrafiltration with a Vivaspin 10 K MWCO filter to concentrate the solution to a volume of approximately 100 μL. The following procedures follow the steps described in the fifth to seventh process of the Subheading 3.2.

4 Notes

1. The optimal timing for passaging HEK293T cells is when the cells have reached approximately 80% confluence. It is important to avoid over-confluent culture, as this may lead to a detrimental impact on cell viability. Furthermore, it is crucial to ensure uniform distribution of the cells in the culture dish. Uneven cell distribution is one of the main causes of low transfection efficiency.

2. The pCAGkGP4.1R and pCAG-RTR2 plasmids are the gag-pol and rev-tat helper vectors, respectively [36, 37]. The pCL20c-MSCV-EGFP plasmid is the HIV-1-based transfer vector [36, 37]. The pCAGGS-VSV-G plasmid is used to produce the VSV-G vector [36, 37]. The pCAGGS-FuG-B or -B2 plasmid is used to produce the HiRet vector [22, 28]. The pCAGGS-FuG-C, -FuG-E, or -FuG-E(P440E) plasmid is used to produce the NeuRet vector [23–25, 30]. To avoid the potential cytotoxicity that may arise at the time of transfection, we recommend that plasmids be purified by using an endotoxin-free plasmid DNA purification kit.
3. The packaging capacity of each viral vector enables it to transport transgenes of specific sizes. This capacity is fixed by the size of the viral genome. Lentiviral vectors have a maximum packaging capacity of 8 kb [8]. When the transgene size exceeds 8 kb, the packaging efficiency is reduced. Therefore, transgenes should be 8 kb or less.
4. The yield of viral vectors depends on the transfection efficiency of plasmids, which is affected by various factors, such as the specific batch of cells, the culture medium, and the quality of the transfection reagents. The cells need to be replaced by a new batch at least once in a half year. To ensure that the culture medium is fresh, it should be replaced with new medium every 2–3 weeks. The pH of the 2 × HBSP solution should be adjusted to 7.05–7.10. Thirty minutes after transfection, the precipitates resulting from the salting-out of plasmid DNA should be checked under the microscope.
5. Carefully suspend the viral pellet in PBS to avoid foaming the solution. Lentiviral particles are denatured by foaming the suspension solution.
6. Our preliminary experiments indicate that the yield of VSV-G vectors by ion-exchange column chromatography purification is inferior to that obtained by ultracentrifugation purification. Therefore, although vectors purified by ultracentrifugation may not be as pure as those obtained by column purification, it is recommended that VSV-G vectors be purified by ultracentrifugation.
7. For filter blocking, add 3 mL of 1% BSA solution to the Vivaspin filter and incubate for 1 h at room temperature before use. When using the filter, discard the blocking solution and wash with PBS.
8. The viral titer varies with the transgene construct. For accurate titer measurement, it is essential that viral vector amplification curves fall within the range of standard amplification curves. The titer of the concentrated lentiviral vectors is typically within the range of 10^{10}–10^{12} genome copies/mL.

9. Lentiviral vectors are fragile, and thus cannot be stored for long periods even at −80 °C. Lentiviral vectors remain stable for at least 6 months when preserved at −80 °C (https://en.vectorbuilder.com). Our preliminary experiments indicate that HiRet/NeuRet vectors remain stable for 1 month, although the functional titer is decreased to about 70% in 2 months when stored at −80 °C. Therefore, in our laboratory, we do not use viral vectors that have been in storage for >1 month. Repeated freezing and thawing should be avoided because of the instability of lentiviral vectors.
10. Although a manual is provided for the AKTA prime plus chromatography system, we recommend that users receive training in the use of the AKTA apparatus from a technical service advisor.

Acknowledgments

We thank R. Kageyama for producing the viral vectors.

Conflict of Interest Statement The authors have no conflicts of interest to declare that are relevant to the content of this chapter.

Funding

This study was funded by Japan Society for the Promotion of Science (JSPS), KAKENHI [25560436] and AMED-Brain/MINDS 2.0, Japan [JP23wm0625001].

References

1. Butler SL, Johnson EP, Bushman FD (2002) Human immunodeficiency virus cDNA metabolism: notable stability of two-long terminal repeat circles. J Virol 76:3739–3747
2. Thomas CE, Ehrhardt A, Kay MA (2003) Progress and problems with the use of viral vectors for gene therapy. Nat Rev Genet 4:346–358
3. Kafri T (2004) Gene delivery by lentivirus vectors an overview. Methods Mol Biol 246:367–390
4. Nielsen MH, Pedersen FS, Kjems J (2005) Molecular strategies to inhibit HIV-1 replication. Retrovirology 2:10
5. Wong LF, Goodhead L, Prat C et al (2006) Lentivirus-mediated gene transfer to the central nervous system: therapeutic and research applications. Hum Gene Ther 17:1–9
6. Cockrell AS, Kafri T (2007) Gene delivery by lentivirus vectors. Mol Biotechnol 36:184–204
7. Pluta K, Kacprzak MM (2009) Use of HIV as a gene transfer vector. Acta Biochim Pol 56:531–595
8. Nectow AR, Nestler EJ (2020) Viral tools for neuroscience. Nat Rev Neurosci 21:669–681
9. Gao J, Gunasekar S, Xia ZJ et al (2024) Gene therapy for CNS disorders: modalities, delivery and translational challenges. Nat Rev Neurosci 25:553–572

10. Cronin J, Zhang XY, Reiser J (2005) Altering the tropism of lentiviral vectors through pseudotyping. Curr Gene Ther 5:387–398
11. Kobayashi K, Inoue KI, Tanabe S et al (2017) Pseudotyped lentiviral vectors for retrograde gene delivery into target brain regions. Front Neuroanat 11:65
12. Kobayashi K, Kato S, Kobayashi K (2018) Genetic manipulation of specific neural circuits by use of a viral vector system. J Neural Transm (Vienna) 125:67–75
13. Kato S, Kobayashi K (2020) Pseudotyped lentiviral vectors for tract-targeting and application for the functional control of selective neural circuits. J Neurosci Methods 344: 108854
14. Kordower JH, Emborg ME, Bloch J et al (2000) Neurodegeneration prevented by lentiviral vector delivery of GDNF in primate models of Parkinson's disease. Science 290:767–773
15. Marr RA, Rockenstein E, Mukherjee A et al (2003) Neprilysin gene transfer reduces human amyloid pathology in transgenic mice. J Neurosci 23:1992–1996
16. Kitagawa R, Miyachi S, Hanawa H et al (2007) Differential characteristics of HIV-based versus SIV-based lentiviral vector systems: gene delivery to neurons and axonal transport of expressed gene. Neurosci Res 57:550–558
17. Haery L, Deverman BE, Matho KS et al (2019) Adeno-associated virus technologies and methods for targeted neuronal manipulation. Front Neuroanat 13:93
18. Callaway EM, Luo L (2015) Monosynaptic circuit tracing with glycoprotein-deleted rabies viruses. J Neurosci 35:8979–8985
19. Junyent F, Kremer EJ (2015) CAV-2—why a canine virus is a neurobiologist's best friend. Curr Opin Pharmacol 24:86–93
20. Pomeranz LE, Reynolds AE, Hengartner CJ (2005) Molecular biology of pseudorabies virus: impact on neurovirology and veterinary medicine. Microbiol Mol Biol Rev 69:462–500
21. Fenno LE, Mattis J, Ramakrishnan C et al (2014) Targeting cells with single vectors using multiple-feature Boolean logic. Nat Methods 11:763–772
22. Kato S, Kobayashi K, Inoue K et al (2011) A lentiviral strategy for highly efficient retrograde gene transfer by pseudotyping with fusion envelope glycoprotein. Hum Gene Ther 22: 197–206
23. Kato S, Kuramochi M, Takasumi K et al (2011) Neuron-specific gene transfer through retrograde transport of lentiviral vector pseudotyped with a novel type of fusion envelope glycoprotein. Hum Gene Ther 22:1511–1523
24. Kato S, Kobayashi K, Kobayashi K (2014) Improved transduction efficiency of a lentiviral vector for neuron-specific retrograde gene transfer by optimizing the junction of fusion envelope glycoprotein. J Neurosci Methods 227:151–158
25. Kato S, Sugawara M, Kobayashi K et al (2019) Enhancement of the transduction efficiency of a lentiviral vector for neuron-specific retrograde gene delivery through the point mutation of fusion glycoprotein type E. J Neurosci Methods 311:147–155
26. Tanabe S, Inoue KI, Tsuge H et al (2017) The use of an optimized chimeric envelope glycoprotein enhances the efficiency of retrograde gene transfer of a pseudotyped lentiviral vector in the primate brain. Neurosci Res 120:45–52
27. Tanabe S, Uezono S, Tsuge H et al (2019) A note on retrograde gene transfer efficiency and inflammatory response of lentiviral vectors pseudotyped with FuG-E vs. FuG-B2 glycoproteins. Sci Rep 9:3567
28. Kato S, Kuramochi M, Kobayashi K et al (2011) Selective neural pathway targeting reveals key roles of thalamostriatal projection in the control of visual discrimination. J Neurosci 31:17169–17179
29. Kinoshita M, Matsui R, Kato S et al (2012) Genetic dissection of the circuit for hand dexterity in primates. Nature 487:235–238
30. Kobayashi K, Sano H, Kato S et al (2016) Survival of corticostriatal neurons by Rho/Rho-kinase signaling pathway. Neurosci Lett 630:45–52
31. Matsuda T, Hiyama TY, Niimura F et al (2017) Distinct neural mechanisms for the control of thirst and salt appetite in the subfornical organ. Nat Neurosci 20:230–241
32. Nomura K, Hiyama TY, Sakuta H et al (2019) [Na+] increases in body fluids sensed by central Nax induce sympathetically mediated blood pressure elevations via H+-dependent activation of ASIC1a. Neuron 101:60–75
33. Ishida A, Kobayashi K, Ueda Y et al (2019) Dynamic interaction between cortico-brainstem pathways during training-induced recovery in stroke model rats. J Neurosci 39: 7306–7320

34. Vancraeyenest P, Arsenault JT, Li X et al (2020) Selective mesoaccumbal pathway inactivation affects motivation but not reinforcement-based learning in macaques. Neuron 108:568–581
35. Matsuda T, Kobayashi K, Kobayashi K et al (2024) Two parabrachial Cck neurons involved in the feedback control of thirst or salt appetite. Cell Rep 43:113619
36. Hanawa H, Kelly PF, Nathwani AC, Persons DA et al (2002) Comparison of various envelope proteins for their ability to pseudotype lentiviral vectors and transduce primitive hematopoietic cells from human blood. Mol Ther 5:242–251
37. Hanawa H, Hematti P, Keyvanfar K et al (2004) Efficient gene transfer into rhesus repopulating hematopoietic stem cells using a simian immunodeficiency virus-based lentiviral vector system. Blood 103:4062–4069

immunodeficiency virus type 1 (HIV-1), for highly efficient retrograde gene transfer (HiRet) and neuron-specific retrograde gene transfer (NeuRet) are reported [1–3] with precise targeting and manipulation of neuronal populations based on their synaptic connectivity. Retrograde transport is a natural cellular process by which neurons transport proteins and lipid-coated particles from axon terminals back to the cell soma. Using this transport mechanism, retrograde lentiviral vectors can deliver cargo genetic material from the synaptic terminals to soma, allowing for targeted gene expression in neurons projecting to the specific virus-infected areas. These capabilities are valuable for the experimental characterization of neuronal circuits, including their connectivity and function to other brain regions, and manipulating neuronal activity with high spatiotemporal specificity.

The basal ganglia (BG) is a mammalian brain region hosting a complex set of neuronal circuits that play important roles in motor control, decision making, and reward processing [4–7]. The cerebral cortex sends information to the major input nuclei of the BG, the striatum and the subthalamic nucleus (STN), which are the input nuclei of the BG (Fig. 1). A fundamental component of the BG for the coordination of voluntary movements and motor learning is the corticostriatal pathway, particularly the projection from the motor cortex to the striatum. Despite these critical roles, the physiological mechanisms by which motor cortical inputs regulate striatal function are not well understood in part, because cortical neurons that project to the striatum and STN are in close proximity [8, 9], so experimental electrical stimulation of the cortex induces excitation in both corticostriatal and corticosubthalamic neurons. In recent decades, optogenetics and chemogenetics have revolutionized the study of neural circuits by enabling pathway-specific excitation and inhibition of neuronal activity [10, 11]. These techniques have allowed researchers to dissect the role of specific neural pathways in various behaviors and physiological processes, and have been applied in diverse brain areas, including those for motor control, memory, emotion, and disease models, to significantly advance our understanding of complex neural networks.

In this chapter, we present a method for injecting lentiviral vectors into the mouse brain using the VSV-G lentiviral vector and a double-infection strategy combining the retrograde lentiviral NeuRet and anterograde AAV DJ vectors for stimulation of cortical or corticostriatal neurons using optogenetics (Fig. 1).In our strategy, the orofacial and forelimb motor regions in the cortex and the striatum are first identified by electrophysiological mapping (*see* **Note 1**). The VSV-G lentiviral vector, VSVG-CaMKII-ChR2/Venus, is injected into the motor regions of the cortex. Or the

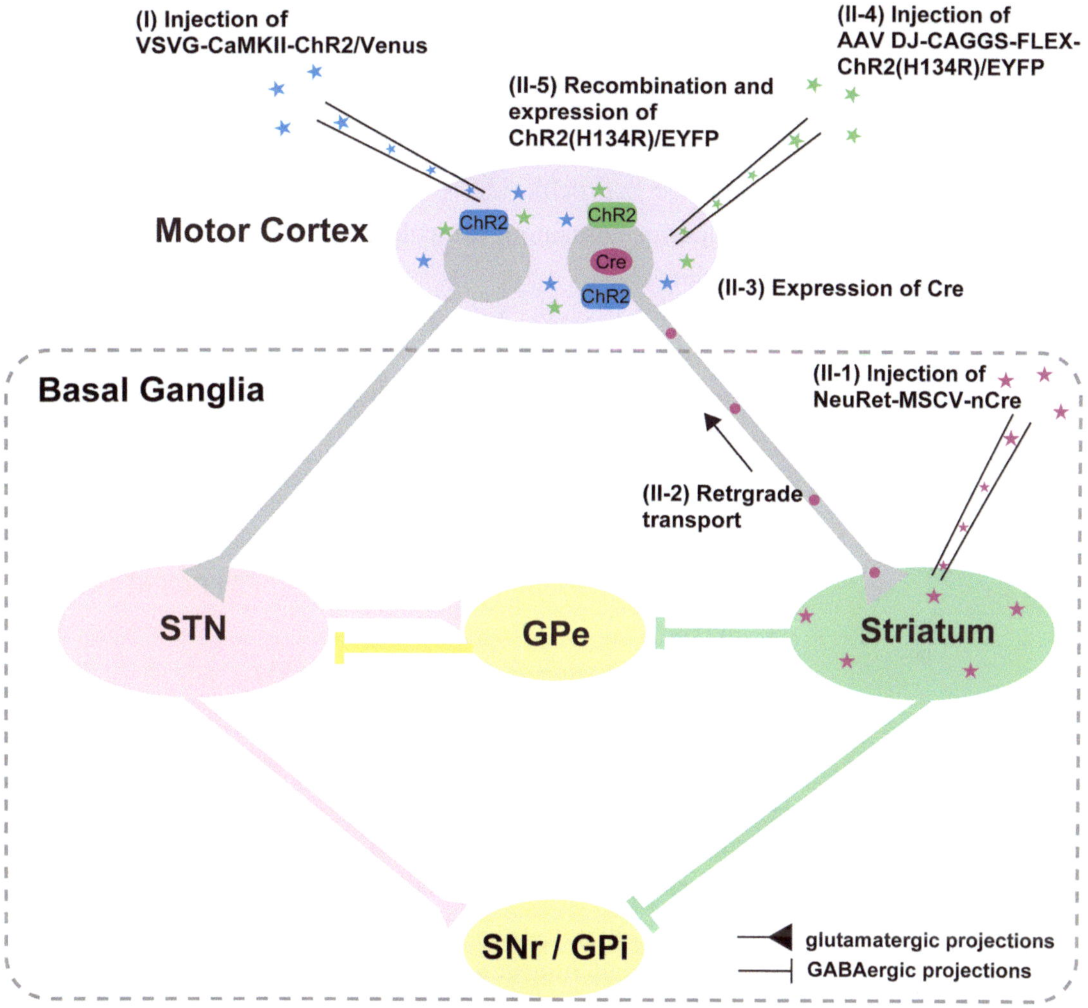

Fig. 1 Neuronal circuitry of the basal ganglia (BG) and the infection strategy. The striatum and subthalamic nucleus (STN) are input nuclei of the BG and receive motor cortical inputs. The VSVG-CaMKII-ChR2/Venus lentiviral vector is injected into the motor cortex to express ChR2 in both corticostriatal and corticosubthalamic neurons (I). To specifically express ChR2 in the corticostriatal neurons (II), first, corticostriatal neurons were labeled by the injection of the NeuRet-MSCV-nCre vector into the striatum (II-1). Second, NeuRet vector-mediated gene transfer was observed via retrograde transport and the Cre transgene was expressed in corticostriatal neurons (II-2 and 3). Then, an AAV DJ-CAGGS-FLEX-ChR2(H134R)/EYFP vector was injected into the motor cortex (II-4). Finally, Cre-mediated recombination occurred and ChR2 was specifically expressed in corticostriatal neurons (II-5). *GPe* external segment of the globus pallidus, *GPi* internal segment of the globus pallidus, *SNr* substantia nigra pars reticulata

retrograde lentiviral vector, NeuRet-MSCV-nCre, is injected into the motor regions of the striatum, and then, an anterograde vector, AAV DJ-CAGGS-FLEX ChR2(H134R/EYFP), is injected into the orofacial and forelimb encoding areas in the motor cortex (*see* **Note 2**).

2 Materials

1. Male C57BL/6 J mice (8–12 weeks).
2. Ketamine hydrochloride.
3. Xylazine hydrochloride.
4. Lidocaine containing epinephrine.
5. Isoflurane.
6. All-In-One Small Animal Anesthetizer (Muromachi Kikai, MK-AT210D).
7. Stereotaxic instruments (Narishige, SR-6 M-HT) with manipulator (Narishige, SM-11, SM-15) and oil hydraulic micromanipulator (Narishige, MO-10).
8. Anesthetic gas mask for mice (Narishige, GM-4).
9. Small U-shaped head holder (Handmade).
10. Bone adhesive resin (Tokuyama Dental, ESTECEM II).
11. Acrylic resin (GC, Unifast II).
12. Hand drill (Minitor, C10M, KM11H35).
13. Microsyringe (Hamilton, 1702 N).
14. Microinfusion pump (WPI, Miro4).
15. Teflon tube (Eicom, JT-10).
16. Glass capillary (Narishige, G-1).
17. Fluorinert (3 M, FC-3283).
18. Teflon-coated tungsten wire (A-M Systems, 795500).
19. Glass-coated tungsten electrode (Alpha Omega, 0.5 MΩ at 1 kHz).
20. Antibiotics ointment.
21. VSV-G vector.
22. NeuRet vector.
23. AAV DJ vector.

3 Methods

3.1 Surgery for Fixing Mice Head Painlessly

1. House the mice under a 12/12 h light/dark cycle with free access to water and food. Handle the mice daily before conducting experiments to accustom them to the experimental manipulations.
2. Anesthetize each mouse with ketamine hydrochloride (100 mg/kg, *i.p.*), xylazine hydrochloride (5 mg/kg, *i.p.*), and isoflurane (1.0%). Restrain the mouse with the stereotaxic apparatus using an auxiliary ear bar, ear bars, and a mouthpiece.

(a)

(b)

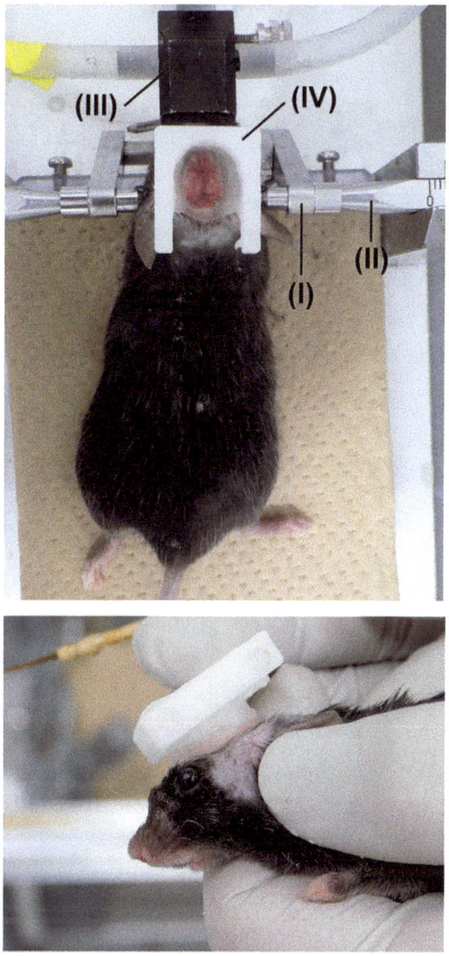

Fig. 2 A head holder for painless attachment of mice to the stereotaxic apparatus. (**a**) The mouse was restrained by the auxiliary ear bars (I), ear bars (II) of the stereotaxic frame and the gas anesthetic mask (III) under isoflurane anesthesia. Then a small U-shaped head holder (IV) was attached to the mouse's head with acrylic resin. Bregma and lambda were marked by India ink (arrow heads) (**b**) A side view of the head of the mouse with the U-shaped head holder

Expose the skull widely and completely remove the periosteum and blood on the skull with saline and 0.3% sodium hypochlorite. Mark the bregma with India ink and completely cover the exposed skull with bone adhesive resin and acrylic resin. Attach a small U-shaped head holder to the skull with acrylic resin (Fig. 2) [12–14].

3.2 Mapping of Target Areas in the Striatum

1. Hold the mouse with its head restrained painlessly by the U-shaped head holder several days after the surgery. Anesthetize the mouse with isoflurane (1.0–1.5%). Shave a portion of the right hemisphere of the skull thin and remove it with a hand drill, leaving the dura mater intact.

2. Insert two pairs of bipolar stimulating electrode (50-μm diameter Teflon-coated tungsten wires, 300–400 μm tip spacing) into the motor cortex. Confirm the orofacial and forelimb regions by intracortical microstimulation (ICMS; one train of ten pulses at 333 Hz, 200 μs duration, ≤ 30 μA) under awake conditions. Temporally fix the bipolar stimulating electrodes with acrylic resin.
3. Insert a glass-coated tungsten electrode vertically into the striatum using an oil hydraulic micromanipulator. Record orthodromic responses to the motor cortical stimulation (200 μs duration, monophasic single pulse at 0.7 Hz, 50 μA power) through the stimulating electrode implanted in the motor cortex to locate the orofacial and forelimb regions in the striatum.
4. Remove the bipolar stimulating electrodes, apply ointment containing antibiotics to the dura, and return the mouse to its home cage.

3.3 Injection of Viral Vector Solutions

1. Hold the mouse in the stereotaxic apparatus by the U-shaped head holder the next day. Anesthetize the mouse with isoflurane (1.0–1.5%). Remove the ointment on the dura mater with cotton balls and saline, exposing the dura mater.
2. Connect the glass capillary (the tip diameter was about 50 ~ 100 μm), Teflon-tubing, and microsyringe and fill the Fluorinert without bubbles (Fig. 3a). Fill the tip of the capillary with an appropriate volume of the VSV-G or NeuRet lentiviral vector solution, ensuring that no air bubbles entered. Attach a microsyringe to the microinfusion pump to precisely control the flow rate of the injection (Fig. 3b).
3. Place the tip of the glass capillary just above the bregma, which is marked with India ink, and then move it to the orofacial and forelimb areas of the motor cortex or the striatum that have already been mapped. First, lower the tip of the glass capillary until it adhered to the dura mater. Then, insert it with the hydraulic micromanipulator to a depth where the orofacial and forelimb regions in the motor cortex or the striatum (*see* **Note 3**).
4. Wait 3–5 min before injection. Inject 0.1 μl or 0.5 μl of the vector solution at a rate of 100 nl/min and wait again 5 min after injection. Slowly remove the glass capillary with the hydraulic micromanipulator. Apply antibiotic ointment to the dura and return the mouse to its home cage.
5. After 1–2 weeks, hold the mouse that injected with NeuRet lentiviral vector in the stereotaxic apparatus by the U-shaped head holder and anesthetize it with isoflurane (1.0–1.5%).

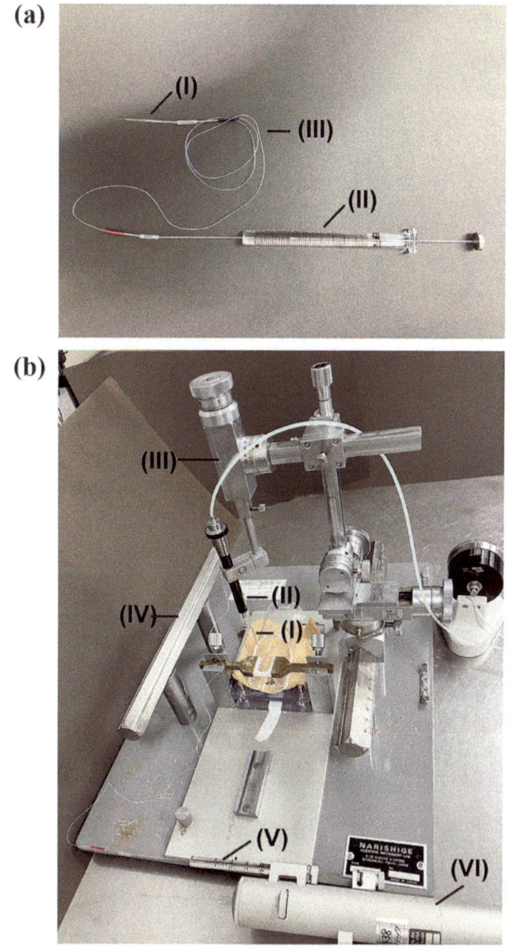

Fig. 3 Equipment setup for stereotaxic intracerebral viral vector delivery. (**a**) The glass capillary (I) was connected to a 25 μl microsyringe (II) via Teflon-tubing (III). (**b**) A picture of the stereotaxic injection system. The glass capillary (I) was held with the hydraulic micromanipulator (II) and the micromanipulator (III) attached to the stereotaxic frame (IV). The microsyringe (V) is attached to the microinfusion pump (VI)

Remove antibiotic ointment on the dura mater, exposing the dura mater.

6. Fill the tip of the capillary with the AAV vector solution as described above. Insert the glass capillary into the orofacial and forelimb regions of the motor cortex that have already been mapped. After waiting 3–5 min, inject 0.1 μl of the vector solution at a rate of 100 nl/min and wait again 5 min after injection.
7. Apply antibiotic ointment to the dura and return the mouse to its home cage.

3.4 Optogenetic Stimulation of the Corticostriatal Neurons

After 3 or 4 weeks, ensure the level of expression of ChR2/Venus or ChR2(H134R)/EYFP has reached the sufficient level to respond to photo-stimulation (Fig. 4a, b). Insert an optrode to the orofacial and/or forelimb regions of the motor cortex, and examine responses to the photo-stimulation (Fig. 4c) (*see* **Note 4**). Change the antibiotic ointment every 2 or 3 days during these 3 or 4 weeks.

4 Notes

1. In order to ensure the injection of viral vectors into the correct target regions of the brain, this chapter has shown a method of mapping the motor cortex using ICMS and the recording orthodromic responses in the striatum. It is important to identify the connections between the brain regions using electrophysiological methods or anterograde and retrograde tracers to provide comprehensive insights into the brain connectivity and functional pathways.
2. In this chapter, we present the double-infection lentiviral vector strategy as an example of a method for introducing viral vectors into specific neuronal circuits in the mouse brain. It is also possible to express the transgene using only retrograde vectors such as the HiRet or NeuRet lentiviral vectors. We first used only the NeuRet lentiviral vector to express channelrhodopsin-2 (ChR2) in corticostriatal neurons. The expression of ChR2 was detected by histology, but we could not obtain photo-stimulated responses. Therefore, we tried the FLEX system using Cre-loxP transgenes. It is critical to consider the tropism and efficiency of viral vectors in the neuronal population of interest. Different viral vectors may have different levels of transduction efficiency and specificity, which can significantly affect the outcome of your experiments. In addition, the promoter used may also affect the expression level and distribution of the transgene. We recommended performing pilot experiments to optimize these parameters before proceeding with large-scale studies. Careful selection and optimization of viral vectors and experimental conditions can improve the reliability and reproducibility of your results.
3. Multiple sites may need to be injected in close proximity. They can be injected in one penetration if they are identical anteroposteriorly and mediolaterally and differ only dorsoventrally. However, if they differ anteroposteriorly or mediolaterally and the distance between them is less than 500 μm, they should be injected on two separate days to avoid injury.

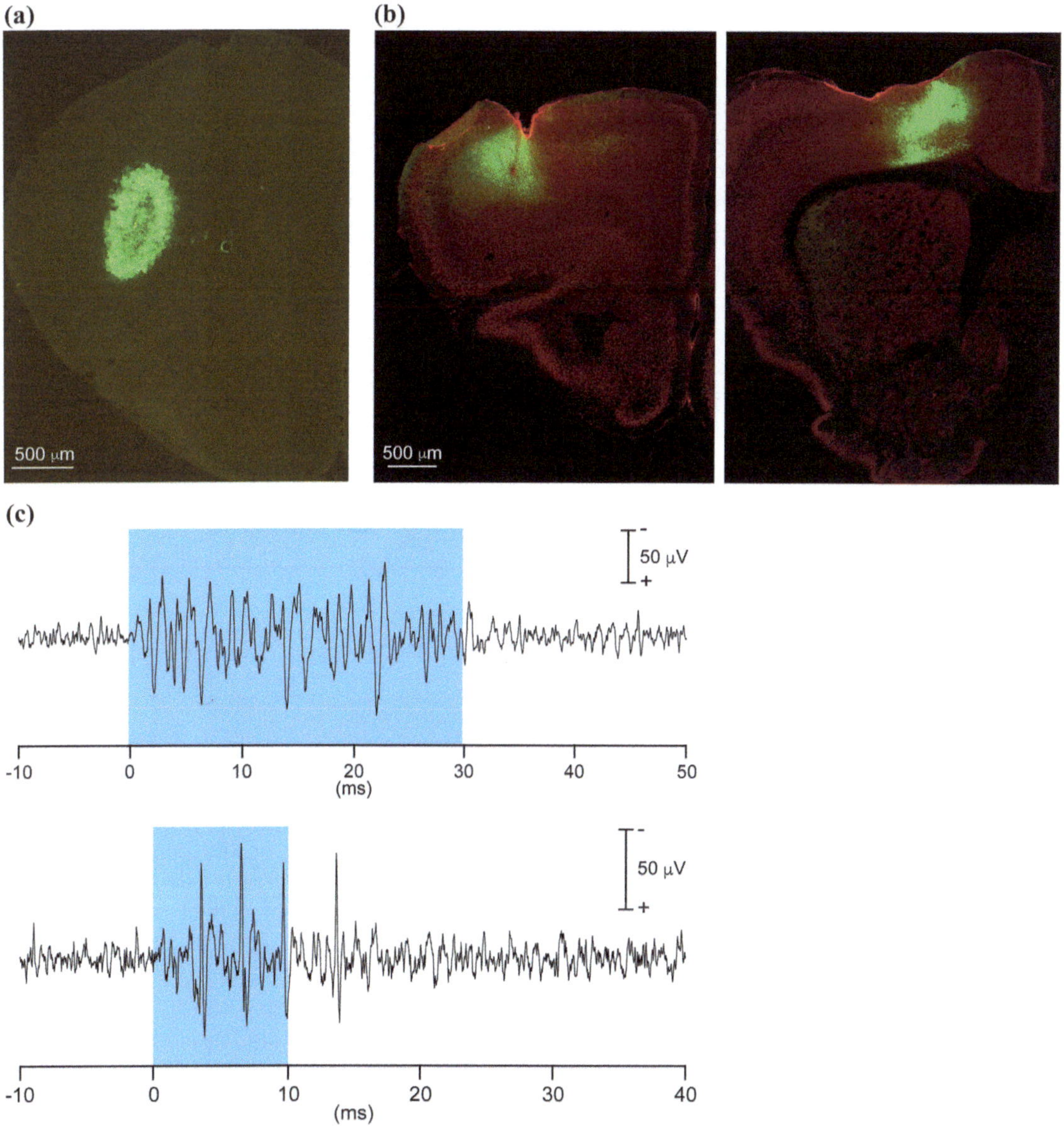

Fig. 4 Neuronal expression of ChR2 or the variant of ChR2(H134R) and responses to photo-stimulation. (**a**) Expression of ChR2 in both corticostriatal and corticosubthalamic neurons in the orofacial area of the motor cortex, as visualized by Venus signals in green. (**b**) Expression of ChR2(H134R) in corticostriatal neurons, as visualized by enhanced yellow fluorescent protein (EYFP) signals in green. Neurons are shown in red, as visualized by an anti-NeuN antibody. The left and right panels show the ChR2(H134R)/EYFP signals in the orofacial and forelimb areas of the motor cortex, respectively. (**c**) An optrode [17], or the combination of an electrode and an optical fiber (diameter in 50 μm), was inserted into the motor cortex, and neuronal activity was recorded. Upper and lower panels show neuronal activity in the motor cortex recorded from mice injected with the VSV-G vector and the NeuRet/AAV DJ vectors, respectively. Photo-stimulation (30 or 10-ms duration single pulse of blue laser, represented by a blue rectangle) from the optrode evoked neuronal excitation

4. We have presented here an example of the local VSV-G lentiviral vector or a double-infection strategy combining the NeuRet retrograde lentiviral vector and optogenetics. These methods can also be applied to chemogenetics using DREADD, calcium imaging using GCaMP, and GRAB sensors such as GRABDA [15, 16]. In addition, these techniques for mapping target areas and injecting viral vectors into the mammalian brain can be adapted not only to other rodents such as rats, but also to nonhuman primates such as common marmosets and macaque monkeys.

Conflict of Interest Statement

The authors have no conflicts of interest to declare that are relevant to the content of this chapter.

Funding

This study was funded by JSPS KAKENHI (Grant Number 21K07257).

References

1. Kato S, Kobayashi K, Inoue K et al (2011) A lentiviral strategy for highly efficient retrograde gene transfer by pseudotyping with fusion envelope glycoprotein. Hum Gene Ther 22(2):197–206. https://doi.org/10.1089/hum.2009.179
2. Kato S, Kuramochi M, Takasumi K et al (2011) Neuron-specific gene transfer through retrograde transport of lentiviral vector pseudotyped with a novel type of fusion envelope glycoprotein. Hum Gene Ther 22(12):1511–1523. https://doi.org/10.1089/hum.2011.111
3. Kobayashi K, Inoue K, Tanabe S et al (2017) Pseudotyped Lentiviral Vectors for Retrograde Gene Delivery into Target Brain Regions. Front Neuroanat 11:65. https://doi.org/10.3389/fnana.2017.00065
4. Cox J, Witten IB (2019) Striatal circuits for reward learning and decision-making. Nat Rev Neurosci 20(8):482–494. https://doi.org/10.1038/s41583-019-0189-2
5. Hikosaka O, Kim HF, Yasuda M et al (2014) Basal ganglia circuits for reward value-guided behavior. Annu Rev Neurosci 37:289–306. https://doi.org/10.1146/annurev-neuro-071013-013924
6. Roth RH, Ding JB (2024) Cortico-basal ganglia plasticity in motor learning. Neuron 112(15):2486–2502. https://doi.org/10.1016/j.neuron.2024.06.014
7. Schultz W (2016) Reward functions of the basal ganglia. J Neural Transm (Vienna) 123(7):679–693. https://doi.org/10.1007/s00702-016-1510-0
8. Nambu A (2011) Somatotopic organization of the primate Basal Ganglia. Front Neuroanat 5:26. https://doi.org/10.3389/fnana.2011.00026
9. Mathai A, Smith Y (2011) The corticostriatal and corticosubthalamic pathways: two entries, one target. So what? Front Syst Neurosci 5:64. https://doi.org/10.3389/fnsys.2011.00064
10. Roth BL (2016) DREADDs for Neuroscientists. Neuron 89(4):683–694. https://doi.org/10.1016/j.neuron.2016.01.040
11. Emiliani V, Entcheva E, Hedrich R et al (2022) Optogenetics for light control of biological systems. Nat Rev Methods Primers 2. https://doi.org/10.1038/s43586-022-00136-4
12. Sano H, Chiken S, Hikida T et al (2013) Signals through the striatopallidal indirect pathway stop movements by phasic excitation in the substantia nigra. J Neurosci 33(17):

7583–7594. https://doi.org/10.1523/JNEUROSCI.4932-12.2013
13. Chiken S, Sato A, Ohta C et al (2015) Dopamine D1 Receptor-Mediated Transmission Maintains Information Flow Through the Cortico-Striato-Entopeduncular Direct Pathway to Release Movements. Cereb Cortex 25(12):4885–4897. https://doi.org/10.1093/cercor/bhv209
14. Dwi Wahyu I, Chiekn S, Hasegawa T et al (2021) Abnormal Cortico-Basal Ganglia Neurotransmission in a Mouse Model of l-DOPA-Induced Dyskinesia. J Neurosci 41(12): 2668–2683. https://doi.org/10.1523/JNEUROSCI.0267-20.2020
15. Nectow AR, Nestler EJ (2020) Viral tools for neuroscience. Nat Rev Neurosci 21(12): 669–681. https://doi.org/10.1038/s41583-020-00382-z
16. Swanson JL, Chin PS, Romero JM et al (2022) Advancements in the Quest to Map, Monitor, and Manipulate Neural Circuitry. Front Neural Circuits 16:886302. https://doi.org/10.3389/fncir.2022.886302
17. Watanabe H, Sano H, Chiken S et al (2020) Forelimb movements evoked by optogenetic stimulation of the macaque motor cortex. Nat Commun 11(1):3253. https://doi.org/10.1038/s41467-020-16883-5

Chapter 4

Stereotaxic Injection of Lentiviral Vectors into the Primate Brain

Ken-ichi Inoue and Masahiko Takada

Abstract

Nonhuman primates (NHPs) have been widely and crucially utilized as model animals for understanding mechanisms of various higher brain functions and pathomechanisms of neurological and psychiatric disorders, since their behaviors mimic those of humans in both normal and pathological states. To know how complex neural networks are involved in such behavioral states, it is essential to define the causal roles of specific pathways and neuron types that constitute these networks. Considering that genetically modified animals to be created through reproductive engineering are not commonly or readily available in NHPs, the gene transduction strategy via viral vectors is the only effective approach to the development of primate models. For this purpose, lentiviral vectors and adeno-associated virus vectors are mainly used. The ability of lentiviral vectors to insert longer sequences is advantageous for achieving cell-type-specific gene expression with the aid of promoters and enhancers. In addition, the use of envelopes that are derived from viruses exhibiting peculiar infection patterns or those being artificially modified is a powerful method for introducing foreign genes into specific cell populations. Furthermore, semipermanent gene expression via lentiviral vectors is an attractive feature for translational gene therapy research. This chapter will outline the stereotaxic injection of lentiviral vectors into the primate brain, primarily for neural network analysis.

Key words Lentiviral vector, Stereotaxic injection, Nonhuman primate

1 Introduction

Nonhuman primates (NHPs) play a vital role as model animals for medical and life science research, because they are evolutionarily closer to humans than other experimental animals, and their body structure and function are similar to those of humans. Since large NHPs, especially macaques, have the excellent ability to learn and perform various motor and cognitive tasks, the use of such monkeys is essential not only for investigating higher brain functions, but also for exploring the pathophysiology of neurological and psychiatric disorders and developing innovative therapeutic approaches to these disorders. So far, considerable amounts of anatomical and physiological findings, mainly on macaques, have been

Kazuto Kobayashi et al. (eds.), *Lentiviral Vectors: Applications for Basic and Clinical Neuroscience*, Methods in Molecular Biology, vol. 2974, https://doi.org/10.1007/978-1-0716-4807-0_4,

accumulated regarding higher brain functions and the underlying neural networks. To understand the central mechanisms for behavioral actions achieved by the complex brain functions, it is of particular importance to elucidate the details of individual pathways and neuron types that constitute each neural network, and to verify which pathway is responsible for symptoms characteristic of each disorder. To address these issues, experimental approaches are required to allow activity measurement and functional intervention in a given pathway by introducing functional probes into a specific neuronal population. At present, when it is not easy or realistic to produce and utilize genetically engineered primate models, the gene transduction strategy via viral vectors is the only effective methodology that permits such analysis in NHPs [1, 2].

Currently, two kinds of viral vectors are mainly used for neural activity manipulation and activity imaging in the primate brain. Lentiviral vectors, first used for optogenic manipulation, allow semipermanent expression of the introduced gene, because they integrate part of their genome into the host genome for expressing the inserted gene. Other kinds of vectors, such as adeno-associated virus (AAV) vectors, are more physically and chemically stable than lentiviral vectors, because it is encircled by a capsid as its outer shell. Both viral vectors are efficient in infecting neurons at their injection sites, but in recent years, vectors with altered infection characteristics have also been used frequently for neural network analysis. A major example is a retrograde viral vector characterized by infection from axon terminals of neurons, and lentiviral vectors can have such an ability by pseudotyping with modified envelope proteins including FuG-E [3, 4]. These lentiviral vectors permitting retrograde gene transfer are effective in the primate brain, and, unlike retrograde AAV vectors such as AAV2-retro [5], they show high retrograde infectivity to subcortical neurons (Fig. 1; [6, 7]). The earlier chapter should be referred to for more detailed information on retrograde lentiviral vectors.

The cell-type-specific gene expression is also useful for functional analysis of neural networks. In rodents, it has become possible to measure and manipulate the activity of a given cell population by using various cell-type-specific recombinase-expressing mice and viral vectors with a recombinase-dependent expression of functional probes. In NHPs, however, it is unlikely that cell-type specific Cre recombinase-expressing animal resources will be established due to the labor and cost involved in their development and maintenance. Therefore, cell-type-specific gene expression in NHPs would largely rely on the expression control by sequences inserted into viral vectors. To date, most of the successful neuron-type-specific activity manipulation studies in NHPs have been carried out by using promoters of which effectiveness had already been proven in mice, such as the CaMKIIα promoter that is specific to excitatory neurons in the cerebral cortex, and the L7 promoter that

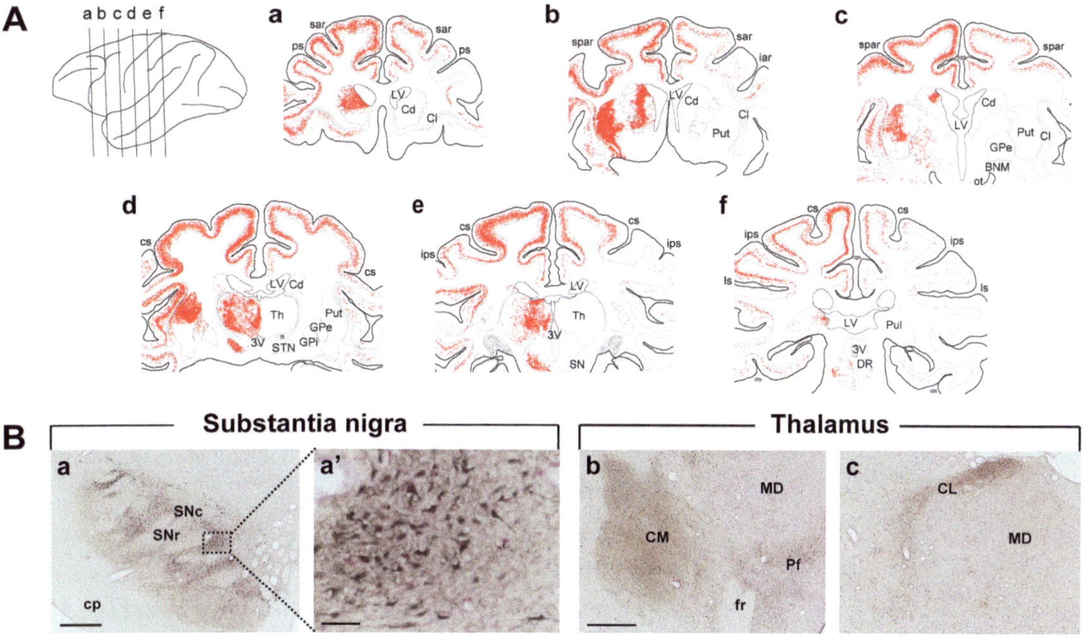

Fig. 1 Retrograde gene transfer of a pseudotyped lentiviral vector in the primate brain. The FuG-E pseudotyped lentiviral vectors (7.0 × 10e10 copies/ml) were injected into the caudate nucleus (Cd) and the putamen (Put) of the macaque monkey, and histological analyses were performed 4 weeks after the injection. (**a**) Patterns of distribution of transgene expression throughout the brain. Each dot corresponds to one transduced neuron. *BNM* basal nucleus of Meynert, *Cd* caudate nucleus, *cgs* cingulate sulcus, *Cl* claustrum, *cs* central sulcus, *DR* dorsal raphe nucleus, *GPe* external segment of the globus pallidus, *GPi* internal segment of the globus pallidus, *iar* inferior limb of the arcuate sulcus, *ips* intraparietal sulcus, *ls* lateral sulcus, *ps* principal sulcus, *Pul* pulvinar, *Put* putamen, *sar* superior limb of the arcuate sulcus, *SN* substantia nigra, *spcd* superior precentral dimple, spur of the arcuate sulcus, *STN* subthalamic nucleus, *Th* thalamus. (**b**) Transgene (GFP) expression revealed by immunostaining in the substantia nigra pars compacta (SNc, **a**) and thalamus (**b, c**). *CL* centrolateral nucleus, *CM* centromedian nucleus, *cp* cerebral peduncle, *fr* fasciculus retroflexus, *MD* mediodorsal nucleus, *Pf* parafascicular nucleus, *SNr* substantia nigra pars reticulata. Scale bar, 1 mm (Adapted from Ref. [6])

is specific to Purkinje cells in the cerebellum [8, 9]. The only exception is a report on the TH (tyrosine hydroxylase) promoter to enable dopaminergic neuron-specific gene expression [10, 11]. In recent years, however, various expression control sequences have been reported. Notably, as the remarkable development in genome analysis, many enhancer sequences have been identified and proven to achieve neuron-type-specific gene expression [12–14]. This leads us to challenge cell-type-specific gene expression in NHPs, although gene expression-regulatory sequences that function in mice may not function in NHPs even if they are derived from the NHP brain. Therefore, it is expected that individual gene expression regulatory sequences will be tested in NHPs, and that the results will be made open through databases or similar media.

Since the macaque monkey brain is about 200 times larger than the mouse brain, it is necessary to introduce more viral vectors for gene transfer into specific brain regions (*see* **Note 1**). Moreover, individual differences exist in brain shape and size, because there are no inbred lines of macaques. Thus, vector injections relying only on standard brain atlases are not recommended. Their procedures should be followed as described below to avoid any failure in precious experiments using valuable animals.

1. Using brain coordinates obtained from MRI imaging.
   (a) Using individual brain coordinates calculated from MRI brain structure images.
   (b) Using a navigation system that uses MRI and fiducial markers to estimate the position of the needle tip inside the brain (Fig. 2a; [7]).
   (c) Combing CT real-time imaging with MRI structural information to perform the injection [15].
2. Injecting while performing MRI real-time imaging [16, 17].

   It is necessary to prepare the MR-compatible injection setup (i.e., cylinder and cannula grid) inside the MRI machine.

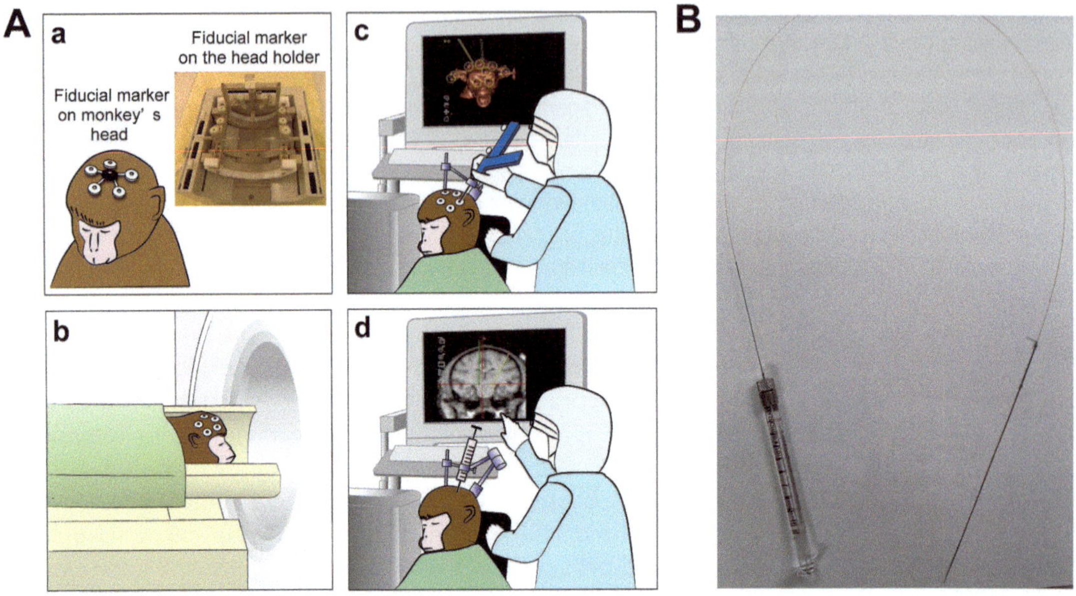

Fig. 2 Injection method for viral vectors into the primate brains. (**a**) Injection method using an MRI navigation system (1-b). (1) Fiducial markers are fixed to the skull or MR-compatible head holder and (2) MRI imaging is performed. (3) In the operating room, registration is performed between the position of the marker on the MRI and the position of the marker in real space using a 3D camera. (4) After the calibration, estimating the position of the target site and syringe needle tip in real space becomes possible, which helps the accurate injection. (**b**) An example of an injectrode. A silica tube connected to a micorsyringe and a tungsten electrode is placed inside a polyimide tube and fixed with epoxy resin

By using contrast agents such as gadolinium and magnesium, it is possible to inject while visualizing the spread of the virus solution as well as the injection needle.

3. Using injectrode, a combination of an electrode and an injection cannula (Fig. 2b).

   It allows the virus solution to be introduced accurately into the site with the known neuronal activity [18].

4. Performing brain area mapping by electrophysiological experiment before the injection [19].

   For details, see the previous chapter.

5. Injecting under visual control after peeling the dura [20].

   After injection, replacing the dura with an artificial dura is necessary, and managing the artificial dura is important for long-term experiments.

In this chapter, we will introduce the method of local intracerebral injection of lentiviral vectors through a microsyringe with the aid of MRI structural images (1-a).

2 Materials

1. Anesthetics (i.e., ketamine hydrochloride and propofol) and animal anesthesia apparatus.
2. Analgesics (including lidocaine).
3. Other pharmaceuticals including antibiotics.
4. Sterile distilled water and sterile saline solution.
5. Sterilized surgical instruments.
6. A sufficient number of sterilized hygiene products (i.e., gauze, cotton balls, drapes).
7. Sterilized hygiene products for the surgeon (i.e., gowns, gloves).
8. Ventilator.
9. MR-compatible head holder (Fig. 2a).
10. Stereotaxic frames for primates.
11. Manipulator.
12. Syringe holder or syringe pump.
13. Microsyringe.
14. Dental hand drill.

3 Methods

3.1 MRI Imaging and Determination of Target Coordinates

1. Under deep anesthesia with an appropriate anesthetic regent, fix the monkey to an MR-compatible head holder (the ear bars contain contrast medium), and take MRI structural brain images.
2. Using an MRI image viewer, check that the line connecting the center of the ear bars (interaural line) is perpendicular to the midline of the brain (if this is not the case, the monkey is not properly fixed in the head holder), and record the XYZ coordinates of the intersection (zero point of brain coordinates).
3. Read the XYZ coordinates of the injection target point and calculate the XYZ distance from the zero point. Determine the angle of the syringe needle insertion and calculate the distance between the point to touch the brain surface and the injection target point.

   * These processes are usually carried out before the date of surgery (for example, when the head holder used during MRI differs from the one used during surgery). However, we sometimes carry out MRI imaging just before the surgery and perform the injection experiment without changing the head holder to avoid any slight misalignment.

3.2 Preparation for Surgery

1. Attach a syringe-equipped manipulator to the stereotaxic frame and record the manipulator coordinates when the syringe tip is at the center of the ear bars of the surgical head holder. We use unique coordinate indicator parts to prevent the tip of the syringe from being damaged.
2. Calculate the manipulator coordinates for the injection target. Check that the syringe tip can reach the target point without problems and that when the syringe is pulled up, the tip is outside the brain and does not cause problems in viral vector loading.
3. After recording the manipulator position so that it can be returned to the exact location, remove the manipulator from the stereotaxic frame and place it so as not to interfere with the surgical operation.

3.3 Surgical Procedures

Compared to rodents, monkeys are more susceptible to infection. Therefore, surgical procedures on monkeys require more rigorous aseptic techniques. *See also* **Note 2** for points to note during the surgery.

- Disinfect the operating table and equipment.
- Have a sufficient number of high-quality sterilized instruments.
- Prepare sterilized reagents such as distilled water and saline solution and appropriate medical reagents such as antibiotics.

1. Shave the surgical field and surrounding area of the monkey with a razor or electric clippers. Use hair removal cream to remove the hair altogether. We perform this process before MRI imaging.
2. Under deep anesthesia and respiratory management, attach the monkey to a stereotaxic frame, disinfect the surgical field using iodine and other agents, dry lightly, and then wipe clean with disinfectant ethanol cotton. Place a sterile drape over the head and body surface, and cut a hole so that only the surgical field is visible. A drape with a window is often convenient.
3. Prepare the surgeon's equipment. The surgeon should wear a surgical globe, mask, cap, sterile gown, foot covers, etc. Wash and disinfect their hands and arms with medicated soap before wearing surgical sterile gloves. The surgeon should not touch anything other than the sterilized instruments and surgical field.
4. Under hygienic control, the skin is incised to expose the skull. After the periosteum and other tissues on the skull are carefully peeled away using hydrogen peroxide, etc., make a hole including the predetermined coordinates of the insertion site in the skull using a hand drill so that the syringe needle can be inserted into the target area of the brain.

3.4 Injecting the Viral Vector Solution

1. Attach the manipulator with the syringe to the stereotaxic frame. Move the tip of the syringe needle just above the insertion point and check that there are no bones or other objects in the way of the insertion.
2. Fill the syringe with the viral vector solution stored at 4 °C. The viral vector solution can be loaded into the syringe from the tip or the back end, so choose the appropriate method. Filling the syringe with liquid in advance is advisable to prevent air bubbles from entering the syringe. In the case of lentiviral vectors, reducing the time the vector solution is at room temperature as much as possible is desirable.
3. Lower the tip of the syringe until it adheres to the dura mater at the insertion point, and then penetrate the dura mater. Move the syringe up and down several times to prevent the brain from moving down during insertion.
4. Move the tip of the syringe needle to the target site, referring to the XYZ coordinates of the target site and the distance from the brain surface decided in advance.
5. Inject the viral vector at an appropriate speed and volume (*see* **Note 3**). After leaving it for 5–20 min after injection, slowly withdraw the syringe needle.
6. After necessary procedures, such as installing artificial bone according to the size of the bone opening, suture the skin.

3.5 Postoperative Management

When returning the monkey to the recovery cage after surgery, position the monkey's head at a suitable distance from the walls. It is essential to keep checking the monkey's condition, such as body temperature and breathing. If necessary, give the monkey an infusion of electrolytes, glucose, amino acids, and vitamins. Do not give water until the animal is standing up continuously. It is also a basic rule to fast on the day of the surgery. In addition, antibiotics will be administered appropriately (i.e., from 2 days before to 2 days after the surgery).

3.6 Confirmation of Virus Vector Introduction and Expression of Introduced Genes

In primate experiments, which often take a long time, it is preferable to be able to confirm that sufficient gene expression is appropriately occurring in the target area before carrying out stimulation or measurement experiments.

1. Visualization of the introduced virus solution.

   It is possible to visualize the contrast agent in the injection solution, such as gadolinium or manganese, by MRI immediately after injection, which makes it possible to visualize the spread of the viral solution and obtain important information on whether the injection site was appropriate. We mainly use manganese chloride at a final concentration of 0.2 mM for this purpose (Fig. 3a). However, note that this method does not guarantee gene expression.

2. Confirming expression of the transgene using fluorescent proteins.

   If the target is in the cortex near the brain surface, the gene expression can be confirmed in the target area by checking the fluorescence of the fluorescent protein expressed with a functional probe using a transparent artificial dura [16, 20, 21]. If the target is in the deep brain, the amount of fluorescence can be checked by detecting fluorescence using fiber photometry [22]. These procedures provide valuable information for selecting the target for stimulation and measurement.

3. Visualization of Transgene Expression Using Brain Imaging.

   In the chemogenetic approach using DREADDs (Designer Receptors Exclusively Activated by Designer Drugs), the recent development of PET ligands that recognize DREADD receptors has made it possible to perform in vivo imaging of the expression kinetics of DREADD receptors (Fig. 3b; [15, 23]). This technique is beneficial because it allows noninvasive confirmation of the expression status in the whole brain before the stimulation experiments, and additional injections can be selected as necessary. It has also become possible to identify the projection targets of the

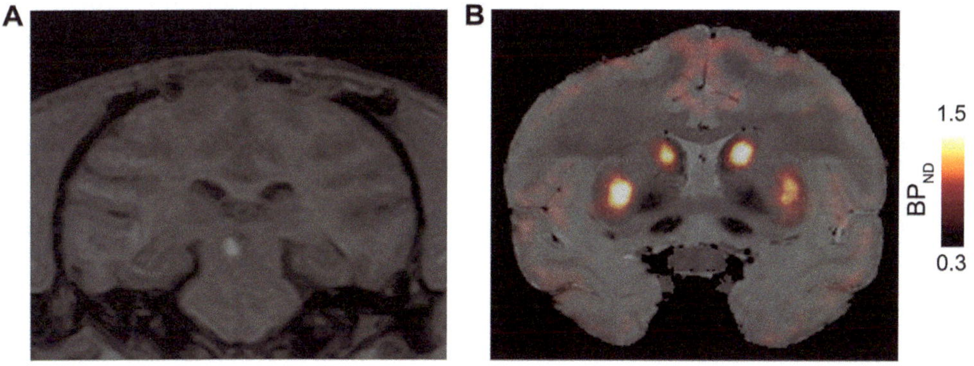

Fig. 3 Confirmation of virus vector introduction and expression of introduced genes. (**a**) Visualization of the administrated viral solution using a contrast agent. The virus vector solution-added manganese chloride solution to achieve a final concentration of 0.2 mM was injected into the macaque raphe nucleus, and MRI imaging (0.3 T) was performed immediately after injection. (**b**) Visualization of the expression of DREADD receptors using PET scan. PET images overlaid on MR images showing [11C]DCZ-specific binding taken 2 months after the DREADD expressing vector injection into the macaque striatum. The value of binding potential is expressed as the regional binding potential relative to a non-displaceable radioligand in tissue (BPND)

gene transfer site by visualizing the DREADD receptors expressed on the axons of the transduced neurons. These improvements are promising for analyzing the neural circuits of primates, and it is hoped that similar methods will be applied to other probes in the future. In addition, the development of a method for detecting gene expression using MRI is also eagerly awaited.

4 Notes

1. Reducing the variation in expression efficiency and side effects for each experiment /individual is important in primate research, where the number of individuals is limited and experiments often take a long time. Due to the primate's larger brain volume, requiring a large amount of viral vector in a single injection, there are fewer cases where the same lot can be used for multiple monkeys. Therefore, the difference in quality between vector lots and the accuracy of the titer calculation becomes more serious in monkey research. It is recommended that a quality check, such as protein staining and confirmation of in vitro infectivity, be performed on the vector for injection.

2. The following points should be noted during surgery.
   - It is important to closely monitor the monkey's various physiological functions under anesthesia.
   - Use a heating pad or something similar to prevent the monkey's body temperature from dropping due to anesthesia.
   - Use eye ointment to prevent the eyes from drying out.
   - Prevent the surgical field from drying out by covering the area with gauze moistened with sterile saline solution or dripping the saline solution.
3. There is currently little consensus regarding the number of injection points, the amount of virus injected per point, and the concentration of the vector solution. In general, however, it is better to inject a small amount of the viral vector at many points to make the vector concentration as uniform as possible because injecting a large amount of the virus at a single point will cause unexpected damage, such as by neuroinflammation from a high concentration of the vector at the injection center [7]. We generally inject about 1 μl of virus into the cortex and 1-5ul into the subcortex, depending on the target area. Because of the size of the target area, we generally set multiple injection points in one track. The distance between injection tracks is generally 1.5 mm or more. On the other hand, it has recently been reported that injecting a large amount of AAV vector solution at high pressure into a single point by convention-enhanced delivery (CED) technique enables the introduction of genes relatively uniformly over an extensive area [16, 17]. Although this method is unsuitable for precise control of the transduced area, it is promising for introducing genes into a vast area in large primate brains.

Conflict of Interest Statement

The authors have no conflicts of interest to declare that are relevant to the content of this chapter.

Funding

This work was supported by KAKENHI Grant Numbers 22H05154 and 22H05154 (K.I.) and AMED Grant Number wm0625103 (K.I.).

References

1. Chan KY, Jang MJ, Yoo BB et al (2017) Engineered AAVs for efficient noninvasive gene delivery to the central and peripheral nervous systems. Nat Neurosci 20:1172–1179
2. Liu Z, Li X, Zhang JT et al (2016) Autism-like behaviours and germline transmission in transgenic monkeys overexpressing MeCP2. Nature 530:98–102
3. Kato S, Inoue K, Kobayashi K et al (2007) Efficient gene transfer via retrograde transport in rodent and primate brains by an HIV-based vector pseudotyped with rabies virus glycoprotein. Hum Gene Ther 18:1141–1151
4. Kato S, Kuramochi M, Takasumi K et al (2011) Neuron-specific gene transfer through retrograde transport of lentiviral vector pseudotyped with a novel type of fusion envelope glycoprotein. Hum Gene Ther 22:1511–1523
5. Tervo DGR, Hwang BY, Viswanathan S et al (2016) A designer AAV variant permits efficient retrograde access to projection neurons. Neuron 92:372–382
6. Tanabe S, Inoue K, Tsuge H et al (2017) The use of an optimized chimeric envelope glycoprotein enhances the efficiency of retrograde gene transfer of a pseudotyped lentiviral vector in the primate brain. Neurosci Res 120:45–52
7. Tanabe S, Uezono S, Tsuge H et al (2019) A note on retrograde gene transfer efficiency and inflammatory response of lentiviral vectors pseudotyped with FuG-E vs. FuG-B2 glycoproteins. Sci Rep 9:3567
8. Klein C, Evrard HC, Shapcott KA et al (2016) Cell-targeted optogenetics and electrical microstimulation reveal the primate koniocellular projection to supra-granular visual cortex. Neuron 90:143–151
9. El-Shamayleh Y, Horwitz GD (2019) Primate optogenetics: Progress and prognosis. Proc Natl Acad Sci USA 116:26195–26203
10. Stauffer WR, Lak A, Yang A et al (2016) Dopamine neuron-specific optogenetic stimulation in rhesus macaques. Cell 166:1564–1571
11. Mimura K, Nagai Y, Inoue K et al (2022) Chemogenetic activation of nigrostriatal dopamine neurons in freely moving common marmosets. iScience 24:103066
12. Hrvatin S, Tzeng CP, Nagy MA et al (2019) A scalable platform for the development of cell-type-specific viral drivers. elife 8:e48089
13. Nair RR, Blankvoort S, Lagartos MJ et al (2020) Enhancer-driven gene expression (EDGE) enables the generation of viral vectors specific to neuronal subtypes. iScience 13: 100888
14. Graybuck LT, Daigle TL, Sederio-Cortes AE et al (2021) Enhancer viruses for combinatorial cell-subclass-specific labeling. Neuron 109: 1449–1464
15. Nagai Y, Miyakawa N, Takuwa H et al (2020) Deschloroclozapine: a potent and selective chemogenetic actuator enables rapid neuronal and behavioral modulations in mice and monkeys. Nat Neurosci 23:1157–1167
16. Yazdan-Shahmorad A, Diaz-Botia C, Hanson TL et al (2016) A large-scale interface for optogenetic stimulation and recording in non-human primates. Neuron 89:927–939
17. Yazdan-Shahmorad A, Tian N, Kharazia V et al (2018) Widespread optogenetic expression in macaque cortex obtained with MR-guided, convection enhanced delivery (CED) of AAV vector to the thalamus. J Neurosci Meth 293: 347–358
18. Maeda K, Inoue K, Takada M et al (2023) Environmental context-dependent activation of dopamine neurons via putative amygdala-nigra pathway in macaques. Nat Commun 14: 2282
19. Inoue K, Takada M, Matsumoto M (2015) Neuronal and behavioral modulations by pathway-selective optogenetic stimulation of the primate oculomotor system. Nat Commun 6:8378
20. Ruiz O, Lustig BR, Nassi JJ et al (2013) Optogenetics through windows on the brain in the non-human primate. J Neurophysiol 110: 1455–1467
21. Ju N, Jiang R, Macknik SL et al (2018) Long-term all-optical interrogation of cortical neurons in awake-behaving non-human primates. PLoS Biol 16:e2005839
22. Tamura K, Takeda M, Setsuie R et al (2017) Conversion of object identity to object-general semantic value in the primate temporal cortex. Science 357:687–692
23. Oyama K, Majima K, Nagai Y et al (2017) Distinct roles of monkey OFC-subcortical pathways in adaptive behavior. Nat Commun 15:6487

Chapter 5

Optimized Lentiviral Vector Production Using Dual-Pseudotyping with VSV-G and Sendai Virus HN Glycoproteins for Enhanced Gene Delivery in Diverse Cell Types

Bat-Erdene Jargalsaikhan, Masanaga Muto, and Masatsugu Ema

Abstract

Highly efficient gene delivery using lentiviral vectors is beneficial not only for basic cell science research but also for human gene therapy applications. Lentiviral pseudotyping with vesicular stomatitis virus G (VSV-G) is a commonly used method that enables the transduction of various cell types. However, the low expression of low-density lipoprotein-receptor, the primary receptor of VSV-G, limits the efficiency of lentiviral vector transduction. Sendai virus hemagglutinin-neuraminidase glycoproteins recognize terminal sialic acids on the host cell plasma membrane, facilitating viral entry. Here, we describe methods for lentiviral dual-pseudotyping with VSV-G and Sendai virus hemagglutinin-neuraminidase to broaden viral tropism.

Key words Lentiviral vector, Dual-pseudotyping, VSV-G, Sendai virus, Hemagglutinin-neuraminidase

1 Introduction

One of the most promising approaches in gene therapy is transduction with lentiviral vectors (LVs) to deliver transgenes into host cells [1, 2]. Although clinical applications using LVs targeting hematopoietic stem cells (HSCs) and T cells have been successful recently, further improvement in the infectious capacity of LV preparations is needed to expand clinical applications [3–5]. The efficiency of lentiviral transduction depends on the initial interaction between the viral envelope proteins and receptors on the cell plasma membrane [6]. Lentiviral pseudotyping, which incorporates different viral envelope proteins into a single viral particle, is an advanced approach for expanding cell tropism and improving the use of these vectors in gene therapy [7]. Pseudotyped LVs with

Kazuto Kobayashi et al. (eds.), *Lentiviral Vectors: Applications for Basic and Clinical Neuroscience*, Methods in Molecular Biology, vol. 2974, https://doi.org/10.1007/978-1-0716-4807-0_5,

vesicular stomatitis virus glycoprotein G (VSV-G) are commonly used in lentivirus production both ex vivo and in vivo across various fields [3, 8, 9]. Envelope proteins derived from members of the *Paramyxoviridae*, including Sendai virus (SeV), have been used in viral vectors to change viral tropism [10–12]. The hemagglutinin-neuraminidase (HN) and fusion (F) glycoproteins are enveloped in SeV. For the initiation of virus entry into host cells, SeV-HN recognizes and binds to sialylated receptors on the host cell membrane. The SeV-F protein is essential for the fusion process between the viral envelope and the host cell membrane [13–15]. We recently found that the infection efficiency of LVs pseudotyped with VSV-G and SeV-HN (V/HN-LV) was significantly increased in various cell types, including primary cultured cells from humans, cynomolgus monkeys, and mice [16]. The molecular mechanism underlying this improved transduction efficiency involves SeV-HN-mediated recognition of sialic acid on the target cell membrane and the SeV-HN-mediated desialylation of VSV-G on the virus particle, which reduces repulsive electrostatic effects, prevents viral self-aggregation, and enhances the association between the virus and the target cell. Importantly, V/HN-LV efficiently transduces primary human HSCs, indicating the potential therapeutic applications of these LV vectors [16]. Here, we describe the methods for the production and transduction of V/HN-LV. We believe that V/HN dual-pseudotyped LVs have the potential to expand the range of human gene therapy applications, as well as having various applications in the field of basic cell biology.

2 Materials

2.1 Cell Culture

1. 293FT cell line (*see* **Note 1**).
2. Basal medium: Dulbecco's modified Eagle's medium (DMEM) with high glucose levels and phenol red, and without L-glutamine, sodium pyruvate, and HEPES.
3. Serum and supplements: Heat-inactivated fetal bovine serum (FBS), L-glutamine 200 mM (100×), sodium pyruvate 100 mM, and nonessential amino acids solution (NEAA, 100×).
4. Complete medium: DMEM supplemented with 10% FBS, 6 mM L-glutamine, 1 mM sodium pyruvate, and 1× NEAA as shown in Table 1. Stored at 4 °C (*see* **Note 2**).
5. Trypsin-EDTA (0.25%) with phenol red.
6. 1× Phosphate-buffered saline (1× PBS) without Ca and Mg.
7. Others: 10-cm or 15-cm dish, 24-well plate, serological pipette, and pipette tips.

Table 1
Composition, concentration, and volume of each of the components in the complete medium (500 ml)

Component	Stock concentration (if applicable)	Required volume (ml)	Final concentration (in medium)
DMEM	Stock solution	425.0 ml	85%
Heat-inactivated FBS	Stock solution	50.0 ml	10%
L-glutamine	200 mM solution	15.0 ml	6 mM
Sodium pyruvate	100 mM solution	5.0 ml	1 mM
NEAA	100× solution	5.0 ml	1×

2.2 Plasmids

1. CS-CA-NLS-GFP: Transfer plasmid vector expressing NLS-GFP protein originating from CS-CA-GFP (RDB05964) plasmid vector (*see* **Note 3**).
2. pCAG-HIVgp (RDB04394): Plasmid vector encoding the HIV-1 *gag* and *pol* genes.
3. pRSV-Rev (RDB08121): Plasmid encoding the HIV-1 Rev. protein.
4. pCMV-VSV-G (RDB04392): Plasmid expressing vesicular stomatitis virus G glycoprotein.
5. pCAG-khcHN: Plasmid vector carrying the gene encoding human codon-optimized SeV-HN with the Kozak sequence.

2.3 Transfection Reagents

1. 25 kDa linear polyethylenimine (25 kDa L-PEI): Aqueous solution with a concentration of 1.0 μg/μl (*see* **Note 4**).
2. 1× Hanks' balanced salt solution (1× HBSS) with Ca and Mg, but without phenol red.
3. 1.5-ml or 5.0-ml polypropylene tubes.

2.4 LV Collection and Concentration

1. 0.22-μm or 0.45-μm PVDF syringe filter (e.g., Millipore Sigma) and 10–50 ml syringes.
2. Cryovials and ultra-deep freezer (−80 °C or −150 °C).
3. Ultracentrifuge (e.g., Beckman Coulter) with the desired rotor (e.g., SW28, SW32, or SW32Ti) and ultracentrifugation tubes (e.g., Beckman Coulter).
4. 20% sucrose solution prepared in 1× PBS and passed through a 0.22-μm PVDF syringe filter. Stored at 4 °C.

2.5 LV Particle Quantification and Biological Titration

1. HIV-1 p24 ELISA kit (e.g., Lenti-X p24 Rapid Titer Kit).
2. 293FT cell line.
3. 35-μm mesh filter.
4. 2% FBS in 1× PBS.
5. 1× TrypLE select without phenol red.
6. Flow cytometry instruments (e.g., Becton Dickinson).

3 Methods

3.1 LV Production

1. Seed 293FT cells at a density of approximately 33,000 cells/cm^2 in a culture dish with complete medium, and incubate the cells at 37 °C in a 5% CO_2 incubator (*see* **Note 5**).
2. Check the confluency of the cells under a microscope. Ideally, the cells should reach approximately 80% confluency 44–48 h after seeding (Fig. 1), with an estimated density of approximately 250,000 cells/cm^2.
3. Remove the old medium by aspiration. Add exactly 15 ml of fresh complete medium to a 10-cm dish, or 45 ml to a 15-cm dish (*see* **Note 6**).
4. At 2.5–4.0 h after medium replacement, prepare the transfection complex and conduct transfection as follows. Add transfer (CS-CA-NLS-GFP), packaging (pCAG-HIVgp), regulator (pRSV-Rev), VSV-G envelope protein (pCMV-VSV-G), and SeV-HN envelope protein (pCAG-khcHN) plasmid vectors in a ratio of 4:2:1:1:0.1 to a polypropylene tube containing HBSS. Subsequently, add L-PEI solution at a 5:1 ratio of PEI:DNA to the tube (Table 2). Then, immediately vortex the PEI/DNA complex for 10–15 s at full speed and leave to stand at room temperature for 10 min.

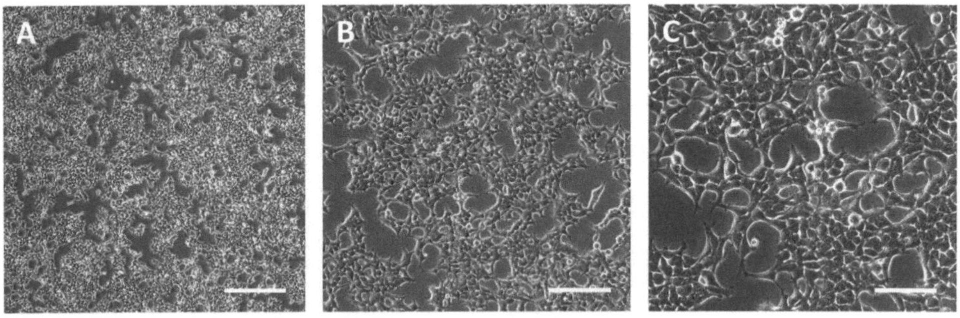

Fig. 1 Representative image of 293FT cells at approximately 80% confluence. Cells were observed under 4× (a), 10× (b), and 20× (c) magnifications. Scale bar: 500 μm (**a**); 200 μm (**b**); and 100 μm (**c**)

Table 2
Components included in the transfection complex for 10-cm and 15-cm dishes

Plasmids and transfection reagents		For 10-cm dish	For 15-cm dish
Plasmids	CS-CA-NLS-GFP (1 μg/μl)	8.0 μg	24.0 μg
	pCAG-HIVgp (1 μg/μl)	4.0 μg	12.0 μg
	pRSV-rev (1 μg/μl)	2.0 μg	6.0 μg
	pCMV-VSV-G (1 μg/μl)	2.0 μg	6.0 μg
	pCAG-khcHN (1 μg/μl)	0.2 μg	0.6 μg
	Total plasmid	16.2 μg	48.6 μg
Buffer	HBSS	900 μl	2700 μl
Transfection reagent	25 kDa L-PEI (1 μg/μl)	81 μl	243 μl
Total transfection complex		997.2 μl	2991.6 μl

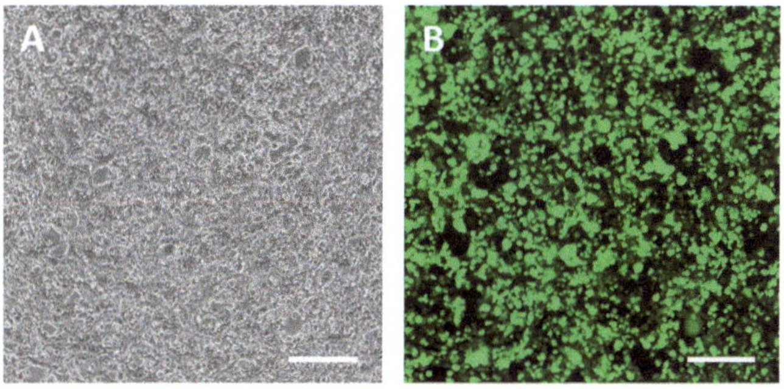

Fig. 2 Representative image of 293FT cells at 48 h post-transfection. (a) Phase-contrast image showing cells at high density. (**b**) Corresponding fluorescence image showing GFP expression Scale bar: 200 μm

5. Pipet the PEI/DNA complex gently, add dropwise onto the medium, and mix by gentle shaking. Incubate at 37 °C with 5% CO_2 overnight.
6. At 18–19 h after transfection, when the cells typically reach near 100% confluency, remove the medium completely by aspiration. Add 10 ml of fresh complete medium to a 10-cm dish, or 25 ml to a 15-cm dish (*see* **Note 6**). Incubate at 37 °C with 5% CO_2 prior to harvesting.
7. Examine the transfected cells using fluorescence microscopy if a visible marker (such as GFP or RFP) is present. Usually, more than 80% of the cells should fluoresce (Fig. 2) 48 h after transfection.
8. At 48–56 h after transfection, carefully collect the supernatant containing recombinant LVs (*see* **Note 7**).

9. Filter the supernatant through a 0.45-μm (or 0.22-μm) PVDF syringe filter under aseptic conditions (*see* **Note 8**).
10. Aliquot the filtered lentiviral supernatant on ice and immediately store at −80 °C (or −150 °C).
11. Alternatively, the filtered lentiviral supernatant can be stored on ice at 4 °C for several hours prior to ultracentrifugation.

3.2 LV Concentration by Ultracentrifugation

1. Add 10 ml of 20% sucrose solution to the ultracentrifugation tube.
2. Carefully overlay the supernatant (up to 20 ml, depending on tube capacity) onto the sucrose, avoiding mixing the two layers.
3. Balance the tubes precisely and place them into the ultracentrifuge rotor.
4. Centrifuge at 50000× *g* for 2 h at 4 °C.
5. After centrifugation, carefully remove the tubes from the rotor and aspirate the supernatant without disturbing the pellet. Discard the supernatant in a biosafety container.
6. Leave the tubes on a paper towel in an inverted position for 3–5 min.
7. Aspirate the remaining supernatant drops.
8. Add 50–500 μl of sterile PBS to the pellet.
9. Cover the tubes with Parafilm and place on ice for 3–4 h.
10. Gently resuspend the pellet by carefully pipetting the solution up and down, taking care to prevent bubble formation.
11. Aliquot the concentrated LVs on ice and immediately store at −80 °C (or −150 °C).

3.3 HIV-1 p24-Based Titration Assay for LV Particle Quantification

1. Prepare serial dilutions of LV samples: 1:100–1:10,000 for unconcentrated and 1:10,000–1:1,000,000 for concentrated samples.
2. Determine p24 levels in the diluted samples using an HIV-1 p24 antigen ELISA kit, according to the manufacturer's instructions (*see* **Note 9**).

3.4 Biological Titer Determination by Flow Cytometry

1. Trypsinize the 293FT cells, then resuspend in complete culture medium. Prepare a suspension of typically 2.5×10^5 cells in 0.5 ml of medium.
2. Add 1–5 μl of LV supernatant to the cells. For concentrated samples, use 10–1000-fold diluted samples.
3. Mix the cells gently and seed into a 24-well plate.
4. Incubate the cells for 48 h.

5. After 48 h, confirm the expression of the fluorescence markers under fluorescence microscopy.
6. Wash the cells with 1× PBS. Add 400 μl of 1× TrypLE select without phenol red and incubate the cells for 2 min at 37 °C.
7. Neutralize the TrypLE by adding 600 μl of 2% FBS in 1× PBS. Immediately, resuspend and filter the suspension through a 35-μm cell strainer.
8. Analyze the cells using flow cytometry.
9. Calculate the biological titer (*see* **Note 10**).

4 Notes

1. 293FT cells are considered potentially biohazardous and require Biosafety Level 2 or higher protection. For additional information regarding the cell line, refer to the manufacturer's (Thermo Fisher Scientific) instructions.
2. Following strict aseptic techniques when preparing and using antibiotic-free cell culture medium is important. Moreover, adding 1% penicillin-streptomycin to the complete medium is optional, depending on laboratory standards.
3. Other transfer plasmid vectors can be used depending on specific research requirements and applications.
4. Add 25 mg of 25 kDa L-PEI to the beaker with a magnetic stirrer and 20 ml of Milli-Q water. Then, add concentrated 1 M HCl dropwise to reach a pH <2.0, and stir for approximately 2.5 h while maintaining the same pH until the PEI is dissolved. Next, adjust the pH to 7.0 by adding 5 M NaOH and transfer to a 100-ml cylinder. Adjust the volume to 25 ml with Milli-Q water, filter through a 0.22-μm PVDF syringe filter, aliquot, and store the L-PEI solution at −25 °C. Avoid freeze/thawing, only thawing just before use and keeping on ice.
5. Passage the 293FT cells at least twice before LV production to ensure healthy, proliferative cells. To reduce the risk of contamination, avoid prolonged maintenance of the cells (no more than 15 passages). Carefully add the cell suspension at the appropriate density to reach the desired confluence on the day of transfection. Specifically, 1.8×10^6 cells for a 10-cm dish and 5.0×10^6 cells for a 15-cm dish.
6. Warm the medium in a 37 °C water bath in advance. Aspirate old medium completely at the edge of the dish. Then, immediately add fresh complete medium slowly and carefully to the edge of the dish. Please note that dropping the medium onto

cells or adding the medium too quickly may cause detachment of the cells.

7. Transfer the viral supernatant by pipetting carefully from the dish into a 50-ml tube. If many cells detach during harvesting, the collected supernatant can be centrifuged at 300–500× *g* for 5 min at 4 °C.
8. Remove the syringe plunger, and mount the syringe barrel with a 0.45-μm (or 0.22-μm) PVDF filter. Carefully transfer the viral supernatant to the syringe barrel mounted with the PVDF filter. Then, carefully insert the syringe plunger and push it slowly to pass the viral supernatant through the filter. Collect the viral supernatant in a 50-ml tube.
9. Generally, 1 ng p24 is equivalent to 1.25×10^7 LV particles. Typically, the protocol results in titers of at least $1 \text{x} 10^9$ physical particles per ml (PP/ml) for an unconcentrated viral supernatant.
10. This cost-effective protocol routinely yields approximately 1×10^7 transduction units per ml (TU/ml) for an unconcentrated viral supernatant, relying on the dual-pseudotyping VSV-G and SeV-HN envelope proteins, without enhancers such as polybrene or protamine sulfate. When calculating the biological titer use the equation below:

$$\text{TU/ml} = \frac{(\text{Initial number of cells} \times \text{Percentage of fluorescent cells})/100}{\text{Volume of viral medium (ml)}}$$

Acknowledgments

This work was supported by AMED under Grant Number JP233fa627008. We thank Edanz (https://jp.edanz.com/ac) for editing a draft of this manuscript.

Funding This work was supported by AMED Grant Number JP223fa627008.

Conflict of Interest Statement The authors have no conflicts of interest to declare that are relevant to the content of this chapter.

References

1. Ginn SL, Mandwie M, Alexander IE et al (2024) Gene therapy clinical trials worldwide to 2023—an update. J Gene Med 26. https://doi.org/10.1002/jgm.3721
2. Milone MC, O'Doherty U (2018) Clinical use of lentiviral vectors. Leukemia 32:1529–1541. https://doi.org/10.1038/S41375-018-0106-0
3. Wolff JH, Mikkelsen JG (2022) delivering genes with human immunodeficiency virus-derived vehicles: still state-of-the-art after 25 years. J Biomed Sci 291(29):1–22. https://doi.org/10.1186/S12929-022-00865-4
4. Amirache F, Lévy C, Costa C et al (2014) Mystery solved: VSV-G-LVs do not allow efficient gene transfer into unstimulated T cells, B cells, and HSCs because they lack the LDL receptor. Blood 123:1422–1424. https://doi.org/10.1182/BLOOD-2013-11-540641
5. Tucci F, Scaramuzza S, Aiuti A, Mortellaro A (2021) Update on clinical ex vivo hematopoietic stem cell gene therapy for inherited monogenic diseases. Mol Ther 29:489–504. https://doi.org/10.1016/j.ymthe.2020.11.020
6. Morizono K, Chen IS (2011) Receptors and tropisms of envelope viruses. Curr Opin Virol 1:13–18. https://doi.org/10.1016/j.coviro.2011.05.001
7. Duvergé A, Negroni M (2020) Pseudotyping lentiviral vectors: when the clothes make the virus. Viruses 12:1–23. https://doi.org/10.3390/v12111311
8. Ghassemi S, Durgin JS, Nunez-Cruz S et al (2022) Rapid manufacturing of non-activated potent CAR T cells. Nat Biomed Eng 6:118–128. https://doi.org/10.1038/s41551-021-00842-6
9. Finkelshtein D, Werman A, Novick D et al (2013) LDL receptor and its family members serve as the cellular receptors for vesicular stomatitis virus. Proc Natl Acad Sci 110: 7306–7311. https://doi.org/10.1073/pnas.1214441110
10. Kobayashi M, Iida A, Ueda Y, Hasegawa M (2003) Pseudotyped lentivirus vectors derived from simian immunodeficiency virus SIVagm with envelope glycoproteins from paramyxovirus. J Virol 77:2607–2614. https://doi.org/10.1128/jvi.77.4.2607-2614.2003
11. Munday RJ, Coradin T, Nimmo R et al (2022) Sendai F/HN pseudotyped lentiviral vector transduces human ciliated and non-ciliated airway cells using α 2,3 sialylated receptors. Mol Ther- Methods Clin Dev 26:239–252. https://doi.org/10.1016/j.omtm.2022.07.002
12. Griesenbach U, Inoue M, Meng C et al (2012) Assessment of F/HN-Pseudotyped lentivirus as a clinically relevant vector for lung gene therapy. Am J Respir Crit Care Med 186:846–856. https://doi.org/10.1164/rccm.201206-1056OC
13. Iida A, Inoue M (2013) Concept and technology underlying Sendai virus (SeV) vector development. In: Sendai Virus Vector. Springer, Tokyo, pp 69–89
14. Matveeva OV, Shabalina SA (2020) Prospects for using expression patterns of paramyxovirus receptors as biomarkers for oncolytic Virotherapy. Cancers (Basel) 12:3659. https://doi.org/10.3390/cancers12123659
15. Suzuki T, Portner A, Scroggs RA et al (2001) Receptor specificities of human Respiroviruses. J Virol 75:4604–4613. https://doi.org/10.1128/JVI.75.10.4604-4613.2001
16. Jargalsaikhan B-E, Muto M, Been Y et al (2024) The dual-Pseudotyped Lentiviral vector with VSV-G and Sendai virus HN enhances infection efficiency through the synergistic effect of the envelope proteins. Viruses 16: 827. https://doi.org/10.3390/v16060827

Chapter 6

Preparation of Lentiviral Vectors for the Generation of Transgenic Cynomolgus Monkeys

Masanaga Muto and Masatsugu Ema

Abstract

Nonhuman primates (NHPs) are invaluable for modeling human diseases because of their high physiological and genetic similarity to humans. Lentiviral vectors, which offer significant advantages in terms of gene delivery and stable in vivo expression, are widely used to generate transgenic NHP models. The protocol detailed herein provides a comprehensive method for preparing lentiviral vectors for the generation of transgenic cynomolgus monkeys. We describe the procedures for packaging lentiviral vectors, including the transfection and enrichment of lentiviral particles. To optimize lentiviral production and ensure high-titer viral preparations, we utilized Sendai virus hemagglutinin-neuraminidase glycoproteins to enhance infection efficiency. This protocol is designed to support the production of high-quality lentiviral vectors, which are essential for achieving robust, uniform transgene expression in transgenic cynomolgus monkeys to advance human disease modeling and therapeutic research.

Key words Lentiviral vector, Cynomolgus monkey, Transgenic animal, Sendai virus hemagglutinin-neuraminidase, Pseudotyping

1 Introduction

The development of transgenic animal models has been a foundation of biomedical research, enabling the study of in vivo gene function and the modeling of human diseases [1]. Rodent models, such as mice and rats, have been extensively used because of the well-established techniques for their genetic manipulation, including DNA microinjection and homologous recombination in embryonic stem cells [2–8]. However, these models are often insufficient in recapitulating human diseases because of the significant evolutionary distance between rodents and humans [9]. This gap can lead to limitations in modeling human-specific diseases, such as AIDS, influenza, and various lung disorders [9]. Moreover, the pathological and behavioral phenotypes observed in rodent

Kazuto Kobayashi et al. (eds.), *Lentiviral Vectors: Applications for Basic and Clinical Neuroscience*, Methods in Molecular Biology, vol. 2974, https://doi.org/10.1007/978-1-0716-4807-0_6,

genetic models frequently differ from those of human conditions, including discrepancies in tumorigenesis, neurodegenerative diseases such as Parkinson's disease [9], and autosomal dominant polycystic kidney disease [10].

Nonhuman primates (NHPs) offer a closer approximation of human physiology and pathology because of their close evolutionary proximity to humans. Among NHPs, Old World monkeys such as cynomolgus monkeys are particularly valuable for disease modeling and therapeutic research [11]. These primates exhibit similarities in organ size and structure and disease progression, making them suitable models for studying the pathogenesis of conditions such as stroke [12, 13], Parkinson's disease [14, 15], and Huntington's disease [16]. Despite their utility, the generation of transgenic cynomolgus monkeys has remained a challenge because of toxicity issues associated with lentiviral transduction.

Lentiviral vectors (LVs) are a powerful tool for generating transgenic animal models because of their ability to stably integrate transgenes into host genomes and their broad tropism [17, 18]. While the use of LVs has been well established in gene therapy and transgenic rodent production [17], their application in nonhuman primates has required further refinement. To improve the efficacy of LVs, it is necessary to optimize their envelope proteins [19]. One approach that can significantly enhance the infectivity and expand the cell tropism of LVs is pseudotyping, which involves the incorporation of different viral envelope proteins into lentiviral particles [19].

A recent paper by our group demonstrated the potential of Sendai virus (SeV) envelope proteins to enhance LV infection efficiency [20]. SeV envelope proteins, including hemagglutinin-neuraminidase (HN), bind to sialylated receptors on target cells and play a critical role in viral entry [21, 22]. The use of LVs pseudotyped with both vesicular stomatitis virus glycoprotein G (VSV-G) and SeV-HN (V/HN-LVs) has been shown to improve transduction efficiency across various cell types, including primary cells from cynomolgus monkeys [20]. By increasing the infection efficiency per viral particle, this approach allows the number of lentiviral particles required for infection to be reduced, thereby minimizing toxicity associated with the transduction process [20].

In this protocol paper, we provide a detailed method for preparing LVs to generate transgenic cynomolgus monkeys. We describe the procedures for packaging LVs, the optimization of the production process using SeV envelope proteins to enhance infection efficiency, and the application of these vectors for the creation of transgenic primates. This protocol aims to support the generation of high-quality transgenic cynomolgus monkeys with robust, uniform transgene expression, thus advancing research involving human disease modeling and therapeutic development.

2 Materials

2.1 Cell Culture

1. Dulbecco's modified Eagle's medium (DMEM).
2. Fetal bovine serum (FBS).
3. Penicillin-streptomycin.
4. DMEM complete medium: DMEM supplemented with 10% FBS and 1X penicillin-streptomycin.
5. 0.25% Trypsin-EDTA.
6. Cell culture plates: 24-well plate, 6-well plate, 10 cm plate, 15 cm plate.
7. Cell line: Lenti-X™ 293 T cell line (632,180, Takara).

2.2 Plasmids

1. CS-CA-NLS-Kusabira Orange (originating from CS-CA-GFP plasmid, RDB05964, RIKEN BRC).
2. pCAG-HIVgp (RDB04394, RIKEN BRC).
3. pCMV-VSV-G-RSV-Rev (RDB04393, RIKEN BRC).
4. pCAG-SeV-HN (wild-type strain of Sendai virus Z hemagglutinin-neuraminidase).

2.3 Transfection

1. TransIT®-293 Transfection Reagent (MIR2700, Mirus Bio).
2. Opti-MEM™ I Reduced Serum Medium (31985–062, Gibco).
3. Hank's balanced salt solution (HBSS).
4. 25 kDa linear polyethylenimine (PEI) (25 kDa L-PEI).

2.4 Collection and First Concentration of Lentiviral Supernatant

1. Corning® 250 mL Vacuum Filter/Storage Bottle System, 0.22 μm Pore.
2. Vivaflow® 50 100,000 MWCO PES (Sartorius).

2.5 Ultracentrifugation

1. Optima L-90 K Ultracentrifuge (Beckman Coulter).
2. SW 32 Ti Rotor (369,650, Beckman Coulter).
3. 20% Sucrose solution/PBS: Add sucrose powder to PBS to make a 20% (w/v) solution.
4. Centrifugation tubes: 31.5 mL Open-Top Thinwall Polypropylene Konical Tube (Beckman Coulter), 25 mm Diameter Conical Tube Adapter (Beckman Coulter).

2.6 Lentiviral Titration (Physical Titer)

1. Lent-X p24 Rapid Titer Kit (TaKaRa Bio).

2.7 Lentiviral Titration (Biological Titer)

1. 2% FBS/PBS: Add FBS to PBS to make a 2% (v/v) solution.
2. 0.25% Trypsin-EDTA.
3. Cell strainer 40 μm.
4. 5 mL polystyrene round tube (Falcon).
5. Flow cytometer (Becton Dickinson).

2.8 Lentiviral Transduction in Cynomolgus Monkey Embryos

1. Leuplin (Takeda Chemical Industries).
2. Micro infusion pump (iPRECIO SMP-200, ALZET Osmotic Pumps Co.).
3. Human follicle-stimulating hormone (hFSH; Asuka Pharmaceutical Co.).
4. Human chorionic gonadotropin (hCG; Puberogen, Nippon Zenyaku Kogyo Co.).
5. Alpha modification of Eagle's medium (MP Biomedicals LLC).
6. Serum Substitute Supplement (SSS; Irvine Scientific).
7. Hyaluronidase (Sigma Chemical Co.).
8. m-TALP: Modified Tyrode solution, with 0.1% lactate, 0.001% pyruvate, 0.3% bovine serum albumin, and 0.24% HEPES.
9. Glass needle for Intracytoplasmic sperm injection (ICSI) (Humagen Fertility Diagnostics).
10. CMRL Medium-1066 (Invitrogen).
11. Bovine serum.
12. 0.25 M Sucrose in m-TALP.

3 Methods

3.1 Lentiviral Production and Concentration

1. Culture Lenti-X™ 293 T cell line in DMEM complete medium at 37 °C in a 5% CO_2 humidified atmosphere.
2. Seed the cells 1 day before transfection to achieve ~80% confluence (seed 15×10^6 cells per 15 cm plate). Prepare five 15 cm plates for transfection (*see* **Note 1**). Use 30 mL of medium per plate.
3. On the day of transfection, add CS-CA-NLS-Kusabira Orange, pCAG-HIVgp, pCMV-VSV-G-RSV-Rev, and pCAG-SeV-HN plasmids to Opti-MEM™ I Reduced Serum Medium as detailed in Table 1 (*see* **Note 2**).
4. Add TransIT®-293 Transfection Reagent to the Opti-MEM/DNA mixture at a 3:1 ratio of TransIT:DNA (*see* **Note 3**). Vortex the solution and incubate at room temperature for 10 min.

Table 1
Amounts of plasmids per 15 cm plate (using TransIT®-293 Transfection Reagent)

Plasmids and transfection reagents		Amounts for one 15 cm plate
Plasmids	CS-CA-NLS-Kusabira Orange	14 μg
	pCAG-HIVgp	7 μg
	pCMV-VSV-G-RSV-Rev	7 μg
	pCAG-SeV-HN	7 μg
	Total plasmids	35 μg
Buffer	Opti-MEM	1500 μL
Transfection reagent	TransIT	105 μL

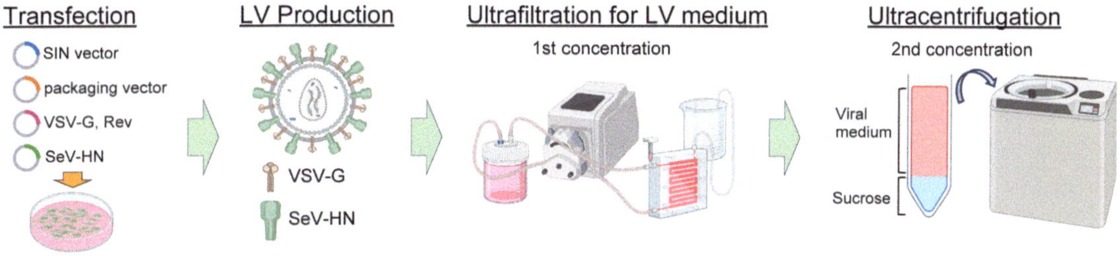

Fig. 1 Scheme for preparing lentiviral vectors. Plasmids for producing LVs are transfected into Lenti-X 293 T cells. The envelope proteins VSV-G and SeV-HN are incorporated into the viral particles. The collected LV medium is then concentrated using ultrafiltration (first concentration) and ultracentrifugation (second concentration) systems

5. Pipette the Opti-MEM/DNA/TransIT mixture, add dropwise onto the cultured cells, and then gently swirl to mix (Fig. 1).
6. After 18 h, replace the medium with 30 mL fresh medium. Collect the medium 48 h after the medium change (Fig. 1), filter it through a 0.22 μm filter, and reserve ~100 μL for biological titer assessment.
7. On the day of the medium collection, concentrate the 150 mL viral medium (30 mL medium for five plates) using a Vivaflow filtration system to reduce the volume to ~20 mL (Fig. 1) (*see* **Note 4**). Reserve ~100 μL of the concentrated medium for biological titer assessment.
8. Add 10 mL of 20% sucrose solution to an Open-Top Thinwall Polypropylene Konical Tube (358,126, Beckman Coulter). Overlay ~20 mL concentrated viral medium on the sucrose solution, and place the tubes with the adapters in centrifuge buckets (Fig. 1).
9. Centrifuge at 17,200 rpm at 4 °C for 2 h.

10. After centrifugation, carefully remove the viral medium and sucrose solution. Invert the tube on tissue paper for 1–2 min to ensure all liquid is removed.
11. Resuspend the semi-dried virus pellet in 35 μL of CMRL medium by thorough pipetting. Aliquot 3 μL of the concentrated virus per tube, and store at −150 °C. Reserve 3 μL of the concentrated medium for biological titer assessment.

3.2 Assessment of Biological Titer and Transduction of Produced LVs Into Cynomolgus Monkey Embryos

1. Dilute the LV medium in Lenti-X™ 293 T cell culture medium as follows: 1:10 to 1:100 for the original LV medium (from **Step 6** of Subheading 3.1, **Step 1** or **Step** 7 of Subheading 3.1, **Step 2**); 1:1000 to 1:4000 for the concentrated LV medium (from **Step 1** of Subheading 3.2, **Step 1** or **Step 5** of Subheading 3.2, **Step 2**); and 1:50,000 to 1:400,000 for the ultracentrifuged LV medium (from **Step 5** of Subheading 3.2, **Step 1** or **Step 9** of Subheading 3.2, **Step 2**).
2. Trypsinize Lenti-X™ 293 T cells and resuspend them in culture medium. Mix 2.5×10^5 cells with 500 μL diluted LV medium in each well of a 24-well plate.
3. After 48 h, remove the medium and add 50 μL Trypsin to each well, followed by incubation for 4 min at 37 °C. Add 450 μL 2% FBS/PBS, thoroughly resuspend the cells, filter through a 40 μm cell strainer, and collect the cell suspension in a 5 mL round polystyrene tube (352,052, Falcon).
4. Quantify the fluorescent protein-positive cells by flow cytometry, and calculate the biological titer (transduction units per milliliter; TU/mL) of LVs using the following equation:

$$\mathrm{TU/mL} = \frac{(\text{initial number of transduced cells} \times \text{percentage fluorescence}^{+}\ \text{cells})/100}{\text{Volume of viral medium (mL)}}$$

5. The biological titer of the concentrated LVs should exceed 1.0×10^{10} TU/mL (*see* **Note 5**).
6. Perform cynomolgus monkey ovarian stimulation and oocyte collection as previously described [23–26]. Collect oocytes 40 h post-hCG treatment via laparoscopic follicular aspiration. Incubate oocytes in Alpha modification of Eagle's medium with 10% SSS at 38 °C in 5% CO_2 for 1–2 h. Transfer the oocytes to m-TALP without hyaluronidase, and store at 38 °C in 5% CO_2 until further use. Classify oocytes by stage as follows: germinal vesicle, metaphase I, metaphase II, or degenerate.
7. Thaw 3 μL of concentrated LVs stored at −150 °C. Pipette the LV medium 150 to 200 times to dissociate viral particles. Centrifuge at ~12,000 rpm for a few seconds, then transfer

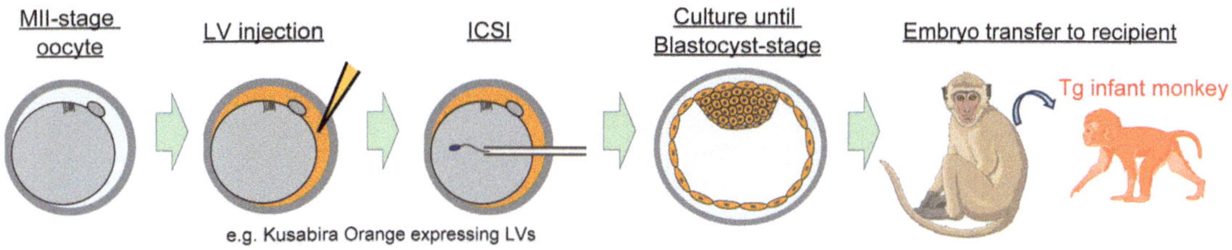

Fig. 2 Scheme for lentiviral vector injection for the generation of transgenic cynomolgus monkeys
The concentrated LVs are injected into the perivitelline space of cynomolgus monkey MII-stage oocytes using a micromanipulator. Four hours after LV injection, ICSI is performed (embryonic day 0; E0), and the embryos are cultured until the blastocyst stage (E8). The lentivirus-transduced E8 blastocysts are then transferred to a pseudopregnant monkey (recipient) to obtain transgenic (Tg) infant monkeys

1 μL of supernatant to a new tube. Dilute the LV medium to 1.0×10^9 TU/mL with CMRL Medium-1066 with 20% bovine serum (see Note 6).

8. Select metaphase II-stage oocytes with the first polar body, and transfer them to 0.25 M sucrose/m-TALP-HEPES. Inject the diluted LVs into the perivitelline space of oocytes using a micromanipulator (Fig. 2). After 4 h, perform intracytoplasmic sperm injection (ICSI).
9. Perform Intracytoplasmic sperm injection (ICSI) in metaphase II-stage oocytes in mTALP-HEPES as previously described [23–26]. Inject fresh sperm into the oocyte cytoplasm, avoiding the polar body. After ICSI, culture embryos in CMRL Medium-1066 with 20% bovine serum at 38 °C in 5% CO_2 and 5% O2 until the blastocyst stage (Fig. 2).
10. Once embryos reach the expanded blastocyst stage, transfer one or two embryos into each female recipient (Fig. 2). Confirm pregnancy via ultrasonography 30 days after ICSI.
11. Assess fluorescent protein expression levels in placental tissues, hair follicles, or other tissues obtained after birth.

4 Notes

1. While five 15 cm plates are specified in this protocol, the number of plates can be increased to obtain high-titer LVs.
2. While a lentiviral production using TransIT®-293 Transfection Reagent is specified here, PEI can also be employed as a cost-effective alternative. If using PEI, follow these steps: Prepare 25 to 30 plates, each with 25 mL of medium. To prepare the PEI solution, dissolve 25 kDa linear PEI in deionized water to a concentration of 1 mg/mL. On the day of transfection, add CS-CA-NLS-Kusabira Orange, pCAG-HIVgp, pCMV-VSV-G-RSV-Rev, and pCAG-SeV-HN plasmids to HBSS as detailed in Table 2. Add PEI solution to the HBSS/DNA mixture at a

Table 2
Amounts of plasmids per 15 cm plate (using PEI solution for transfection)

Plasmids and transfection reagents		Amounts for one 15 cm plate
Plasmids	CS-CA-NLS-Kusabira Orange pCAG-HIVgp pCMV-VSV-G-RSV-Rev pCAG-SeV-HN Total plasmids	14 μg 7 μg 7 μg 7 μg 35 μg
Buffer	HBSS	1500 μL
Transfection reagent	25 kDa linear PEI	105 μL

3:1 ratio of PEI:DNA. Vortex the solution, and incubate at room temperature for 10 min. Pipette the HBSS/DNA/PEI mixture, add it dropwise onto the cultured cells, and then gently swirl to mix. After 18 h, replace the medium with 30 mL fresh medium. Collect the medium 48 h after the medium change, filter it through a 0.22 μm filter, and reserve ~100 μL for biological titer assessment. On the day of medium collection, concentrate ~625 mL viral medium (25 mL medium for 25 plates) via ultracentrifugation. Transfer ~38 mL of viral medium to an Open-Top Thinwall Ultra-Clear Tube (344,058, Beckman Coulter), and place the tubes into the centrifuge buckets. Six tubes can be processed per ultracentrifugation run. Centrifuge at 17,200 rpm at 4 °C for 2 h. After centrifugation, carefully remove the supernatant. Invert the tube on tissue paper for 1–2 min to ensure that all liquid is removed. Resuspend the semi-dried virus pellet in 500 μL CMRL medium by thorough pipetting. Use the same CMRL medium to resuspend the virus pellets from all tubes. Repeat Steps 1 to 4 to concentrate the entire volume of viral medium. Reserve ~10 μL of the concentrated medium for biological titer assessment. Add 10 mL of 20% sucrose solution to an Open-Top Thinwall Polypropylene Konical Tube (358,126, Beckman Coulter). Overlay 500 ~ μL concentrated viral medium onto the sucrose solution, and place the tubes with the adapters in centrifuge buckets. Centrifuge at 17,200 rpm for 2 h at 4 °C. After centrifugation, carefully remove the viral medium and sucrose solution. Invert the tube on tissue paper for 1–2 min to ensure that all liquid is removed. Resuspend the semi-dried virus pellet in 35 μL CMRL medium by thorough pipetting. Aliquot 3 μL of the concentrated virus per tube and store at −150 °C. Reserve 3 μL of the concentrated medium for biological titer assessment.

3. Determine the appropriate TransIT:DNA ratio in your laboratory to achieve maximum transfection efficiency.

4. While a Vivaflow filtration system is specified here, ultracentrifugation can also be employed for initial LV concentration. If using ultracentrifugation, follow these steps: Add LV medium to Open-Top Thinwall Ultra-Clear Tube (344,058, Beckman Coulter), and centrifuge at 17,200 rpm for 2 h at 4 °C. After centrifugation, remove the viral medium carefully, and invert the tube on the tissue paper for 1–2 min to remove the liquid completely. Add ~500 μL CMRL medium to the semi-dried virus pellet in the tube, and suspend by pipetting well. Overlay the first concentrated viral medium on 10 mL sucrose solution. Place the tube adapters and the tubes containing the viral medium and sucrose solution in centrifuge buckets. Centrifuge at 17,200 rpm for 2 h at 4 °C. After centrifugation, remove the viral medium carefully, and invert the tube on the tissue paper for 1–2 min to remove the liquid completely. Add 35 μL CMRL medium to the semi-dried virus pellet in the tube, and suspend by pipetting well. Aliquot 3 μL concentrated virus into each tube, and store at −150 °C.
5. To generate high-quality transgenic cynomolgus monkeys with robust transgene expression, ensure that the biological titer of injected LV medium exceeds 1.0×10^9 TU/mL.
6. To reduce toxicity during the LV infection of cynomolgus monkey embryos, dilute the concentrated LV medium 1:10 or more.

Acknowledgments

This work was supported by AMED under Grant Number JP233fa627008. Figures were created at BioRender.com (Fig. 1: BioRender.com/s71g898, Fig. 2: BioRender.com/p89m642). We thank Edanz (https://jp.edanz.com/ac) for editing a draft of this manuscript.

Funding This work was supported by AMED Grant Number JP223fa627008.

Conflict of Interest Statement The authors have no conflicts of interest to declare that are relevant to the content of this chapter.

References

1. Shakweer WME, Krivoruchko AY, Dessouki SM, Khattab AA (2023) A review of transgenic animal techniques and their applications. J Genet Eng Biotechnol 21:55. https://doi.org/10.1186/s43141-023-00502-z
2. Jaenisch R, Mintz B (1974) Simian virus 40 DNA sequences in DNA of healthy adult mice derived from preimplantation blastocysts injected with viral DNA. Proc Natl Acad Sci USA 71:1250–1254. https://doi.org/10.1073/pnas.71.4.1250
3. Gordon JW, Scangos GA, Plotkin DJ, Barbosa JA, Ruddle FH (1980) Genetic transformation of mouse embryos by microinjection of

purified DNA. Proc Natl Acad Sci USA 77: 7380–7384. https://doi.org/10.1073/pnas.77.12.7380
4. Bachmann S, Peters J, Engler E, Ganten D, Mullins J (1992) Transgenic rats carrying the mouse renin gene–morphological characterization of a low-renin hypertension model. Kidney Int 41:24–36. https://doi.org/10.1038/ki.1992.4
5. Mullins JJ, Peters J, Ganten D (1990) Fulminant hypertension in transgenic rats harbouring the mouse Ren-2 gene. Nature 344:541–544. https://doi.org/10.1038/344541a0
6. Brinster RL, Chen HY, Trumbauer M, Senear AW, Warren R, Palmiter RD (1981) Somatic expression of herpes thymidine kinase in mice following injection of a fusion gene into eggs. Cell 27:223–231. https://doi.org/10.1016/0092-8674(81)90376-7
7. Wagner TE, Hoppe PC, Jollick JD, Scholl DR, Hodinka RL, Gault JB (1981) Microinjection of a rabbit beta-globin gene into zygotes and its subsequent expression in adult mice and their offspring. Proc Natl Acad Sci USA 78: 6376–6380. https://doi.org/10.1073/pnas.78.10.6376
8. Doetschman T, Gregg RG, Maeda N, Hooper ML, Melton DW, Thompson S, Smithies O (1987) Targetted correction of a mutant HPRT gene in mouse embryonic stem cells. Nature *330*:576–578. https://doi.org/10.1038/330576a0
9. Harding JD (2017) Nonhuman primates and translational research: progress, opportunities, and challenges. ILAR J 58:141–150. https://doi.org/10.1093/ilar/ilx033
10. Nagao S, Kugita M, Yoshihara D, Yamaguchi T (2012) Animal models for human polycystic kidney disease. Exp Anim 61:477–488. https://doi.org/10.1538/expanim.61.477
11. Niu Y, Shen B, Cui Y, Chen Y, Wang J, Wang L, Kang Y, Zhao X, Si W, Li W et al (2014) Generation of gene-modified cynomolgus monkey via Cas9/RNA-mediated gene targeting in one-cell embryos. Cell 156:836–843. https://doi.org/10.1016/j.cell.2014.01.027
12. Marshall JWB, Ridley RM, Baker HF, Hall LD, Carpenter TA, Wood NI (2003) Serial MRI, functional recovery, and long-term infarct maturation in a non-human primate model of stroke. Brain Res Bull 61:577–585. https://doi.org/10.1016/s0361-9230(03)00214-4
13. Monsein LH, Mathews VP, Barker PB, Pardo CA, Blackband SJ, Whitlow WD, Wong DF, Bryan RN (1993) Irreversible regional cerebral ischemia: serial MR imaging and proton MR spectroscopy in a nonhuman primate model. AJNR Am J Neuroradiol 14:963–970
14. Soderstrom K, O'Malley J, Steece-Collier K, Kordower JH (2006) Neural repair strategies for Parkinson's disease: insights from primate models. Cell Transplant 15:251–265. https://doi.org/10.3727/000000006783982025
15. Barraud Q, Lambrecq V, Forni C, McGuire S, Hill M, Bioulac B, Balzamo E, Bezard E, Tison F, Ghorayeb I (2009) Sleep disorders in Parkinson's disease: the contribution of the MPTP non-human primate model. Exp Neurol 219:574–582. https://doi.org/10.1016/j.expneurol.2009.07.019
16. Isacson O, Riche D, Hantraye P, Sofroniew MV, Maziere M (1989) A primate model of Huntington's disease: cross-species implantation of striatal precursor cells to the excitotoxically lesioned baboon caudate-putamen. Exp Brain Res 75:213–220. https://doi.org/10.1007/BF00248544
17. Sakuma T, Barry MA, Ikeda Y (2012) Lentiviral vectors: basic to translational. Biochem J 443:603–618. https://doi.org/10.1042/BJ20120146
18. Park F (2007) Lentiviral vectors: are they the future of animal transgenesis? Physiol Genomics 31:159–173. https://doi.org/10.1152/physiolgenomics.00069.2007
19. Cronin J, Zhang X-Y, Reiser J (2005) Altering the tropism of lentiviral vectors through pseudotyping. Curr Gene Ther 5:387–398. https://doi.org/10.2174/1566523054546224
20. Jargalsaikhan BE, Muto M, Been Y, Matsumoto S, Okamura E, Takahashi T, Narimichi Y, Kurebayashi Y, Takeuchi H, Shinohara T et al (2024) The Dual-Pseudotyped Lentiviral Vector with VSV-G and Sendai Virus HN Enhances Infection Efficiency through the Synergistic Effect of the Envelope Proteins. Viruses 16. https://doi.org/10.3390/v16060827
21. Nakanishi M, Otsu M (2012) Development of Sendai virus vectors and their potential applications in gene therapy and regenerative medicine. Curr Gene Ther 12:410–416. https://doi.org/10.2174/156652312802762518
22. Markwell MA, Paulson JC (1980) Sendai virus utilizes specific sialyloligosaccharides as host cell receptor determinants. Proc Natl Acad Sci USA 77:5693–5697. https://doi.org/10.1073/pnas.77.10.5693
23. Seita Y, Tsukiyama T, Iwatani C, Tsuchiya H, Matsushita J, Azami T, Okahara J, Nakamura S, Hayashi Y, Hitoshi S et al (2016) Generation of

transgenic cynomolgus monkeys that express green fluorescent protein throughout the whole body. Sci Rep 6. https://doi.org/10.1038/srep24868

24. Seita Y, Morimura T, Watanabe N, Iwatani C, Tsuchiya H, Nakamura S, Suzuki T, Yanagisawa D, Tsukiyama T, Nakaya M et al (2020) Generation of Transgenic Cynomolgus Monkeys Overexpressing the Gene for Amyloid-β Precursor Protein. J Alzheimer's Dis 75:45–60. https://doi.org/10.3233/jad-191081
25. Seita Y, Tsukiyama T, Azami T, Kobayashi K, Iwatani C, Tsuchiya H, Nakaya M, Tanabe H, Hitoshi S, Miyoshi H et al (2019) Comprehensive evaluation of ubiquitous promoters suitable for the generation of transgenic cynomolgus monkeys. Biol Reprod 100:1440–1452. https://doi.org/10.1093/biolre/ioz040
26. Tsukiyama T, Kobayashi K, Nakaya M, Iwatani C, Seita Y, Tsuchiya H, Matsushita J, Kitajima K, Kawamoto I, Nakagawa T et al (2019) Monkeys mutant for PKD1 recapitulate human autosomal dominant polycystic kidney disease. Nat Commun 10. https://doi.org/10.1038/s41467-019-13398-6

Chapter 7

Ubiquitous and Brain Cell-Type-Specific Promoters and Enhancers in Lentiviral Vectors

Yasunori Matsuzaki, Ayumu Konno, and Hirokazu Hirai

Abstract

This chapter discusses the development and application of constitutive and brain cell-type-specific promoters and enhancers in lentiviral vectors to enable precise and efficient gene delivery. The fundamental mechanisms of transcriptional regulation by promoters, distinguished between constitutive and cell-type-specific promoters, are outlined and the role of enhancers in modulating transcriptional activity is emphasized. Recent breakthroughs in single-cell transcriptomics and assay for transposase-accessible chromatin (ATAC) sequencing have led to the identification of novel enhancer regions, thus paving the way for the development of more refined and targeted lentiviral vectors. These advances have the potential to significantly improve the specificity and efficacy of transgene expression in distinct brain cell populations, including excitatory and inhibitory neurons, astrocytes, and microglia.

Key words Viral vector, Gene therapy, Promoter, Enhancer, eRNA, Single-cell analysis, Neuron, Astrocyte, Microglia, Oligodendrocyte

1 Introduction

Viral vectors, including lentiviral vectors, have emerged as powerful tools for delivering genes of interest (GOI) to target cells. In recent years, rapid progress has been made in the development of these vectors as agents for genetic therapy [1, 2]. Simply incorporating GOIs into viral vectors does not lead to gene expression; transcription initiation requires sequences known as "promoters." Lentiviral vectors pseudotyped with vesicular stomatitis virus glycoprotein (VSV-G) transduce both dividing and nondividing cells. Expression levels and target cell types were adjusted by modifying the promoters. Although there are some differences, the mechanism of gene expression through promoters is conserved across viruses and multicellular eukaryotes, enabling the spatial, temporal, and quantitative control of gene expression. Retroviruses like human immunodeficiency virus type 1 (HIV-1), from which lentiviral

Kazuto Kobayashi et al. (eds.), *Lentiviral Vectors: Applications for Basic and Clinical Neuroscience*, Methods in Molecular Biology, vol. 2974, https://doi.org/10.1007/978-1-0716-4807-0_7,

vectors are derived, possess a long terminal repeat (LTR) containing a U3 promoter/enhancer in the 5′ LTR region that controls the transcription of viral genes. However, in self-inactivating (SIN) lentiviral vectors, the promoter/enhancer function of LTR is eliminated, and GOI expression is driven by an exogenously introduced promoter [3].

The promoters used in viral vectors can be broadly categorized into two types: constitutive promoters, which are active across a wide variety of cell types and tissues, and cell-type-specific promoters, which are active only in certain types of cells. Both promoter types share a common structure: a core promoter that recruits basic transcription factors and RNA polymerase II (Pol II) and an enhancer that binds transcriptional activators (Fig. 1a). The core promoter consists of several elements, including the TFIIB recognition element (BRE), TATA box, initiator (Inr) crucial for transcription initiation, and downstream elements, such as the downstream core element (DCE) and motif ten element (MTE) (Fig. 1a, diagram below). Although the TATA box is a well-known promoter element, most human genes do not contain a TATA box that requires a downstream promoter element (DPE) [4, 5].

When the TATA-binding protein (TBP) binds to the TATA box and Pol II binds to Inr, transcription begins at the transcription start site located near the initiator [6–9]. Many promoters used in viral vectors (isolated from genomes) include sequences that extend beyond the transcription start site to downstream exon 1 sequences, because the core promoter region extends into this range.

When an activator binds to an enhancer, it recruits the mediator complex, basic transcription factors, and Pol II to form the pre-initiation complex (PIC) and initiate transcription (Fig. 1b) [9–11]. In eukaryotic cells, particularly mammalian cells, multiple enhancers exist both upstream and downstream of the core promoter that regulates transcription. In some cases, enhancers can be located more than 1 MB upstream of the transcription start site, as observed in the sonic hedgehog (Shh) gene [12]. Moreover, some enhancers reside within introns downstream of the core promoter, and isolated promoters may be supplemented with introns to enhance their transcriptional activity. Pol II binds not only to the core promoter, but also to the enhancer to produce enhancer RNAs (eRNAs). These eRNAs bridge the enhancer and core promoter regions (Fig. 1b), reinforcing the PIC and promoting transcription [13]. Cell-type-specific gene expression is believed to be regulated by enhancers.

In viral vectors, the enhancer and core promoter induce the expression of the GOI. However, this expression is not always constant, exhibiting on-and-off fluctuations known as "transcriptional bursting" [14, 15]. Although the precise mechanisms are unclear, transcriptional bursts are thought to be influenced by the type of enhancer, structure of the core promoter, and distance

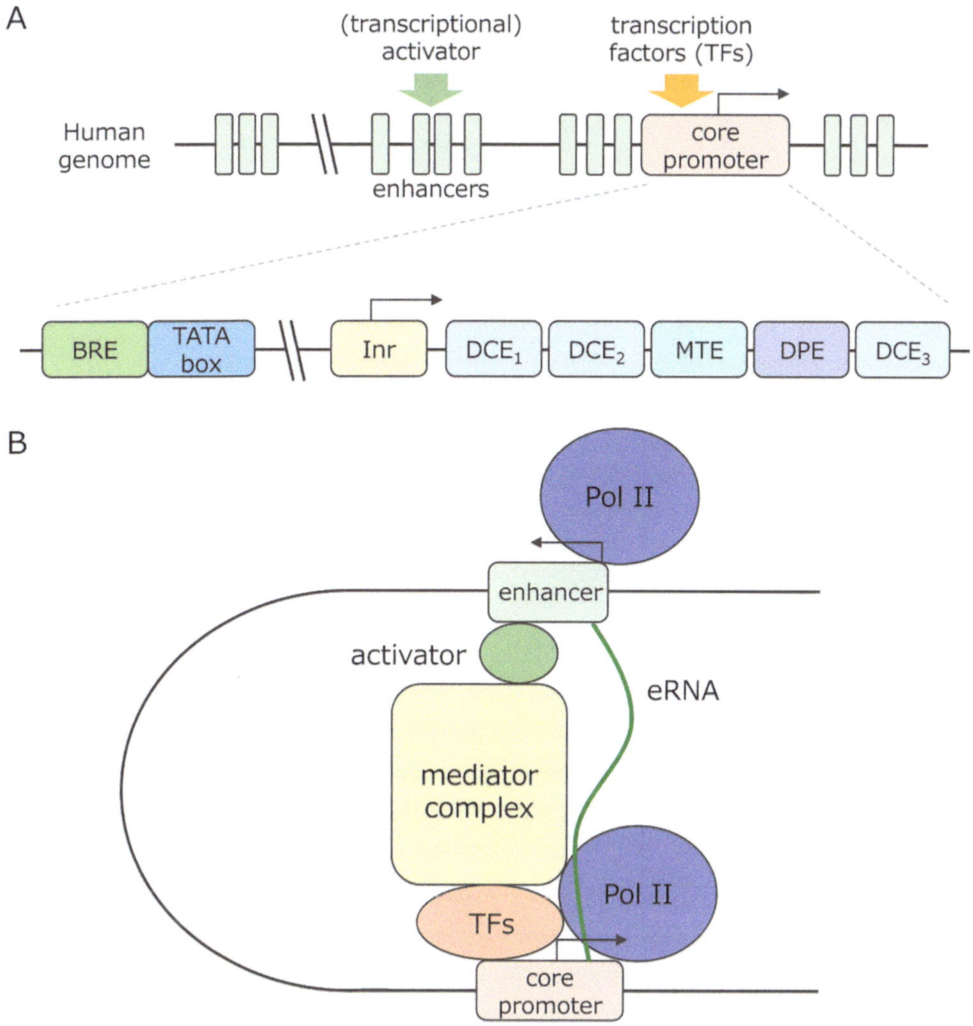

Fig. 1 Schematic representation of the promoter regions in the genome. (**a**) Transcription initiation in the genome involves the core promoter, where transcription factors (TFs) bind, and enhancers, where activators bind. Multiple enhancers are located both upstream and downstream of the core promoter, regulating transcription. As shown in the lower panel, the core promoter is composed of several elements, including the B recognition element (BRE), TATA-box, initiator (Inr), downstream core element (DCE), motif ten element (MTE), and downstream promoter element (DPE). (**b**) Structure of the pre-initiation complex (PIC) required for transcription initiation. PIC formation begins with the binding of an activator to the enhancer. The enhancer, bound to the activator, attracts various proteins, including the mediator complex and transcription factors (TFs). RNA polymerase II (Pol II) is associated with the mediator complex. Transcription is initiated at the Inr element within the core promoter when TFs, the mediator complex, and Pol II are recruited to this region. Pol II also binds to enhancers, producing enhancer RNA (eRNA), a type of noncoding RNA. eRNA bridges the enhancer and core promoter, stabilizing PIC formation and enhancing transcriptional output

between the enhancer and core promoter, contributing to the spatial, temporal, and quantitative control of transcription [16].

To maximize utility, the promoters used in viral vectors should be as short as possible while maintaining high transcriptional activity in target cells. Since the 1990s, the development of cell-type-specific promoters has involved isolating genome sequences, including the transcription start sites of genes specifically expressed in the target cell type, adjusting them to various lengths, and using luciferase or fluorescent proteins as the GOI to analyze enzyme activity, fluorescence intensity, and cell type expression.

In recent years, advances in single-cell analysis and ATAC sequencing techniques have enabled detailed analysis of open chromatin structures around specific gene loci in various cell types. This has enabled the prediction and identification of enhancer regions used in a cell-type-specific manner. It has also been demonstrated that combining these predicted enhancer regions with core promoters enables cell-type-specific functionality [17]. The following sections describe the representative promoters used for the viral vectors.

2 Constitutive Promoters

Promoters expressed in a wide range of cell types and tissues are referred to as constitutive or ubiquitous promoters. These promoters generally drive strong and versatile expression, although they may not be expressed equally across all cell types; therefore, it is important to select a promoter suited to the target cell type. Constitutive promoters originated from viruses and mammals are summarized in Table 1, and the characteristics of each are described below.

2.1 CMV Promoter

The cytomegalovirus (CMV) promoter, derived from *the immediate earlygene 1 (IE)* of human herpesvirus 5 (HHV-5), is one of the most widely used promoters because of its ability to induce strong expression of the GOI across various cell types. Three variations of CMV promoter exist based on differences in the enhancer, core promoter, and intron sequences (Fig. 2a). The original CMV promoter was developed in 1986 and consists of a 0.3 kb CMV genome region (containing the CMV *IE* enhancer and core promoter) linked to a 0.5 kb intron from the human *HBB* intron. In 1994, a modified version of the human *HBB* intron replaced by a 0.6 kb rabbit *HBB* gene intron was developed [18, 19]. Furthermore, in 1996, an intron-free version containing a 0.6 kb segment—double the size of the original CMV genome region—was introduced, and went on to become the standard CMV promoter used today [20]. Notably, CMV promoters can be silenced epigenetically, which may affect long-term expression [21].

Table 1
Constitutive promoters used in viral vectors

Promoter	Length (kb)	Enhancer	Core promoter	Intron	Publication year	Reference (DOI)
CMV	0.8	CMV *IE*	CMV *IE*	Human *HBB*	1986	https://doi.org/10.1016/0378-1119(86)90137-X
CMV	0.9	CMV *IE*	CMV *IE*	Rabbit *HBB*	1994	https://doi.org/10.1073/pnas.91.20.9564
CMV	0.6	CMV *IE*	CMV *IE*	–	1996	https://doi.org/10.1126/science.272.5259.263
CAG/CBA	1.7	CMV *IE*	Chicken *Actb*	Chicken *Actb*/rabbit *HBB*	1991	https://doi.org/10.1016/0378-1119(91)90434-d
CB	0.8	CMV *IE*	Chicken *Actb*	SV40 late 16S	2003	https://doi.org/10.1038/sj.gt.3302133
CBh	0.8	CMV *IE*	Chicken *Actb*	Chicken *Actb*/munute virus of mice VP	2011	https://doi.org/10.1089/hum.2010.245
MSCV	1.4	PCMV LTR	PCMV LTR	(ψ + first ATG deleted gag gene)	1994	(PMID: 7584069)
RSV	0.3	RSV LTR	RSV LTR	–	1980	https://doi.org/10.1016/0092-8674(80)90555-3
EF1α	1.2	Human *EEF1A1*	Human *EEF1A1*	Human *EEF1A1*	1990	https://doi.org/10.1016/0378-1119(90)90091-5
PGK	0.5	Mouse *Pgk-1*	Mouse *Pgk-1*	–	1991	https://doi.org/10.1093/nar/19.20.5755

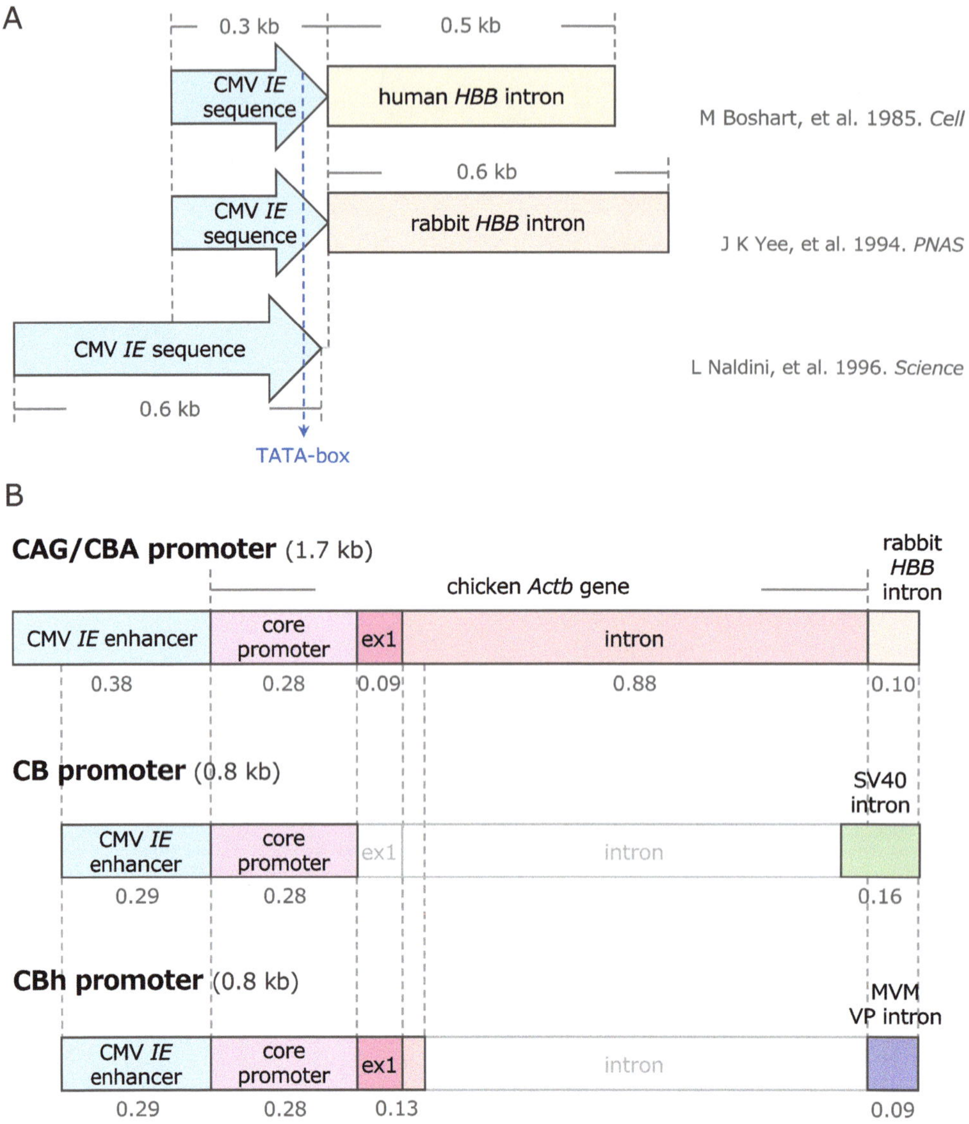

Fig. 2 Structure of constitutive promoters used in viral vectors. (**a**) The structure of three types of CMV promoters. The CMV *IE* sequence (blue arrow) includes the enhancer of the CMV *IE* gene and a core promoter containing the TATA box. At the top, a CMV promoter is shown where a 0.5 kb intron sequence of the human *HBB* gene is connected to a 0.3 kb CMV *IE* sequence. In the middle, the human *HBB* intron is replaced with a 0.6 kb intron from the rabbit *HBB* gene. At the bottom, a CMV promoter with a total length of 0.6 kb, consisting of a 0.3 kb extension upstream of the 0.3 kb CMV *IE* sequence without an intron sequence, is depicted. (**b**) Structure of the CAG/CBA, CB, and CBh promoters. These promoters share a CMV *IE* enhancer and the core promoter of the chicken *Actb* (*β-actin*) gene. The CMV *IE* enhancer of the CAG/CBA promoter is approximately 0.1 kb longer than that of the other two promoters. The key distinguishing feature of these three promoters lies in the sequences downstream of the core promoter. The CAG/CBA promoter includes around 1.0 kb of exon 1 (ex1) and intron sequences from the chicken *Actb* gene, along with a 0.1 kb rabbit *HBB* intron as a splicing acceptor sequence. The downstream sequence of the CB promoter's core promoter consists only of a 0.16 kb intron sequence from simian virus 40 (SV40). Although not illustrated, the CB promoter contains a 0.04-kb spacer sequence between the core promoter and the SV40 intron. The downstream sequence of the CBh promoter's core promoter comprises the chicken *Actb* gene's ex1/very short intron sequence and an additional 0.09 kb VP intron from minute virus of mice (MVM). The gray text below each component indicates the sequence length (kb)

2.2 CAG, CB, and CBh Promoters

The CAG (or CBA), CB, and CBh promoters are constructed by combining the CMV *IE* enhancer with a core promoter derived from the chicken *Actb* (*β-actin*) gene (Fig. 2b). The CAG promoter was designed to achieve efficient expression across a wide variety of cell types by linking the CMV *IE* enhancer to the 5′ side of the AG promoter, which includes the core promoter, exon 1, and intron sequences of the chicken *β-actin* gene, as well as a rabbit *HBB* intron sequence downstream [22]. The rabbit *HBB* intron sequence acts as a splicing acceptor, enabling the splicing out of the chicken *β-actin* gene's intron (0.9 kb) after transcription [23]. The CAG promoter is sometimes referred to as the CBA promoter. However, since the CAG promoter is relatively large (1.7 kb), a truncated version called the CB promoter was developed by replacing the chicken *β-actin* gene's exon 1 and intron (1.0 kb) and the rabbit *HBB* intron (0.1 kb) with a 0.16-kb sequence derived from SV40, reducing the overall length to 0.9 kb [24]. The CB promoter is not well suited for expression in motor neurons, leading to the development of the CBh promoter, in which the SV40 sequence is replaced with a sequence from minute virus of mice (MVM) for motor neuron-specific expression [25]. Unlike the CMV promoter, the CAG and CBh promoters are not prone to silencing.

2.3 MSCV Promoter

The murine stem cell virus (MSCV) promoter was derived from a virus isolated from a myeloproliferative sarcoma virus, a type of Moloney murine sarcoma virus (MoMSV), after the serial passage of the cell surface antigens of a multipotent embryonal carcinoma (EC) line, PCC4. This virus is referred to as PCC4-cell-passaged myeloproliferative sarcoma virus (PCMV) or murine stem cell virus (MSCV) and is used as an MSCV retroviral vector [26, 27]. A portion of the gag gene, which lacks the ATG start codon and is located near the 5′ LTR, where promoter activity resides, was later isolated as the MSCV promoter [28]. This promoter is active in stem cells, where a series of CMV promoters is inactive. Additionally, when directly administered to the cerebellum of mice and marmosets using viral vectors, it exhibits strong promoter activity in Purkinje cells, robustly expressing the transgene [29].

2.4 RSV Promoter

The RSV promoter was isolated from the 5′ LTR region of the Rous sarcoma virus. Although relatively short, it induces strong gene expression [30].

2.5 EF1α Promoter

The EF1α promoter was isolated from the promoter and intron regions of the human *EEF1A1* (*elongation factor 1α*) gene. It operates in a wide variety of cell types and exhibits high gene expression levels. The total length of this promoter is 1.2 kb, with approximately 1.0 kb consisting of an intron between exon 1 and exon 2 [31].

2.6 PGK Promoter

The PGK promoter was isolated from the *Pgk-1* gene of the mouse, which encodes an enzyme involved in glucose metabolism and is expressed throughout the body. Because of its widespread expression, this promoter is expected to be active in a wide variety of cell types. The PGK promoter demonstrates long-term stable activity [32].

3 Cell-Type-Specific Promoters

Cell-type-specific promoters are isolated from the promoter regions of genes that are specifically expressed in target cell types and induce GOI expression in a cell-type-specific manner. Compared to constitutive promoters, cell-type-specific promoters generally exhibit weaker promoter activity, resulting in lower expression levels (Table 2).

3.1 SynI Promoter

The synapsin-I (SynI) promoter is the most widely used pan-neuron-specific promoter. Both human- and rat-derived versions exist. Initially, the lengths of these promoters varied between 2.0 kb (human-derived) and 4.3 kb (rat-derived), but were later shortened to 0.4 kb (human-derived) and 1.0 kb (rat-derived) [33–36]. To improve expression efficiency, the human-derived version was combined with a CMV enhancer (CMVE-hSynI) and the rat-derived version with the minimal CMV core promoter (minCMV) (rSynI-minCMV). However, viral vectors containing CMVE-hSynI often show gene expression in glial cells. For this reason, the hSynI promoter alone is commonly used [37–39].

3.2 CaMKII Promoter

The CaMKII promoter is isolated from the *Camk2a* (*Calmodulin-dependent protein kinase II (CaMKII)* gene in mice and is available at two lengths: 0.4 kb and 1.3 kb [36]. The CaMKII protein is specifically expressed in excitatory neurons, and the 0.4 kb version of the CaMKII promoter functions specifically in excitatory neurons [40]. In contrast, the 1.3 kb CaMKII promoter is active in both excitatory and inhibitory neurons, making it a pan-neuronal promoter, at least in the cerebral cortex. The 1.3 kb promoter exhibits higher promoter activity than the 0.4 kb version and shows greater promoter activity than the rSynI promoter in the adult rat cortex [36].

3.3 NSE Promoter

In 1990, the 1.8-kb rNSE promoter was isolated from the rat *Eno2* (*neuron-specific enolase, NSE*) gene. Although promoter activity decreased, a neuron-specific 1.2 kb rNSE promoter was later developed [41]. In 2004, a 1.1-kb human-derived NSE promoter was also isolated [42, 43]. The NSE promoter is fundamentally a pan-neuronal-specific promoter; however, when an adeno-associated virus (AAV) vector containing the NSE promoter is directly

Table 2
Cell-type-specific promoters used in viral vectors

Target cell type	Promoter	Species	Length (bp)	Publication year	Reference (DOI)
Pan-neuron	(CMVE-)SynI	Human	(368-)401	2007	https://doi.org/10.1038/sj.gt.3302924
Pan-neuron	SynI	Human	448	2014	https://doi.org/10.1038/nature13321
Pan-neuron	SynI	Rat	536, 1106	2004	https://doi.org/10.1073/pnas.0407976101
Pan-neuron	SynI(-minCMV)	Rat	1046(−129)	2013	https://doi.org/10.1016/j.jneumeth.2013.12.004
Excitatory neuron Pan-neuron	CaMKII(0.3) CaMKII(1.3)	Mouse Mouse	364 1293	2004 2004	https://doi.org/10.1073/pnas.0407976101
Pan-neuron (astrocyte)	NSE	Rat	1807	1990	https://doi.org/10.1016/0896-6273(90)90308-3
Pan-neuron (astrocyte)	NSE	Rat	1200	2017	https://doi.org/10.1007/s12311-017-0866-5
Pan-neuron (astrocyte)	NSE	Human	1083	2004	https://doi.org/10.1210/en.2003-0772
Inhibitory neuron	GAD65	Mouse	2542	2021	https://doi.org/10.1186/s13041-021-00746-1
Purkinje cell	L7-6	Mouse	844	2017	https://doi.org/10.1016/j.omtm.2017.07.006
Astrocyte	Gfa(ABC1D)	Human	681	2008	https://doi.org/10.1002/glia.20622
Astrocyte	Gfap	Marmoset	314	2016	https://doi.org/10.1371/journal.pone.0162023
Astrocyte	GfaABC1D	Mouse	582	2018	https://doi.org/10.1007/s12035-018-1366-4
Astrocyte	ALDH1L1	Human	1245	2017	https://doi.org/10.5607/en.2017.26.6.350
Oligodendrocyte	MBP	Mouse	1951	1998	https://doi.org/10.1038/sj.gt.3300547
Oligodendrocyte Oligodendrocyte Oligodendrocyte	MAG(0.3) MAG(1.5) MAG(2.2)	Human Human Human	298 1451 2184	2016 2016 2016	https://doi.org/10.3389/fnmol.2016.00013
Microglia	Iba1 + miR.T.	Mouse	1678 + 204	2022	https://doi.org/10.1038/s42003-022-04200-3

(continued)

Table 2
(continued)

Target cell type	Promoter	Species	Length (bp)	Publication year	Reference (DOI)
Microglia	HexB	Rhesus macaque	190	2022	https://doi.org/10.3389/fncel.2022.808598
Microglia	HEXB	Human	183	2022	
Microglia	CD68	Human	135	2022	
Microglia	CD68	Human	695	2022	
Microglia	Iba1a + miR.T.	Human	466 + 96	2024	https://doi.org/10.1016/j.isci.2024.110706
Microvascular endothelial cell	GLUT1	Human	302	2018	https://doi.org/10.1002/jgm.3013
Microvascular endothelial cell	Cldn5	Mouse	1131	2012	https://doi.org/10.1016/j.cyto.2011.10.016
Microvascular endothelial cell	CLDN5	Human	1245	–	Our unpublished data

injected into brain tissue, the resulting inflammation can activate the promoter specifically in astrocytes [44]. In contrast, when using a blood–brain barrier-permeable AAV variant (AAV-PHP.B) to express a fluorescent protein, the NSE promoter exhibits neuron-specific expression (owing to the absence of inflammation) and demonstrates stronger activity than the SynI and CaMKII promoters [45].

3.4 GAD65(delE1) Promoter

The GAD65(delE1) promoter is a 2.5-kb GABAergic neuron-specific promoter isolated from the murine *Gad2* (*Gad65*) gene [46]. This promoter contains Regions I and II, which are densely packed with enhancers that serve as biding sites for Dlx activators [47]. When administered intravenously in mice using the AAV-PHP.B vector, the GAD65(delE1) promoter drives gene expression not only in the forebrain, such as in the mDlx enhancer, but also in the hindbrain.

3.5 L7-6 Promoter

The L7-6 promoter is a 0.8-kb cerebellar Purkinje cell-specific promoter isolated from the *Pcp2* (*L7*) gene in mice [48]. This promoter maintains high cell type specificity not only in mice but also in nonhuman primates, such as the common marmoset. The transcriptional activity significantly decreases when the RAR-related orphan receptor α (RORα) binding region, which functions as an enhancer, is deleted, suggesting that this enhancer plays a critical role in transcriptional activity [48].

3.6 GFAP Promoter

The first reported human-derived gfa2 promoter was 2.2 kb in size; however, its activity in mouse tissues was insufficient. As a result, the gfa (ABC1D) promoter was developed in 2008 [49, 50]. This promoter was created by dividing the upstream 1.7 kb region of the human *GFAP* gene into four sections (A–D) and further subdividing the longest C region into C1 to C6. The promoter was created by removing the low-transcriptional activity regions from C2 to C6. Initially used to create transgenic mice, it was later found to maintain sufficient astrocyte specificity when used in viral vectors, making it one of the most commonly used astrocyte-specific promoters for viral vectors.

The authors developed a 0.3-kb cjGFAP promoter by combining the 3′-end of the C region with the D region, excluding the AB regions and the 5′-end of the C region from the common marmoset genome [51]. Despite being very short, this promoter exhibited high astrocyte specificity of approximately 90% in the cerebral cortex and cerebellum of common marmosets. However, astrocyte specificity was significantly decreased in the mouse cerebral cortex, suggesting that the cjGFAP promoter may be specific to astrocytes in primates. GFAP expression was significantly upregulated during astroglial activation due to inflammation [52, 53]. Therefore, if a viral vector were directly injected into the brain parenchyma and caused inflammation, the activity of the GFAP promoter would be enhanced, resulting in increased transgene expression.

Moreover, the authors developed a mouse GFAP (mGfaABC1D) promoter based on the human gfa (ABC1D) promoter [54]. The mGfaABC1D promoter exhibited high astrocyte specificity of over 90% in the cerebral cortex of mice using the AAV-PHP.eB vector [55].

3.7 ALDH1L1 Promoter

The *ALDH1L1* gene is specifically expressed in astrocytes. Although the expression level of *Gfap* is very low in normal cerebral cortex gray matter, *Aldh1l1* is strongly expressed throughout the cerebral cortex gray matter. Therefore, *Aldh1l1*-specific transgenic mice were developed using a bacterial artificial chromosome (BAC) system to drive astrocyte-specific gene expression [56, 57]. In addition, a human-derived ALDH1L1 (hALDH1L1) promoter was developed for use in viral vectors [58]. However, the hALDH1L1 promoter drives astrocyte-specific expression only in the mouse thalamus, with promoter activity also detected in neurons of the amygdala, hippocampus, and prefrontal cortex. In marmosets, hALDH1L1 promoter-driven AAV vectors administered to the cerebral cortex induced transgene expression in both astrocytes and neurons.

3.8 MBP and MAG Promoters

In 1998, the myelin basic protein (MBP) promoter derived from mice was developed as an oligodendrocyte-specific promoter that could be incorporated into viral vectors [59]. Although this promoter is relatively long (approximately 2.0 kb), it exhibits high specificity and expression efficiency and has been widely used. When combined with the AAV5 serotype capsid, the MBP promoter demonstrated approximately 80% oligodendrocyte specificity in the cerebral cortex of marmosets [60]. In 2016, a Myelin-Associated Glycoprotein (MAG) promoter with higher specificity than the MBP promoter was reported. This MAG promoter exists in three versions of different lengths: 0.3, 1.5, and 2.2 kb [61].

3.9 Microglia-Specific Promoters

Aif1 (also known as Iba1, ionized calcium-binding adaptor molecule 1) is a protein expressed in microglia and macrophages and is frequently used as a marker to detect these cells. In 2022, a 1.7 kb region upstream of the ATG start codon within exon 1 of the mouse *Aif1* (*Iba1*) gene was identified as having microglia-specific promoter activity in the mouse brain and was named the Iba1 promoter [62]. However, viral vectors containing the Iba1 promoter also drive the expression in neurons. To overcome this limitation, microRNA target (miR.T) sequences complementary to *miR-9-5p* (*miR-9*) and *miR-129-2-3p*, which are expressed in non-microglial cells, were incorporated downstream of the WPRE in the AAV genome. In non-microglial cells, endogenous *miR-9* and *miR-129-2-3p* bind to the miR.T sequences in viral mRNA, leading to degradation of the mRNA. In contrast, because *miR-9* and *miR-129-2-3p* are not expressed in microglia, the viral mRNA is not degraded, allowing for gene expression in microglia.

Upon the administration of the AAV vector to the cerebellum and striatum, high specificity and expression efficiency were achieved, although gene expression in neurons of the cerebral cortex could not be fully suppressed. In addition to the Iba1 promoter developed by the authors, in 2022, human- and macaque-derived short *HEXB* promoters, as well as a human-derived *CD68* promoter, have been reported to function in cultured microglial cell lines. In 2023, AAV vectors with a human-derived 0.5 kb *Iba1a* promoter and *miR-124* target sequence were also shown to function in mouse microglia in vivo [63, 64]. During proof review of this manuscript, we reported that placing target sequences complementary to *miR-9* and *miR-129-2-3p* on both sides of the WPRE markedly improves microglial specificity and transduction efficiency compared with conventional designs, and have added this information here [76]. Thus, recent advances have enabled gene expression in mouse microglia in vivo and in cultured primate microglia; however, there have been no reports thus far on gene expression in primate microglia in vivo.

3.10 Brain Microvascular Endothelial Cell (BMEC)-Specific Promoter

The blood–brain barrier (BBB) exists to keep the brain safe from external influences, with brain microvascular endothelial cells (BMECs) forming the basis of its vascular structures [65, 66]. To enable gene expression specifically in BMECs, promoters such as human-derived hGLUT1, mouse-derived mCldn5, and human-derived CLDN5 (hCLDN5) have been developed. When these promoters were packaged into a BBB-permeable AAV2-BR1N capsid variant and intravenously injected into mice, the hGLUT1 promoter showed activity in brain parenchymal cells, whereas the hCLDN5 promoter was specifically active in brain vascular endothelial cells. Thus, viral vectors incorporating the hCLDN5 promoter are considered suitable for gene expression restricted to BMECs.

4 Cell-Type Specific Enhancers

Recent advances in nucleic acid analysis techniques, such as single-cell transcriptomics (scRNA-seq) and assays for transposase-accessible chromatin sequencing (ATAC-seq), have enabled the analysis of gene expression and open chromatin regions in specific cell types [67–71]. As a result, enhancer regions involved in the transcriptional regulation of particular genes were identified. Enhancers targeting specific cell types, including astrocytes and oligodendrocytes, have been developed, with a particular focus on GABAergic neurons, which comprise many subtypes (Table 3).

Table 3
Cell-type-specific enhancers used in viral vectors

Target cell type	Enhancer	Species	Length (bp)	Core promoter	Intron	Total length (bp)	Publication year	Reference (DOI)
Oligodendrocyte	eHGT_410m	Mouse	295	Human *HBB*	–	354	2023	https://doi.org/10.1101/2023.09.20.558718
Oligodendrocyte	eHGT_395h	Human	624	Human *HBB*	–	683	2023	
Astrocyte	3xCore2 (390 m)	Mouse	750	Human *HBB*	–	808	2023	
Astrocyte	3xCore2 (390 h)	Human	735	Human *HBB*	–	793	2023	
Inhibitory neuron	Dlx	Mouse	541	Human *HBB*	Human *HBB* + mouse *Ighv*	834	2016	https://doi.org/10.1038/nn.4430
Inhibitory neuron	Dlx	Human	542	Human *HBB*	Human *HBB* + mouse *Ighv*	835	2016	
SST+ neuron	GRE44 (SST44)	Mouse	720	Human *HBB*	Human *HBB* + mouse *Ighv*	1016	2019	https://doi.org/10.7554/eLife.48089
PV+ neuron	E2	Mouse	619	Human *HBB*	Human *HBB* + mouse *Ighv*	903	2020	https://doi.org/10.1038/s41593-020-0692-9
PV+ cortical basket cell	BiPVe3	Mouse	681	Human *HBB*	Human *HBB* + mouse *Ighv*	748	2025	https://doi.org/10.1016/j.neuron.2025.05.002
PV+ cortical chandelier cell	BiPVe4	Mouse	542	Human *HBB*	Human *HBB* + mouse *Ighv*	826	2025	
SST+ non-Martinotti neuron	BiSSTe10	Mouse	775	Human *HBB*	Human *HBB* + mouse *Ighv*	1059	2025	

SST+ Martinotti neuron	BiSSTe4	Mouse	662	Human *HBB*	Human *HBB* + mouse *Ighv*	946	2025	
VIP+ neuron	BiVIPe4	Mouse	416	Human *HBB*	Human *HBB* + mouse *Ighv*	700	2025	
LAMP5+ neuron	BiLAMP5e3	Mouse	646	Human *HBB*	Human *HBB* + mouse *Ighv*	930	2025	
ChAT+ neuron	BiCHATe27	Mouse	854	Human *HBB*	Human *HBB* + mouse *Ighv*	1138	2025	

According to the Allen Institute, GABAergic neurons in the cerebral cortex and hippocampus can be broadly classified as Lhx6-positive and Adarb2-positive. Lhx6-positive cells are further divided into parvalbumin (PV)-positive and somatostatin (SST)-positive neurons. Adarb2-positive cells include Vasoactive Intestinal Peptide (VIP)-positive, Inhibitor of DNA Binding 2 (Id2)-positive, and Meis Homeobox 2 (Meis)-positive neurons [72]. These groups can be subdivided into specific populations revealing a wide variety of subtypes. Several enhancers specific to GABAergic neuronal populations have been developed. In the following sections, we describe the enhancers specific to these GABAergic neuron populations, along with those specific to glial cells. The core promoter used in most of these enhancers is derived from the human *HBB* gene, with some groups utilizing a chimeric intron combining the human *HBB* and mouse *Ighv* (immunoglobulin heavy chain variable region) gene sequences.

4.1 Astrocyte and Oligodendrocyte-Specific Enhancers

In 2023, Mich et al. identified astrocyte- and oligodendrocyte-specific enhancers in human and mouse genomes [73]. Among the astrocyte-specific enhancers, human-derived 3xCore2(390 h) and mouse-derived 3xCore2(390 m) have been reported to be the most efficient. For oligodendrocytes, human-derived eHGT_395h and mouse-derived eHGT_410m are considered as the most effective.

4.2 GABAergic Neuron-Specific Enhancer

The **Dlx enhancer**, derived from both the mouse and human genomes, was the first to target pan-GABAergic neurons [17]. When packaged into an AAV-PHP.B capsid and administered intravenously, the mDlx enhancer limits gene expression in the forebrain. In 2019, a GRE44 (SST44) enhancer specific for somatostatin (SST)-positive neurons was developed, followed by an S5E2 enhancer specific for parvalbumin (PV)-positive neurons in 2020 [74, 75]. In 2024, further development of enhancers targeting even smaller subpopulations of GABAergic neurons, such as Bi series enhancers, was reported, greatly expanding the targeting capabilities for numerous subpopulations of GABAergic neurons. Although not in GABAergic neurons, cholinergic neuron–specific enhancers targeting choline acetyltransferase (ChAT)-positive neurons have subsequently been developed.

Enhancer development has revolutionized cell-type-specific promoter research, which has previously been constrained by limited data. Viral vector-based exploration has enabled the identification of highly specific enhancers by logically identifying enhancer regions that regulate cell-type-specific gene expression. This approach is expected to become a mainstream method for developing cell-type-specific enhancers. However, similar to cell-type-specific promoters, enhancers still face challenges with respect to expression efficiency, which tends to be relatively low. Increasing expression efficiency is crucial as it allows for the reduction of viral vector doses. The development of stronger cell-type-specific promoters and enhancers that can be incorporated into viral vectors will help advance both our understanding of brain pathophysiology and the clinical application of gene therapy.

Conflict of Interest The authors declare that there is no conflict of interest.

Author Contributions Y.M. prepared the original draft. H.H. supervised and completed the manuscript. All the authors have read and approved the final version of this chapter.

Funding

This research was partially supported by the Program for Multidisciplinary Frontier Brain and Neuroscience Discoveries (Brain/MINDS 2.0) of the Japan Agency for Medical Research and Development (AMED) (grant number JP24wm0625103) and JSPS KAKENHI (grant numbers 24 K15743, 24H01221, 23 K27482, and 22 K06454).

References

1. Papaioannou I, Owen JS, Yanez-Munoz RJ (2023) Clinical applications of gene therapy for rare diseases: a review. Int J Exp Pathol 104(4):154–176. https://doi.org/10.1111/iep.12478
2. Perry C, Rayat A (2021) Lentiviral vector bioprocessing. Viruses 13(2):268. https://doi.org/10.3390/v13020268
3. Cockrell AS, Kafri T (2007) Gene delivery by lentivirus vectors. Mol Biotechnol 36(3):184–204. https://doi.org/10.1007/s12033-007-0010-8
4. Burke TW, Kadonaga JT (1997) The downstream core promoter element, DPE, is conserved from drosophila to humans and is recognized by TAFII60 of drosophila. Genes Dev 11(22):3020–3031. https://doi.org/10.1101/gad.11.22.3020
5. Breathnach R, Chambon P (1981) Organization and expression of eucaryotic split genes coding for proteins. Annu Rev Biochem 50:349–383. https://doi.org/10.1146/annurev.bi.50.070181.002025
6. Lagrange T, Kapanidis AN, Tang H et al (1998) New core promoter element in RNA polymerase II-dependent transcription: sequence-specific DNA binding by transcription factor IIB. Genes Dev 12(1):34–44. https://doi.org/10.1101/gad.12.1.34
7. Lim CY, Santoso B, Boulay T et al (2004) The MTE, a new core promoter element for transcription by RNA polymerase II. Genes Dev 18(13):1606–1617. https://doi.org/10.1101/gad.1193404
8. Lewis BA, Kim T-K, Orkin SH (2000) A downstream element in the human β-globin promoter: evidence of extended sequence-specific transcription factor IID contacts. Proc Natl Acad Sci USA 97(13):7172–7177. https://doi.org/10.1073/pnas.120181197
9. Kadonaga JT (2012) Perspectives on the RNA polymerase II core promoter. WIREs Dev Biol 1(1):40–51. https://doi.org/10.1002/wdev.21
10. Khoury G, Gruss P (1983) Enhancer elements. Cell 33(2):313–314. https://doi.org/10.1016/0092-8674(83)90410-5
11. Andersson R, Sandelin A (2020) Determinants of enhancer and promoter activities of regulatory elements. Nat Rev Genet 21(2):71–87. https://doi.org/10.1038/s41576-019-0173-8

12. Lettice LA, Heaney SJ, Purdie LA et al (2003) A long-range Shh enhancer regulates expression in the developing limb and fin and is associated with preaxial polydactyly. Hum Mol Genet 12(14):1725–1735. https://doi.org/10.1093/hmg/ddg180
13. Lam MT, Li W, Rosenfeld MG et al (2014) Enhancer RNAs and regulated transcriptional programs. Trends Biochem Sci 39(4):170–182. https://doi.org/10.1016/j.tibs.2014.02.007
14. So LH, Ghosh A, Zong C et al (2011) General properties of transcriptional time series in Escherichia coli. Nat Genet 43(6):554–560. https://doi.org/10.1038/ng.821
15. Rodriguez J, Larson DR (2020) Transcription in living cells: molecular mechanisms of bursting. Annu Rev Biochem 89:189–212. https://doi.org/10.1146/annurev-biochem-011520-105250
16. Fukaya T, Lim B, Levine M (2016) Enhancer control of transcriptional bursting. Cell 166(2):358–368. https://doi.org/10.1016/j.cell.2016.05.025
17. Dimidschstein J, Chen Q, Tremblay R et al (2016) A viral strategy for targeting and manipulating interneurons across vertebrate species. Nat Neurosci 19(12):1743–1749. https://doi.org/10.1038/nn.4430
18. Foecking MK, Hofstetter H (1986) Powerful and versatile enhancer-promoter unit for mammalian expression vectors. Gene 45(1):101–105. https://doi.org/10.1016/0378-1119(86)90137-X
19. Yee JK, Miyanohara A, LaPorte P et al (1994) A general method for the generation of high-titer, pantropic retroviral vectors: highly efficient infection of primary hepatocytes. Proc Natl Acad Sci USA 91(20):9564–9568. https://doi.org/10.1073/pnas.91.20.9564
20. Naldini L, Blömer U, Gallay P et al (1996) In vivo gene delivery and stable transduction of nondividing cells by a lentiviral vector. Science (New York, NY) 272(5259):263–267. https://doi.org/10.1126/science.272.5259.263
21. Mehta AK, Majumdar SS, Alam P et al (2009) Epigenetic regulation of cytomegalovirus major immediate-early promoter activity in transgenic mice. Gene 428(1):20–24. https://doi.org/10.1016/j.gene.2008.09.033
22. Niwa H, Yamamura K, Miyazaki J (1991) Efficient selection for high-expression transfectants with a novel eukaryotic vector. Gene 108(2):193–199. https://doi.org/10.1016/0378-1119(91)90434-d
23. Miyazaki J, Takaki S, Araki K et al (1989) Expression vector system based on the chicken *β-actin* promoter directs efficient production of interleukin-5. Gene 79(2):269–277. https://doi.org/10.1016/0378-1119(89)90209-6
24. Wang Z, Ma HI, Li J et al (2003) Rapid and highly efficient transduction by double-stranded adeno-associated virus vectors in vitro and in vivo. Gene Ther 10(26):2105–2111. https://doi.org/10.1038/sj.gt.3302133
25. Gray SJ, Foti SB, Schwartz JW et al (2011) Optimizing promoters for recombinant adeno-associated virus-mediated gene expression in the peripheral and central nervous system using self-complementary vectors. Hum Gene Ther 22(9):1143–1153. https://doi.org/10.1089/hum.2010.245
26. Grez M, Akgün E, Hilberg F et al (1990) Embryonic stem cell virus, a recombinant murine retrovirus with expression in embryonic stem cells. Proc Natl Acad Sci USA 87(23):9202–9206. https://doi.org/10.1073/pnas.87.23.9202
27. Hawley RG, Lieu FH, Fong AZ et al (1994) Versatile retroviral vectors for potential use in gene therapy. Gene Ther 1(2):136–138
28. Persons DA, Allay JA, Allay ER et al (1997) Retroviral-mediated transfer of the green fluorescent protein gene into murine hematopoietic cells facilitates scoring and selection of transduced progenitors in vitro and identification of genetically modified cells in vivo. Blood 90(5):1777–1786. https://doi.org/10.1182/blood.V90.5.1777
29. Oue M, Handa H, Matsuzaki Y et al (2012) The murine stem cell virus promoter drives correlated transgene expression in the leukocytes and cerebellar Purkinje cells of transgenic mice. PLoS One 7(11):e51015. https://doi.org/10.1371/journal.pone.0051015
30. Yamamoto T, de Crombrugghe B, Pastan I (1980) Identification of a functional promoter in the long terminal repeat of Rous sarcoma virus. Cell 22(3):787–797. https://doi.org/10.1016/0092-8674(80)90555-3
31. Kim DW, Uetsuki T, Kaziro Y et al (1990) Use of the human elongation factor 1α promoter as a versatile and efficient expression system. Gene 91(2):217–223. https://doi.org/10.1016/0378-1119(90)90091-5
32. McBurney MW, Sutherland LC, Adra CN et al (1991) The mouse Pgk-1 gene promoter contains an upstream activator sequence. Nucleic Acids Res 19(20):5755–5761. https://doi.org/10.1093/nar/19.20.5755
33. Thiel G, Greengard P, Sudhof TC (1991) Characterization of tissue-specific transcription

by the human synapsin I gene promoter. Proc Natl Acad Sci USA 88(8):3431–3435

34. Sauerwald A, Hoesche C, Oschwald R et al (1990) The 5′-flanking region of the synapsin I gene. A G+C-rich, TATA- and CAAT-less, phylogenetically conserved sequence with cell-type-specific promoter function. J Biol Chem 265(25):14932–14937
35. Hoesche C, Sauerwald A, Veh RW et al (1993) The 5′-flanking region of the rat synapsin I gene directs neuron-specific and developmentally regulated reporter gene expression in transgenic mice. J Biol Chem 268(35): 26494–26502
36. Dittgen T, Nimmerjahn A, Komai S et al (2004) Lentivirus-based genetic manipulations of cortical neurons and their optical and electrophysiological monitoring in vivo. Proc Natl Acad Sci USA 101(52):18206–18211. https://doi.org/10.1073/pnas.0407976101
37. Hioki H, Kameda H, Nakamura H et al (2007) Efficient gene transduction of neurons by lentivirus with enhanced neuron-specific promoters. Gene Ther 14(11):872–882. https://doi.org/10.1038/sj.gt.3302924
38. Matsuzaki Y, Oue M, Hirai H (2013) Generation of a neurodegenerative disease mouse model using lentiviral vectors carrying an enhanced synapsin I promoter. J Neurosci Methods 223C:133–143. https://doi.org/10.1016/j.jneumeth.2013.12.004
39. Xue M, Atallah BV, Scanziani M (2014) Equalizing excitation-inhibition ratios across visual cortical neurons. Nature 511(7511):596–600. https://doi.org/10.1038/nature13321
40. Scheyltjens I, Laramée ME, Van den Haute C et al (2015) Evaluation of the expression pattern of rAAV2/1, 2/5, 2/7, 2/8, and 2/9 serotypes with different promoters in the mouse visual cortex. J Comp Neurol 523(14): 2019–2042. https://doi.org/10.1002/cne.23819
41. Shinohara Y, Ohtani T, Konno A et al (2017) Viral vector-based evaluation of regulatory regions in the neuron-specific enolase (NSE) promoter in mouse cerebellum in vivo. Cerebellum 16(5–6):913–922. https://doi.org/10.1007/s12311-017-0866-5
42. Forss-Petter S, Danielson PE, Catsicas S et al (1990) Transgenic mice expressing beta-galactosidase in mature neurons under neuron-specific enolase promoter control. Neuron 5(2):187–197. https://doi.org/10.1016/0896-6273(90)90308-3
43. Kim J, Adam RM, Solomon KR et al (2004) Involvement of cholesterol-rich lipid rafts in Interleukin-6-induced neuroendocrine differentiation of LNCaP prostate cancer cells. Endocrinology 145(2):613–619. https://doi.org/10.1210/en.2003-0772
44. Sawada Y, Konno A, Nagaoka J et al (2016) Inflammation-induced reversible switch of the neuron-specific enolase promoter from Purkinje neurons to Bergmann glia. Sci Rep 6: 27758. https://doi.org/10.1038/srep27758
45. Radhiyanti PT, Konno A, Matsuzaki Y et al (2021) Comparative study of neuron-specific promoters in mouse brain transduced by intravenously administered AAV-PHP.eB. Neurosci Lett 756:135956. https://doi.org/10.1016/j.neulet.2021.135956
46. Hoshino C, Konno A, Hosoi N et al (2021) GABAergic neuron-specific whole-brain transduction by AAV-PHP.B incorporated with a new GAD65 promoter. Mol. Brain 14(1):33. https://doi.org/10.1186/s13041-021-00746-1
47. Le TN, Zhou QP, Cobos I et al (2017) GABAergic interneuron differentiation in the basal forebrain is mediated through direct regulation of glutamic acid decarboxylase isoforms by dlx Homeobox transcription factors. J Neurosci 37(36):8816–8829. https://doi.org/10.1523/JNEUROSCI.2125-16.2017
48. Nitta K, Matsuzaki Y, Konno A et al (2017) Minimal Purkinje cell-specific PCP2/L7 promoter virally available for rodents and non-human primates. Mol Ther Methods Clin Dev 6:159–170. https://doi.org/10.1016/j.omtm.2017.07.006
49. Lee Y, Messing A, Su M et al (2008) GFAP promoter elements required for region-specific and astrocyte-specific expression. Glia 56(5): 481–493. https://doi.org/10.1002/glia.20622
50. Su M, Hu H, Lee Y et al (2004) Expression specificity of GFAP transgenes. Neurochem Res 29(11):2075–2093. https://doi.org/10.1007/s11064-004-6881-1
51. Shinohara Y, Konno A, Takahashi N et al (2016) Viral vector-based dissection of marmoset GFAP promoter in mouse and marmoset brains. PLoS One 11(8):e0162023. https://doi.org/10.1371/journal.pone.0162023
52. Eng LF, Ghirnikar RS (1994) GFAP and Astrogliosis. Brain Pathol 4(3):229–237. https://doi.org/10.1111/j.1750-3639.1994.tb00838.x
53. Pekny M, Nilsson M (2005) Astrocyte activation and reactive gliosis. Glia 50(4):427–434. https://doi.org/10.1002/glia.20207

54. Shinohara Y, Konno A, Nitta K et al (2018) Effects of neutralizing antibody production on AAV-PHP.B-mediated transduction of the mouse central nervous system. Mol Neurobiol 56(6):4203–4214. https://doi.org/10.1007/s12035-018-1366-4
55. Kawabata H, Konno A, Matsuzaki Y et al (2024) Improving cell-specific recombination using AAV vectors in the murine CNS by capsid and expression cassette optimization. Mol Ther Methods Clin Dev 32(1):101185. https://doi.org/10.1016/j.omtm.2024.101185
56. Yang Y, Vidensky S, Jin L et al (2011) Molecular comparison of GLT1+ and ALDH1L1+ astrocytes in vivo in astroglial reporter mice. Glia 59(2):200–207. https://doi.org/10.1002/glia.21089
57. Bayraktar OA, Bartels T, Holmqvist S et al (2020) Astrocyte layers in the mammalian cerebral cortex revealed by a single-cell in situ transcriptomic map. Nat Neurosci 23(4):500–509. https://doi.org/10.1038/s41593-020-0602-1
58. Koh W, Park YM, Lee SE et al (2017) AAV-mediated astrocyte-specific gene expression under human ALDH1L1 promoter in mouse thalamus. Exp Neurobiol 26(6):350–361. https://doi.org/10.5607/en.2017.26.6.350
59. Chen H, McCarty DM, Bruce AT et al (1998) Gene transfer and expression in oligodendrocytes under the control of myelin basic protein transcriptional control region mediated by adeno-associated virus. Gene Ther 5(1):50–58. https://doi.org/10.1038/sj.gt.3300547
60. Matsuzaki Y, Fukai Y, Konno A et al (2024) Optimal different adeno-associated virus capsid/promoter combinations to target specific cell types in the common marmoset cerebral cortex. Mol Ther Methods Clin Dev 32(4):101337. https://doi.org/10.1016/j.omtm.2024.101337
61. von Jonquieres G, Fröhlich D, Klugmann CB et al (2016) Recombinant human myelin-associated glycoprotein promoter drives selective AAV-mediated transgene expression in oligodendrocytes. Front Mol Neurosci 9:13. https://doi.org/10.3389/fnmol.2016.00013
62. Okada Y, Hosoi N, Matsuzaki Y et al (2022) Development of microglia-targeting adeno-associated viral vectors as tools to study microglial behavior in vivo. Commun Biol 5(1):1224. https://doi.org/10.1038/s42003-022-04200-3
63. Shah S, Wong LM, Ellis K et al (2022) Microglia-specific promoter activities of HEXB gene. Front Cell Neurosci 16:808598. https://doi.org/10.3389/fncel.2022.808598
64. Serrano C, Cananzi S, Shen T et al (2024) Simple and highly specific targeting of resident microglia with adeno-associated virus. iScience 27(9):110706. https://doi.org/10.1016/j.isci.2024.110706
65. Hwang DY, Ismail-Beigi F (2006) Control of Glut1 promoter activity under basal conditions and in response to hyperosmolarity: role of Sp1. Am J Physiol Cell Physiol 290(2):C337–C344. https://doi.org/10.1152/ajpcell.00089.2005
66. Burek M, Forster CY (2009) Cloning and characterization of the murine claudin-5 promoter. Mol Cell Endocrinol 298(1–2):19–24. https://doi.org/10.1016/j.mce.2008.09.041
67. Mereu E, Lafzi A, Moutinho C et al (2020) Benchmarking single-cell RNA-sequencing protocols for cell atlas projects. Nature Biotechnol 38(6):747–755. https://doi.org/10.1038/s41587-020-0469-4
68. Tang F, Barbacioru C, Wang Y et al (2009) mRNA-Seq whole-transcriptome analysis of a single cell. Nat Methods 6(5):377–382. https://doi.org/10.1038/nmeth.1315
69. Buenrostro JD, Wu B, Litzenburger UM et al (2015) Single-cell chromatin accessibility reveals principles of regulatory variation. Nature 523(7561):486–490. https://doi.org/10.1038/nature14590
70. Cusanovich DA, Daza R, Adey A et al (2015) Multiplex single cell profiling of chromatin accessibility by combinatorial cellular indexing. Science (New York, NY) 348(6237):910–914. https://doi.org/10.1126/science.aab1601
71. Preissl S, Fang R, Huang H et al (2018) Single-nucleus analysis of accessible chromatin in developing mouse forebrain reveals cell-type-specific transcriptional regulation. Nat Neurosci 21(3):432–439. https://doi.org/10.1038/s41593-018-0079-3
72. Machold R, Rudy B (2024) Genetic approaches to elucidating cortical and hippocampal GABAergic interneuron diversity. Front Cell Neurosci 18:1414955. https://doi.org/10.3389/fncel.2024.1414955
73. Mich JK, Sunil S, Johansen N et al (2023) Enhancer-AAVs allow genetic access to oligodendrocytes and diverse populations of astrocytes across species. bioRxiv [Preprint]. https://doi.org/10.1101/2023.09.20.558718

74. Hrvatin S, Tzeng CP, Nagy MA et al (2019) A scalable platform for the development of cell-type-specific viral drivers. eLife 8:e48089. https://doi.org/10.7554/eLife.48089

75. Vormstein-Schneider D, Lin JD, Pelkey KA et al (2020) Viral manipulation of functionally distinct interneurons in mice, non-human primates and humans. Nat Neurosci 23(12): 1629–1636. https://doi.org/10.1038/s41593-020-0692-9

76. Aoki R, Konno A, Hosoi N, Kawabata H, Hirai H (2025) AAV vectors for specific and efficient gene expression in microglia. Cell Rep Methods 5(8):101116. https://doi.org/10.1016/j.crmeth.2025.101116

Chapter 8

Efficient Labeling of Somatodendritic Membranes of the Central Neurons

Hiroshi Kameda, Shinichiro Okamoto, Masato Koike, Takahiro Furuta, and Hiroyuki Hioki

Abstract

Neuronal cells possess dendritic and axonal processes, and through these processes they form neuronal circuits in the brain. To elucidate the neuronal circuits, it is indispensable to visualize both the cell bodies of neurons as well as their processes. Although there are several methods for visualizing these processes, such as immunostaining and intracellular staining, labeling with recombinant virus vectors has become a powerful tool because reporter proteins can be expressed in infected cells. Among the available vectors, the lentivirus vector shows low cytotoxicity, and its genome can be integrated into the genome of infected cells, resulting in stable transgene expression. Using this vector, we previously developed a reporter protein anchored to the somatodendritic membranes of neurons. The reporter protein clearly visualizes the contours of dendritic shafts and spines, which preferentially receive excitatory synaptic inputs and show morphological changes associated with synaptic activity. In this chapter, the detailed protocol for the production, purification, and injection of lentivirus vectors is described. Subsequently, the clear visualization of dendritic spines of the neurons in the neocortex and caudate-putamen is demonstrated.

Key words Membrane-targeting signal, Dendrite-targeting signal, Spine, Neuron-specific promoter, Pyramidal cell, Medium-sized spiny neuron

1 Introduction

Visualization of dendritic and axonal processes is indispensable for elucidating neuronal circuits in the brain because these processes are the input and output sites of neuronal information, respectively. To date, several methods have been reported for visualizing these processes. Immunodetection of some proteins, such as parvalbumin and somatostatin, can visualize proximal dendrites and axon terminals, which are the final output sites in axons [1, 2]. In contrast, the Golgi method and intracellular staining are suitable for the visualization of distal dendrites and dendritic protrusions called spines [3, 4], which preferentially receive excitatory synaptic inputs and show morphological changes associated with synaptic activity

Kazuto Kobayashi et al. (eds.), *Lentiviral Vectors: Applications for Basic and Clinical Neuroscience*, Methods in Molecular Biology, vol. 2974, https://doi.org/10.1007/978-1-0716-4807-0_8,

and brain disorders [5]. Thus, although these techniques have significantly contributed to the elucidation of neuronal circuits, labeling with recombinant virus vectors has been developed with the advantage of expressing reporter proteins in infected cells.

Gene transfer via recombinant virus vectors has become a powerful tool for visualizing neurons. Among the available virus vectors, the lentivirus vector enables stable transgene expression because it can integrate its genome into the genome of infected cells [6], and shows low cytotoxicity [7]. Although the lentivirus vector infects not only neurons but also glial cells [6, 8], and transgene expression in neurons by the vector is generally lower than that by other vectors [9, 10], the introduction of neuron-specific promoters and the Tet-Off system enables strong expression of transgenes in neurons, resulting in clear visualization of neuronal processes in vivo [11, 12]. Moreover, a lentivirus vector combined with modified rabies G-protein and the Tet-Off system can strongly retrogradely label dendritic processes [13]. Taken together, the recombinant lentivirus vector is a useful anterograde or retrograde tracer for the visualization of these neuronal processes.

In a previous study, to clearly visualize fine structures of dendritic processes, we first developed a lentivirus vector that expresses the myristoylation/palmitoylation (myr) site of the Fyn N-terminal peptide-attached green fluorescent protein (myrGFP) under the control of a neuron-specific promoter [14]. myrGFP locates just beneath the plasma membrane and allows clear visualization of the contours of the dendritic shafts and spines of neurons in the neocortex and caudate-putamen (CPu) (Fig. 1). To exclusively label the somatodendritic part of neurons, we further investigated the signal polypeptide that can be anchored there and found that myrGFP fused to the C-terminal cytoplasmic domain of low-density lipoprotein receptor (LDLRct), a basolateral sorting signal of polarized epithelial cells [15], was efficiently transported to the somatodendritic part of infected neurons, indicating more intense labeling of dendritic processes (Fig. 2) [14]. In this chapter, we describe the following detailed procedures: (1) preparation and injection of lentivirus particles, (2) fixation and sectioning of rat brains, and (3) visualization of the dendritic spines using immunoperoxidase or immunofluorescence staining.

2 Materials

2.1 Production and Concentration of Lentivirus Particles

1. ViraPower™ Lentiviral Expression System (Thermo Fisher Scientific).
2. pLenti expression plasmid: The genomic plasmids encoding the GFP, palGFP, myrGFP, myrGFP-LDLRct between SYN

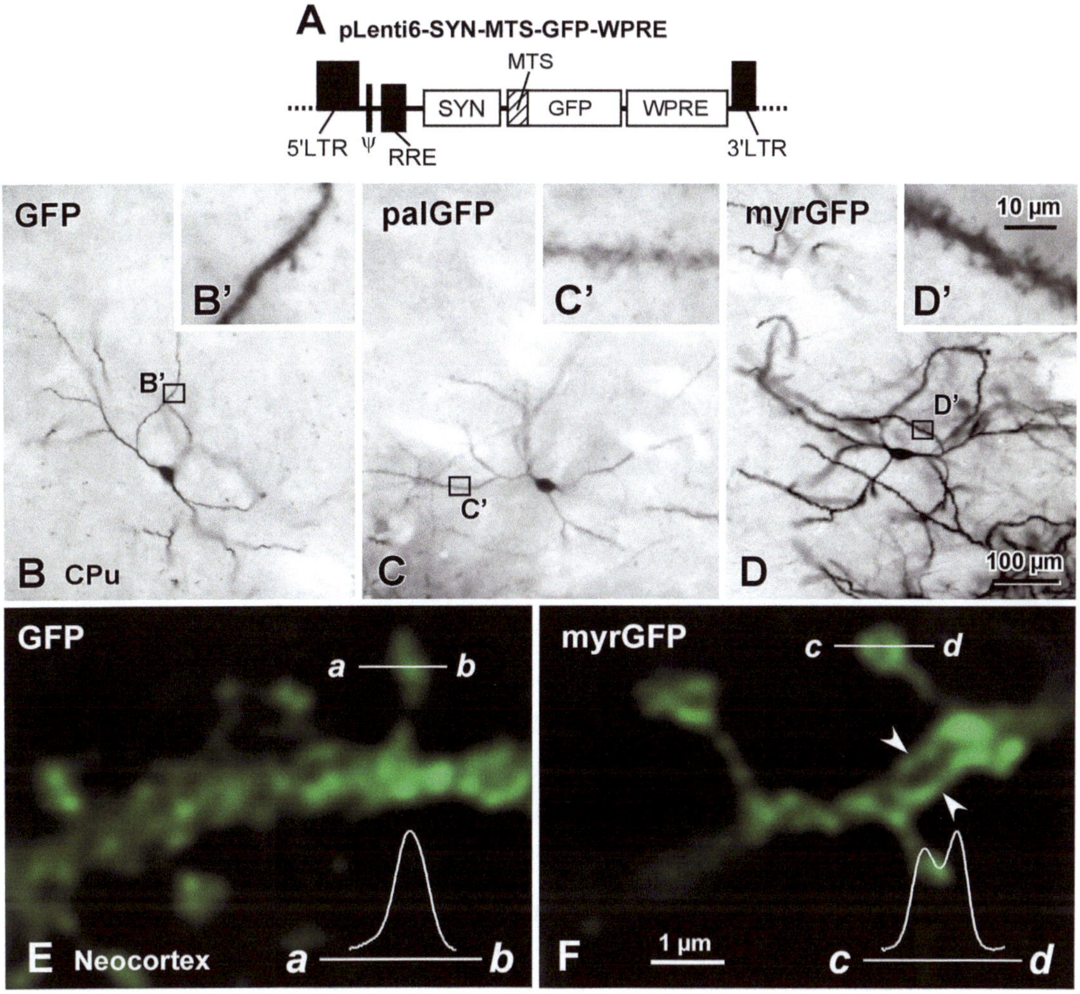

Fig. 1 Visualization of dendritic membranes of neurons in the caudate-putamen (CPu) and neocortex. (**A**) Construction of the lentivirus vector that expresses green fluorescent protein (GFP) fused with a membrane-targeting signal (MTS), the N-terminal palmitoylation (pal) site of growth-associated protein-43 (GAP-43) or myr, under the control of a neuron-specific promoter, human synapsin I (SYN). (**B–D′**) The lentivirus vector expressing GFP, palGFP, or myrGFP was injected into the CPu. Intensity of GFP immunolabeling in the somatodendritic part of infected striatal neurons was in the order myrGFP, GFP, and palGFP. Dendritic spines were most clearly labeled with myrGFP (**B′–D′**). (**E, F**) The lentivirus vector expressing GFP or myrGFP was injected into the neocortex. Confocal laser scanning microscopic images of GFP-immunopositive apical dendritic tufts were compared between GFP- (**E**) and myrGFP-labeled pyramidal cells (**F**). The lower right graphs in **E** and **F** are fluorescence intensity plots along lines *ab* and *cd*, respectively. Only one peak is observed in **E**, whereas two peaks are observed in **F**, suggesting that myrGFP was located just beneath the plasma membrane of the spine. Arrowheads in **F** indicate the two walls of the dendritic shaft, also indicating that myrGFP was concentrated at the membrane. LTR, long terminal repeat of lentivirus; RRE, rev responsive element; WPRE, woodchuck hepatitis virus posttranscriptional regulatory element; Ψ, packaging signal. Bar in **D** applies to **B–D**, that in **D′** to **B′–D′**, and that in **F** to **E** and **F**

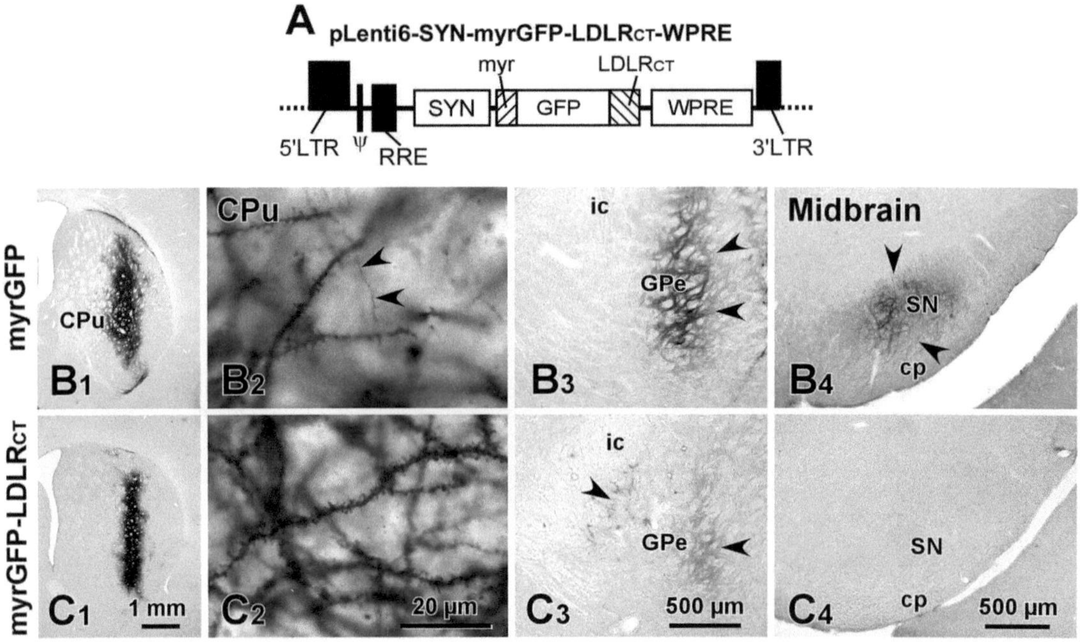

Fig. 2 Effects of low-density lipoprotein receptor (LDLRcT) as a dendrite-targeting signal. (**A**) Construction of the lentivirus vector, which expresses myrGFP fused with LDLRcT under the control of SYN promoter. (**B**) When lentivirus vector that expresses myrGFP was injected in the CPu (**B$_1$**), anterogradely transported GFP was detected in the external segment of the globus pallidus (GPe) (arrowheads in **B$_3$**) and substantia nigra (SN) (arrowheads in **B$_4$**), which are the target regions of striatal projection neurons. GFP-immunopositive axon collaterals were also often observed in the injected site of the vector (arrowheads in **B$_2$**). (**C**) With myrGFP-LDLRcT virus, very weak GFP immunoreactivity was observed in the GPe (arrowheads in **C$_3$**); however, no immunoreactivity was found in the SN (**C$_4$**). GFP-immunopositive axon collaterals were rarely observed (**C$_2$**). cp, cerebral peduncle; ic, internal capsule. Bars in **C*n*** apply to **B*n*–C*n***

promoter and WPRE sequence, termed pLenti6-SYN-GFP-WPRE, pLenti6-SYN-palGFP-WPRE, pLenti6-SYN-myrGFP-WPRE, or pLenti6-SYN-myrGFP-LDLRcT-WPRE, respectively (*see* **Note 1**).

3. Helper plasmids: ViraPower™ Packaging Mix (Thermo Fisher Scientific). The mixture contains pLP1, pLP2, and pLP/VSVG plasmids.
4. 293FT cells (Thermo Fisher Scientific).
5. Phosphate-buffered saline (PBS)(−).
6. Opti-MEM™ (Thermo Fisher Scientific).
7. Opti-MEM™(+): Add fetal bovine serum (FBS) to a final concentration of 2%.
8. Trypsin solution.
9. Lipofectamine® 2000 (Thermo Fisher Scientific).

10. Complete medium: Dulbecco's Modified Eagle Medium (Thermo Fisher Scientific) containing 0.1% Minimum Essential Medium Non-Essential Amino Acids, 2 mM L-glutamine, 1 mM Sodium pyruvate, 500 μg/mL G418, 1% penicillin-streptomycin solution, and 10% FBS.
11. 100 mm tissue culture plastic dish.
12. BioCoat™ poly-D-lysine 100 mm dish (Corning).
13. Round-bottom polystyrene 5 mL tubes.
14. Polypropylene 50 mL conical tubes.
15. Centrifuge.
16. CO_2 incubator.
17. 0.45 μm filter.
18. Polypropylene 50 mL conical tubes.
19. Centricon® Plus-20 (Millipore).
20. Centrifuge.

2.2 Injection

1. Anesthesia: All animal care and use should be in accordance with the National Institutes of Health Guide for the Care and Use of Laboratory Animals, and the experiments should be approved by the Committee for Animal Care and Use and the Committee for Recombinant DNA Study at each institute.
2. Lentivirus expressing GFP, palGFP, myrGFP, or myrGFP-LDLRCT.
3. PBS: 5 mM PBS (0.9%, w/v, pH 7.4).
4. Bovine serum albumin (BSA).
5. Stereotaxic instrument (Narishige or David Kopf).
6. Surgical scissors.
7. Superfine tweezers.
8. Picospritzer® III (General Valve Corporation).
9. N_2 gas.
10. Hand drill.
11. Steel hole cutter.
12. Clip applier.
13. Glass micropipette.

2.3 Fixation

1. Anesthesia.
2. PBS.
3. Fixative solution: 3% (v/v) formaldehyde, 75% (v/v) saturated picric acid (equal to 0.95% (w/v) picric acid), and 0.1 M Na_2HPO_4 (adjusted to pH 7.0 using NaOH) (*see* **Note 2**).
4. Needle (18 gauge).

5. Tweezers.
6. Surgical scissors.
7. Bone forceps.

2.4 Sectioning

1. 0.1 M sodium phosphate buffer (PB) (pH 7.4) (*see* **Note 3**).
2. 30% (w/v) sucrose in 0.1 M PB.
3. Optimal cutting temperature (O.C.T.) compound.
4. Dry ice powder.
5. Freezing microtome.
6. 24-Well cell culture plates.
7. Sodium azide.

2.5 Immuno-peroxidase Staining

1. H_2O_2.
2. PBS.
3. PBS-X: PBS containing 0.3% (v/v) Triton X-100.
4. Affinity-purified rabbit antibody to GFP (*see* **Note 4**).
5. PBS-XCD: PBS-X containing 1% (v/v) normal donkey serum and 0.25% (w/v) λ-carrageenan (*see* **Note 5**).
6. Goat antibody to rabbit IgG (MP Biomedicals).
7. Rabbit peroxidase-anti-peroxidase (Jackson Immunoresearch Laboratories).
8. Diaminobenzidine-4HCl (DAB).
9. 0.5 M Tris–HCl, pH 7.6.
10. Gelatinized glass slides.
11. Graded series of ethanol in water (30, 50, 70, 95, and 100%).
12. Xylene.
13. NEW MX: Mounting medium (Matsunami Glass).
14. Microphot-FXA (Nikon).
15. 1× objective lens.
16. 4× objective lens.
17. 40× objective lens.
18. 100× objective lens.
19. Digital monochrome camera.

2.6 Immuno-fluorescence Staining

1. PBS-X.
2. PBS-XCD.
3. Affinity-purified rabbit antibody to GFP (*see* **Note 4**).
4. AlexaFluor 488-conjugated anti-rabbit IgG goat antibody (A-11008; Thermo Fisher Scientific).

5. Gelatinized glass slides.
6. Fluorescence mounting medium: 90% (v/v) glycerol and 2.5% (w/v) triethylenediamine in 20 mM Tris–HCl (pH 7.4) (*see* **Note 6**).
7. TCS SP2 confocal laser scanning microscope (Leica microsystems).
8. 63× objective lens.
9. ImageJ.

3 Methods

3.1 Production and Concentration of Lentivirus Vectors

1. The day before transfection, plate 293FT cells in a 100 mm tissue culture dish to get 90–95% confluency the next day.
2. Pre-warm 7 mL of Opti-MEM™(+) in the poly-D-lysine coated 100 mm dish.
3. For every transfection sample, prepare DNA-Lipofectamine® 2000 complexes as follows:
   (a) In a 5 mL round-bottom polystyrene tube, mix 9 μg of ViraPower™ Packaging Mix and 3 μg of pLenti expression plasmid, pLenti6-SYN-GFP-WPRE, pLenti6-SYN-palGFP-WPRE, pLenti6-SYN-myrGFP-WPRE, or pLenti6-SYN-myrGFP-LDLRct-WPRE, with 1.5 mL of Opti-MEM™. Mix gently and incubate at RT for 5 min.
   (b) In a 5 mL round-bottom polystyrene tube, dilute 36 μL of Lipofectamine® 2000 with 1.5 mL of Opti-MEM™. Mix gently and incubate at RT for 5 min.
   (c) After incubation for 5 min, combine the diluted DNA with diluted Lipofectamine® 2000 and mix gently.
   (d) Incubate at RT for 20 min with gentle shaking to allow for the formation of DNA-Lipofectamine® 2000 complexes.
4. In the meantime, prepare 293FT cells for virus production as follows:
   (a) Remove the medium and wash the cells once with 5 mL of PBS(−).
   (b) Add 2 mL of trypsin solution to the monolayer and incubate for 3 min in a CO_2 incubator.
   (c) Briefly pipet the solution to obtain a single cell suspension.
   (d) Transfer the cells to a 50 mL conical tube and add 2 mL of Opti-MEM™(+).
   (e) Centrifuge the cells at 200× *g* for 3 min at 4 °C.

(f) Aspirate the medium and resuspend the cells in 150 μL of Opti-MEM™(+).

5. Add the resuspended cells into 3 mL of DNA-Lipofectamine® 2000 complexes.
6. Incubate them at RT for 5 min.
7. Add the mixture of the cells and DNA-Lipofectamine® 2000 complexes to the poly-D-lysine coated 100 mm dish containing pre-warmed Opti-MEM™(+).
8. Mix gently by rocking the dish back and forth.
9. Incubate the cells in a CO_2 incubator.
10. After 8–12 h, remove the medium containing the DNA-Lipofectamine® 2000 complexes and replace with 12 mL of "complete medium."
11. Harvest virus-containing supernatant 48–72 h (60 h recommended) post-transfection by removing medium in a 50 mL conical tube.
12. To separate the cell debris, centrifuge at 2000× *g* for 15 min at 4 °C.
13. Collect the supernatant, and filter it through a sterile 0.45 μm low protein-binding filter.
14. Concentrate the virus solution using a Centricon® Plus-20 Centrifugal Filter Concentrator as follows:
    (a) Add virus solution to the Centricon® Plus-20 filter device.
    (b) Centrifuge at 3500× *g* for 10 min at 4 °C.
    (c) Place sample reservoir upside down in a new vial.
    (d) Centrifuge at 200× *g* for 2 min at 4 °C to transfer the concentrate to the vial.
15. Divide the solution into 10 μL aliquots and store at −80 °C.
16. Determine the actual virus titer by adding serial dilutions of the virus solution to 293FT cells.

3.2 Injection

1. Dilute the lentivirus vector to 1.0×10^6 infectious units (IU) in PBS containing 2% (w/v) BSA (*see* **Note** 7).
2. Anesthetize the rats following the regulations of the institution.
3. Place the anesthetized rats in a stereotaxic instrument, and adjust the incisor bar until the heights of the bregma and lambda are equal to achieve a flat skull position.
4. Make an incision in the scalp along the midline and expose the skull.
5. Thin the skull above the CPu (0.3 mm posterior to the bregma and 3.5 mm lateral to the midline) or primary somatosensory

cortex (S1) of rats (2.6 mm posterior to the bregma and 4.0 mm lateral to the midline) using a hand drill with a steel hole cutter, and carefully remove the remaining bones using superfine tweezers to expose the brain surface (*see* **Note 8**).

6. Load 1.0 μL of the lentivirus vector into a glass micropipette, and attach the micropipette to a Picospritzer® III.
7. Move the tip of the micropipette to the CPu (0.3 mm posterior to the bregma, 3.5 mm lateral to the midline, and 4.7 mm deep from the brain surface) or S1 of rats (2.6 mm posterior to the bregma, 4.0 mm lateral to the midline, and 1.5 mm deep from the brain surface) (*see* **Note 9**).
8. Inject 1.0 μL of the virus into the CPu or S1 using pressure for 5 min.
9. Slowly remove the micropipette from the brain, and suture the skin incision.
10. Fix the rats 7–28 days after the injection.

3.3 Fixation

1. Anesthetize the rats following the regulations of the institution.
2. Open the thoracic cavity with surgical scissors, insert an 18-gauge needle from the left ventricle of the heart into the aorta, and cut the right atrial appendage (*see* **Note 10**).
3. Perfuse with 200 mL of PBS for 10 min to remove the blood from the circulatory system.
4. Perfuse with 200 mL of fixative solution for 10 min.
5. Remove the scalp, skull, and dura mater using bone forceps and tweezers.
6. Remove the brain and place it in a fixative solution for 4 h at RT or overnight at 4 °C.

3.4 Sectioning

1. Replace the fixative with 30% (w/v) sucrose in 0.1 M PB and gently shake it for 1–2 days at 4 °C to cryoprotect the brain tissue.
2. Place the brain on the stage of a freezing microtome with a 2:1 mixture of 30% (w/v) sucrose:O.C.T. compound and briefly freeze them using dry ice powder (*see* **Note 11**).
3. Adjust the angle of the stage to keep the coronal plane of the brain horizontal.
4. Layer 30% (w/v) sucrose in 0.1 M PB on the surface of the brain, and freeze the sucrose solution and brain completely by embedding in dry ice powder on the stage for at least 5 min (*see* **Note 12**).

5. Cut the brain into 40-μm-thick coronal sections on a freezing microtome, and collect the sections serially in 24-well cell culture plate with 0.1 M PB containing 0.02% sodium azide.
6. Store at 4 °C until further use.

3.5 Immunoperoxidase Staining

1. Pick the sections containing the CPu, the external segment of the globus pallidus (GPe), or the substantia nigra (SN) from the serial sections.
2. Wash the sections with PBS twice at RT for 10 min.
3. Incubate the sections at RT for 30 min with 1% (v/v) H_2O_2 in PBS to block endogenous peroxidase activity.
4. Wash the sections with PBS-X twice at RT for 10 min.
5. Incubate the sections at RT overnight with 1.0 μg/mL affinity-purified rabbit antibody to GFP in PBS-XCD.
6. Wash the sections with PBS-X twice at RT for 10 min.
7. Incubate the sections at RT for 2 h with 60 μg/mL anti-rabbit IgG goat antibody in PBS-XCD.
8. Wash the sections with PBS-X twice at RT for 10 min.
9. Incubate the sections at RT for 1 h with 50 μg/mL rabbit peroxidase-anti-peroxidase in PBS-X.
10. Wash the sections with PBS twice at RT for 10 min.
11. Incubate the sections at RT for 20–40 min with 0.02% (w/v) DAB and 0.001% (v/v) H_2O_2 in 50 mM Tris–HCl (pH 7.6).
12. Wash the sections with PBS twice at RT for 10 min.
13. Mount all the stained sections onto gelatinized glass slides.
14. Dry the sections.
15. Soak the glass slides in ethanol-water mixtures of 30%, 50%, and 70% (v/v) at RT for 30 min each, and dip the slides into 95% (v/v) ethanol in water.
16. Soak the glass slides in 100% ethanol overnight at RT.
17. Dip the glass slides in 50% xylene-ethanol mixture (v/v), and soak the slides in xylene overnight at RT.
18. Cover the sections with the mounting medium NEW MX and place the coverslips.
19. Capture digital images with a digital monochrome camera using 1×, 4×, 40×, and 100× objective lenses.

3.6 Immunofluorescence Staining

1. Pick the cortical sections containing the cell bodies and dendrites of the infected neurons from the serial sections.
2. Wash the sections with PBS-X twice at RT for 10 min.
3. Incubate the sections at RT overnight with 1.0 μg/mL affinity-purified rabbit antibody to GFP in PBS-XCD.

4. Wash the sections with PBS-X twice at RT for 10 min.
5. Incubate the sections at RT for 1 h with 5 μg/mL AlexaFluor 488-conjugated anti-rabbit IgG goat antibody in PBS-XCD.
6. Wash the sections with PBS-X twice at RT for 10 min.
7. Mount all the stained sections onto gelatinized glass slides.
8. Cover the sections with the fluorescence mounting medium and place the coverslips.
9. Examine the samples under confocal laser scanning microscope TCS SP2 with a 488-nm laser beam and a 510–530 nm emission prism window.
10. Capture digital fluorescence images with the optical slice thickness of 81 nm (pinhole corresponding to 1 Airy unit), using a 63× objective lens.
11. Measure the fluorescence intensity of confocal laser scanning images using the ImageJ software.

4 Notes

1. The plasmid DNA used for transfection should be clean and should contain low endotoxin levels. Usually, plasmid DNA are purified using PureLink™ HiPure Plasmid Kits (Thermo Fisher Scientific). Instead, you can purify the plasmid using cesium chloride equilibrium centrifugation. The detailed information of the plasmids is shown in Kameda et al. [14].
2. After preparing the fixation solution, leave it at least overnight at RT to completely hydrate formaldehyde. You can also use 4% paraformaldehyde in 0.1 M PB as a fixative solution.
3. To prevent the blade from rusting, PB but not PBS should be used for sectioning. In addition, after sectioning, the blades should be washed thoroughly with deionized water and ethanol.
4. This antibody is self-made and affinity-purified by column [16].
5. λ-Carrageenan makes the incubation medium highly viscous and contributes to avoiding uneven binding of antibodies to sections [17].
6. Triethylenediamine is used as an anti-fading reagent.
7. BSA blocks the nonspecific binding of virus particles to the surfaces of pipette tips, tubes, and glass micropipettes, thus contributing to a linear decrease in the virus titer with dilution.
8. The stereotaxic coordinates for the CPu and S1 are determined according to the rat brain atlas [18].

9. When the tip of the glass micropipette reaches the target point, move the micropipette 0.05 mm upward to create space for injection of the virus solution.
10. Before perfusion, cut and smooth the tip of the 18-gauge needle to avoid stabbing the wall of the blood vessel.
11. The mixture of sucrose solution and the O.C.T. compound enables smooth slicing of the brain tissue compared with using the O.C.T. compound alone.
12. Layering sucrose solution on the surface of the brain protects the brain from melting during sectioning.

Acknowledgments

The authors thank Editage (www.editage.com) for English language editing.

Conflict of Interest Statement The authors have no conflicts of interest to declare that are relevant to the content of this chapter.

Funding

This work was supported by KAKENHI (JP23K20044 and JP25K02371 to H.H.) from the Japan Society for the Promotion of Science (JSPS), the Japan Agency for Medical Research and Development (AMED) (JP21dm0207112 and JP24wm0625103 to H.H.), Moonshot R&D from the Japan Science and Technology Agency (JST) (JPMJMS2024 to H.H.), and Fusion Oriented Research for disruptive Science and Technology (FOREST) from JST (JPMJFR204D to H.H.).

References

1. DeFelipe J, Hendry SHC, Jones EG (1989) Visualization of chandelier cell axons by parvalbumin immunoreactivity in monkey cerebral cortex. Proc Natl Acad Sci USA 86:2093–2097
2. Milner TA, Bacon CE (1989) Ultrastructural localization of somatostatin-like immunoreactivity in the rat dentate gyrus. J Comp Neurol 290:544–560
3. Westman J (1968) The lateral cervical nucleus in the cat. I. A Golgi study. Brain Res 10:352–368
4. Takato M, Goldring S (1979) Intracellular marking with Lucifer Yellow CH and horseradish peroxidase of cells electrophysiologically characterized as glia in the cerebral cortex of the cat. J Comp Neurol 186:173–188
5. Nimchinsky EA, Sabatini BL, Svoboda K (2002) Structure and function of dendritic spines. Annu Rev Physiol 64:313–353
6. Naldini L, Blömer U, Gallay P, Ory D, Mulligan R, Gage FH, Verma IM, Trono D (1996) In vivo gene delivery and stable transduction of nondividing cells by a lentiviral vector. Science 272:263–267
7. Pan D, Gunther R, Duan W, Wendell S, Kaemmerer W, Kafri T, Verma IM, Whitley CB (2002) Biodistribution and toxicity studies of VSVG-pseudotyped lentiviral vector after

intravenous administration in mice with the observation of in vivo transduction of bone marrow. Mol Ther 6:19–29

8. Dittgen T, Nimmerjahn A, Komai S, Licznerski P, Waters J, Margrie TW, Helmchen F, Denk W, Brecht M, Osten P (2004) Lentivirus-based genetic manipulations of cortical neurons and their optical and electrophysiological monitoring in vivo. Proc Natl Acad Sci USA 101:18206–18211
9. Kanter-Schlifke I, Georgievska B, Kirik D, Kokaia M (2007) Brain area, age and viral vector-specific glial cell-line-derived neurotrophic factor expression and transport in rat. Neuroreport 18:845–850
10. Wickersham IR, Finke S, Conzelmann KK, Callaway EM (2007) Retrograde neuronal tracing with a deletion-mutant rabies virus. Nat Methods 4:47–49
11. Hioki H, Kameda H, Nakamura H, Okunomiya T, Ohira K, Nakamura K, Kuroda M, Furuta T, Kaneko T (2007) Efficient gene transduction of neurons by lentivirus with enhanced neuron-specific promoters. Gene Ther 14:872–882
12. Hioki H, Kuramoto E, Konno M, Kameda H, Takahashi Y, Nakano T, Nakamura KC, Kaneko T (2009) High-level transgene expression in neurons by lentivirus with Tet-off system. Neurosci Res 63:149–154
13. Watakabe A, Kato S, Kobayashi K, Takaji M, Nakagami Y, Sadakane O, Ohtsuka M, Hioki H, Kaneko T, Okuno H, Kawashima T, Bito H, Kitamura Y, Yamamori T (2012) Visualization of cortical projection neurons with retrograde TET-off lentiviral vector. PLoS One 7:e46157
14. Kameda H, Furuta T, Matsuda W, Ohira K, Nakamura K, Hioki H, Kaneko T (2008) Targeting green fluorescent protein to dendritic membrane in central neurons. Neurosci Res 61:79–91
15. Matter K, Hunziker W, Mellman I (1992) Basolateral sorting of LDL receptor in MDCK cells: the cytoplasmic domain contains two tyrosine-dependent targeting determinants. Cell 71:741–753
16. Tamamaki N, Nakamura K, Furuta T, Asamoto K, Kaneko T (2000) Neurons in Golgi-stain-like images revealed by GFP-adenovirus infection in vivo. Neurosci Res 38:231–236
17. Hioki H, Okamoto S, Konno M, Kameda H, Sohn J, Kuramoto E, Fujiyama F, Kaneko T (2013) Cell type-specific inhibitory inputs to dendritic and somatic compartments of parvalbumin-expressing neocortical interneuron. J Neurosci 33:544–555
18. Paxinos G, Watson C (2007) The rat brain in stereotaxic coordinates. Academic Press, San Diego

Chapter 9

Visualization of Selective Neural Pathways by Double Viral Infection Method

Akiya Watakabe

Abstract

Application of viral vector technology for neural tracing enabled the visualization of neural pathways in a manner that had been impossible by classic techniques. The development of highly efficient retrograde vector played a pivotal role in its success. This chapter explains the strategy, protocol, and the expected results for visualization of selective neural pathways by double viral vector infection method. This method utilizes the combination of lentiviral vector showing retrograde gene transfer and adeno-associated viral (AAV) vector to split the TET-Off/TET-On system to these vectors. By co-injection of these vectors at the target and the origins of the pathway, it is possible to label a specific population of neurons that connect the two injection sites. The enhancement of transgene expression by TET system results in Golgi-like visualization of the cells from their dendrites to axon terminals. Using this technique, the corticothalamic cells are specifically labeled in their entirety including their collateral axons.

Key words Lentiviral vector, Retrograde gene transfer, NeuRet vector, Adeno-associated viral vector, Axon collateral, Corticothalamic, Tracer, TET-Off system

1 Introduction

The nervous system is a huge network of various cell types, in which neurons make long- and short distance connections to form functional circuits. Elucidation of neural connections is one of the central questions of neuroscience. Development of anterograde and retrograde tracers for neural connectivity was pivotal in solving this problem [1, 2]. These tracers are either taken up from the cell body or nerve terminals and transported to the other end of the connection. They are then visualized using isotopic (e.g., tritiated amino acids), enzymatic (e.g., HRP: horse radish peroxidase), immunohistochemical (e.g., BDA: biotinylated dextran amine), or fluorescent (e.g., FastBlue) labeling for axonal and/or cellular localization [1, 2]. These tracers are extremely useful to visualize the bulk outputs (for anterograde tracers) and inputs (for

Kazuto Kobayashi et al. (eds.), *Lentiviral Vectors: Applications for Basic and Clinical Neuroscience*, Methods in Molecular Biology, vol. 2974, https://doi.org/10.1007/978-1-0716-4807-0_9,

retrograde tracers) of the neurons around the injection site. Furthermore, by restricted administration, some tracers (e.g., HRP, BDA and biocytin) have been used to visualize fine morphology of isolated axons and neurons [3–5]. A recent addition to this array of neuroanatomical tools is the viral-based tracers.

Rabies virus is a neurotropic virus that moves retrogradely from neuron to neuron via synaptic junctions, accompanied by self-replication during passage. This property has been successfully used to trace a chain of transsynaptic connectivity [6]. Furthermore, genetically engineered rabies viral vectors are now used as monosynaptic retrograde tracer [7] and cell-type specific transsynaptic tracer [8]. Another interesting application of the rabies' ability for retrograde transport is pseudotyping the lentiviral vector with the rabies virus glycoprotein [9]. By fusing the glycoprotein of the rabies and vesicular stomatitis virus, a highly efficient retrograde vector with low toxicity was developed [10–13]. Using this vector for expression by "TET-Off" system, a high level expression was achieved to visualize fine morphology of specific projection neuron subtypes [14]. In this system, a transgene of interest is placed downstream of the tetracycline responsive element (TRE), which becomes active in the presence of the tetracycline transactivator (tTA). By splitting these two components into the retrograde gene transfer vector and local infection vector, it became possible to restrict the transgene expression to neurons that connect the two sites of viral injections (see Fig. 1). In this chapter, the basic methodology and expected result of this double viral vector infection method is explained, taking an example of corticothalamic cells in the somatosensory barrel field (S1BF) in mice.

2 Materials

2.1 Viral Vectors

1. TET-Off system: An example combination is shown in Fig. 1. tTA2 and fluorescent protein (e.g., tRFP) under TRE promoter are encoded in separate vectors (*see* **Note 1**). By simultaneously transducing neurons at the cell bodies and the terminals, TET-Off system can work only in neurons that were doubly infected by these viruses (*see* **Note 2**). These pathway-specific neurons exhibit highly enhanced expression of the fluorescent protein due to the amplification step by tTA [14, 15].
2. Local infection vector: AAV is most suitable. We routinely use AAV1 with success, but other serotypes would also be usable [16, 17] (*see* **Note 3**).
3. Retrograde gene transfer vector: a neuron-specific retrograde gene transfer (NeuRet) vector shows excellent result [12, 13]. It provides highly efficient retrograde transduction

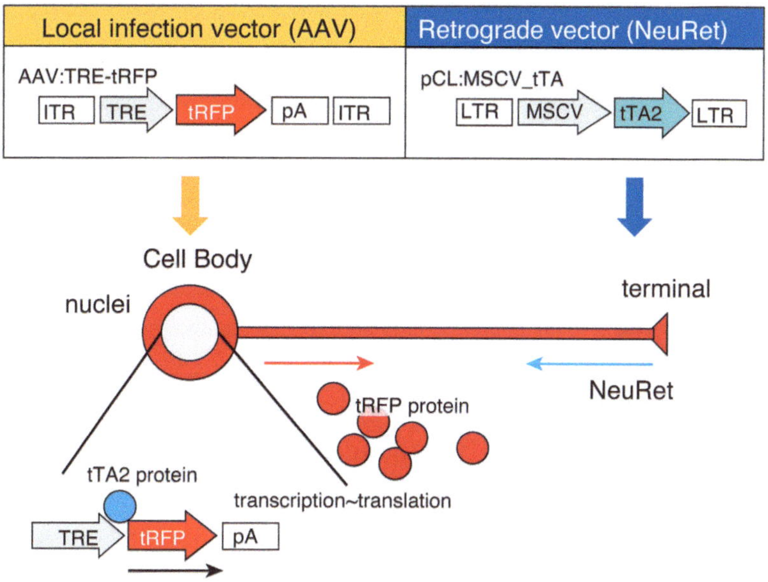

Fig. 1 Double viral infection strategy for pathway-specific labeling. Adeno-associated viral (AAV) vector provides the tRFP (turbo Red Fluorescent Protein) red fluorescein coding sequence under TRE promoter in locally infected cells, whereas NeuRet vector retrogradely infects the same cells from the terminal, which provide the tTA2 transactivator. When the neurons are doubly infected by AAV and NeuRet vectors, tTA2 starts expression from TRE promoter to produce a large amount of tRFP, which fills the entire neuronal cells

with low toxicity. Alternative options include rabies viral vector [18], adenoviral vector [19], and AAV2-retro [20] (*see* **Note 4**).

2.2 Viral Injections

1. Injector: Nanoject 2000 or newer versions of nanoject (Drummond). Programmable injectors are suitable for delicate control of injections.
2. Injection Capillary: Nanoject Glass Capillaries (3–000-203-G/X). Control the puller temperature to make suitable shapes: short taper for cortical injections (outer diameter 30 ~ 40 μm) and long taper for deeper injections (outer diameter 60 ~ 70 μm) (Fig. 2; *see* **Note 5**). It is recommended to sharpen the tip by microgrinder (Narishige EG-45).
3. Retrograde tracer: CTB-Alexa488, final 0.2 mg/ml in PBS (Invitrogen, C22841).

2.3 Immunostaining

1. Dent's solution: 80% methanol/20% dimethyl sulfoxide solution.
2. Tris-buffered saline (TBS)): 25 mM Tris-HCl, pH 7.4, 137 mM NaCl, 2.7 mM KCl.

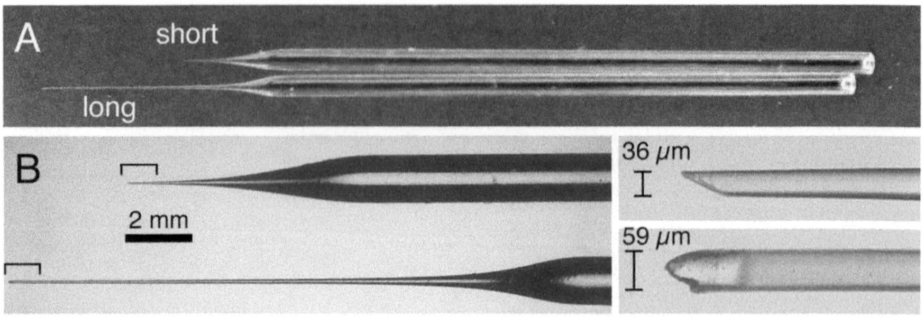

Fig. 2 Examples of short- and long-tapered injection capillaries

3. 4% paraformaldehyde: 4% paraformaldehyde in 0.1 M phosphate buffer or PBS.
4. Immersion buffer: 10% FBS, 2% BSA 0.5% TritonX100 in TBS.
5. TNT: 0.1 M Tris-HCl, pH 7.5, 0.15 M NaCl, 0.05% Tween20.
6. Thirty percent sucrose solution: 30% sucrose in PBS.
7. Hoechst 33342 (Molecular Probes).
8. Fluoromount-G mounting medium (Invitrogen). (*see* **Note 6**).

3 Methods

Preparation of lentiviral vectors is described in Chap. 2. In this chapter, the protocols for vector injection and transgene detection by immunostaining are mentioned.

3.1 Viral Injection

1. Anesthetize the mouse with intraperitoneal injection of ketamine/xylazine mixture (100 and 10 mg/kg, respectively). Alternatively, gas anesthesia by isoflurane provides better control of the anesthesia. Securely fix the mouse in a stereotaxic apparatus. Disinfect the head with Povidone-iodine.
2. Incise the mouse's scalp, scrape the skull surface to expose the cranial bone and make a small hole using a dental drill. Be careful not to damage the dura mater.
3. Wet the dura mater with saline and pierce into the brain tissue with the sharpened injection capillary (*see* **Note 7**). Inject the desired volume of viruses/tracers. The speed should be less than 0.1–0.2 μL/min to reduce back propagation along the injection capillary. Confirm before and during injection that the capillary is not clogged.
4. Suture the incision and warm the mouse for recovery.

3.2 Transgene Detection by Immunostaining (Floating Method)

1. After 3–4 weeks of waiting period, perfusion fix the mouse with 4% paraformaldehyde. Cryoprotect the brain with 30% sucrose.
2. Section the brain on a freezing mircrotome (20–40 μm width).
3. Treat the sections with Dent's solution for 30 min for better penetration of the antibody for axonal staining (*see* **Note 8**).
4. Block the sections with Immersion buffer and incubate the primary antibody in Immersion buffer overnight at 4 °C (*see* **Note 9**).
5. After rinsing the sections three times with TNT, incubate the sections with the secondary fluorescent tagged antibodies in TNT for 3 h at room temperature.
6. After rinsing the sections three times with TNT, counterstain the sections with Hoechst 33342 (1:2000).
7. Mount the sections immersed in TBS onto a slideglass using paintbrushes, absorb excess moisture, and dry. Envelop into Fluoromount-G with a coverslip. Wait until the mounting medium hardens and gets ready for observation. Typical results are shown in Figs. 3 and 4 (*see* **Note 10**).

4 Notes

1. The TRE-fluorescent protein transcription unit and MSCV-tTA2 transcription units can be swapped between local infection vector and retrograde vector. For example, two NeuRet vectors encoding different colors can be injected to different target regions, while a tTA-expressing AAV vector is injected to their common origin, visualizing two different projection cell types in the same area (e.g., corticothalamic vs callosal) [21].
2. The pathway-specific TET system by double viral method has been successfully used for functional studies in macaques [22].
3. The bulk labeling highlights the high degree of layer-specificity of neurite extension as a population [21]. Instead, the neurons can be sparsely labeled for more detailed morphological analyses by reducing the titer of the local infection vector (that is, AAV) [23].
4. AAV2-retro [20] is a superb alternative to label corticocortical cells. Note, however, that AAV2-retro may not be suitable to probe subcortical connections [24].
5. Generally speaking, pulling a capillary that was melted at a lower temperature result in shorter taper.
6. DAPI-Fluoromount-G (SouthernBiotech) can be used to stain nuclei during mounting.

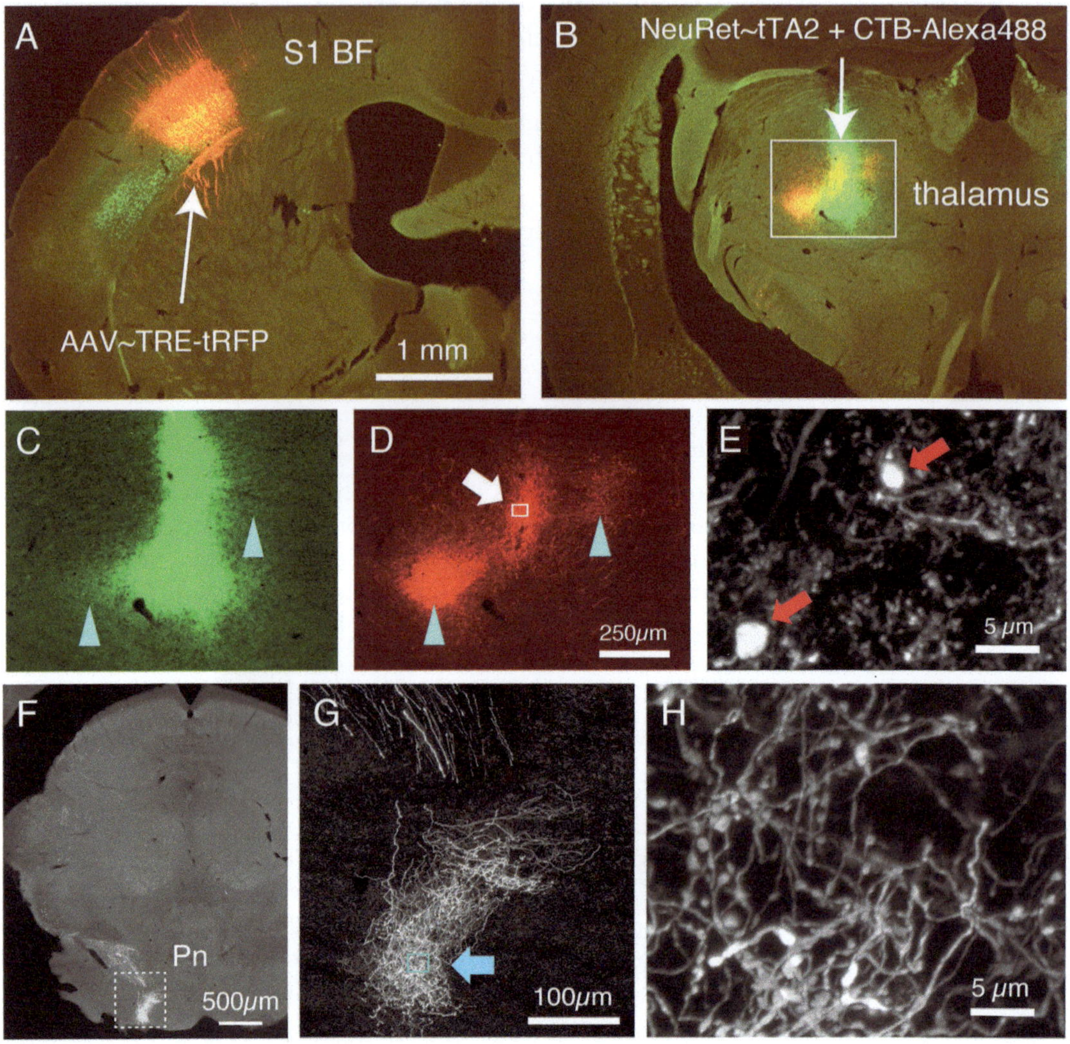

Fig. 3 Visualization of the entirety of the corticothalamic cells by double viral infection method. The AAV carrying tetracycline responsive element (TRE-tRFP was injected into the somatosensory barrel field (panel A; S1BF) and the NeuRet vector encoding tTA2 was injected into the thalamus (panel B). The NeuRet vector was co-injected with CTB-Alexa488, a fluorescent retrograde tracer, to label corticothalamic cells in general (panel C). Note that expression of tRFP in S1BF is restricted to the site of AAV injection as compared to widely spread CTB-labeled green cells. A big advantage of this method is that the terminal structure of the labeled corticocortical cells can be fully visible, which is not possible by retrograde vector alone. In panel D, three clusters of axon terminals are visible, among which the central one (white arrow) is considered to have taken up the NeuRet vector. Two other clusters (cyan arrowhead) are likely to be the axon collaterals of the doubly infected neurons. By confocal microscopic imaging, it is possible to image very detailed structures of the axon terminals (panel E), such as large boutons (red arrows), characteristic of layer 5 corticothalamic terminals in the nucleus posterior (PO) [25]. This labeling also visualized the axon collaterals toward the pontine nucleus (panels F-H). Panels G and H are confocal microscopic imaging of the region shown in panel F

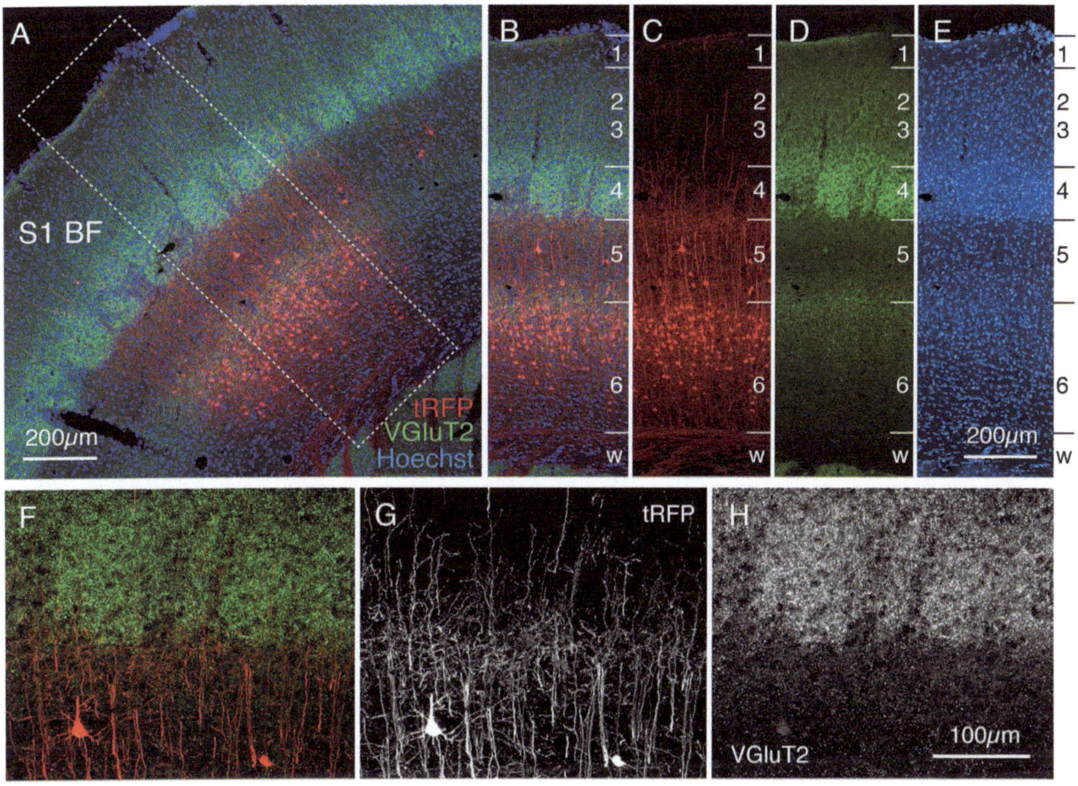

Fig. 4 Immunolocalization of thalamocortical terminals in comparison to the corticothalamic cells. (A-E) Corticothalamic cells were labeled by tRFP (red). This is the same sample as shown in Fig. 3. Thalamocortical terminals were labeled by VGluT2 antibody [26] (green). Nuclear counterstaining was performed using Hoechst33342 (blue). Note high concentration of VGluT2 terminals in layer 4, where the neurites from the corticothalamic cells appear to be excluded. (F-H) High magnification views of the layer 4/5 boundary. Note good segregation of the thalamocortical terminals and corticothalamic neurites. All these images were taken by confocal microscopic imaging [21]

7. If the condition is good, the capillary penetrates through the dura with no resistance. If the capillary pushes the dura, you can make a small hole with a tip of a 27G needle.
8. The sections are treated in 12-well plate in a floating condition. Methanol treatment hardens the section. Take care not to let the section attached to each other during treatment, which will lead to uneven staining.
9. Although fluorescent proteins are visible with no further treatment, antibody staining improves visibility of weak signals, which may be otherwise obscured by tissue autofluorescence or nearby strong signals (Fig. 3).
10. When you fail to observe high-level expression of transgene, either of the two injections may have failed. It is possible to confirm the successful injection of the lentiviral vector by

immunostaining tTA, which should be abundantly expressed around the injection site. Furthermore, the leak expression of the TRE-transgene in AAV may be also confirmed by immunostaining. If the transgene of interest is not detectable by immunostaining, you can still visualize the site of injection by in situ hybridization; you are expected to detect abundant hybridization signals to the AAV genome (see Fig. S1D of [21]). These hybridization signals can be detected even with the sense probe and without waiting period after injection (unpublished data).

Acknowledgments

I thank Dr. Shigeki Kato for providing NeuRet Vectors and sharing precious information during vector development. I thank Dr. Hiroaki Mizukami for always supporting our work by generously providing AAV vectors. I thank Drs. Masafumi Takaji, Tetsuo Yamamori, and Ms. Sonoko Ohsawa for advice, technical help, and encouragement.

Funding This study was funded by JSPS KAKENHI Grant Number 24K09678.

Conflict of Interest Statement The authors have no conflicts of interest to declare that are relevant to the content of this chapter.

References

1. Lanciego JL, Wouterlood FG (2011) A half century of experimental neuroanatomical tracing. J Chem Neuroanat 42:157–183
2. Lanciego JL, Wouterlood FG (2020) Neuroanatomical tract-tracing techniques that did go viral. Brain Struct Funct 225:1193–1224
3. Blasdel GG, Lund JS (1983) Termination of afferent axons in macaque striate cortex. J Neurosci 3:1389–1413
4. Rockland KS, Andresen J, Cowie RJ, Robinson DL (1999) Single axon analysis of pulvinocortical connections to several visual areas in the macaque. J Comp Neurol 406:221–250
5. Pinault D, Deschênes M (1998) Projection and innervation patterns of individual thalamic reticular axons in the thalamus of the adult rat: a three-dimensional, graphic, and morphometric analysis. J Comp Neurol 391: 180–203
6. Kelly RM, Strick PL (2000) Rabies as a transneuronal tracer of circuits in the central nervous system. J Neurosci Methods 103:63–71
7. Wickersham IR, Finke S, Conzelmann K-K, Callaway EM (2007) Retrograde neuronal tracing with a deletion-mutant rabies virus. Nat Methods 4:47–49
8. Wickersham IR, Lyon DC, Barnard RJO, Mori T, Finke S, Conzelmann K-K, Young JAT, Callaway EM (2007) Monosynaptic restriction of transsynaptic tracing from single, genetically targeted neurons. Neuron 53:639–647
9. Mazarakis ND, Azzouz M, Rohll JB et al (2001) Rabies virus glycoprotein pseudotyping of lentiviral vectors enables retrograde axonal transport and access to the nervous system after peripheral delivery. Hum Mol Genet 10:2109–2121

10. Kato S, Inoue K, Kobayashi K, Yasoshima Y, Miyachi S, Inoue S, Hanawa H, Shimada T, Takada M, Kobayashi K (2007) Efficient gene transfer via retrograde transport in rodent and primate brains using a human immunodeficiency virus type 1-based vector pseudotyped with rabies virus glycoprotein. Hum Gene Ther 18:1141–1151
11. Kato S, Kobayashi K, Inoue K, Kuramochi M, Okada T, Yaginuma H, Morimoto K, Shimada T, Takada M, Kobayashi K (2011) A lentiviral strategy for highly efficient retrograde gene transfer by pseudotyping with fusion envelope glycoprotein. Hum Gene Ther 22: 197–206
12. Kato S, Kuramochi M, Takasumi K et al (2011) Neuron-specific gene transfer through retrograde transport of lentiviral vector pseudotyped with a novel type of fusion envelope glycoprotein. Hum Gene Ther 22:1511–1523
13. Kato S, Kobayashi K (2020) Pseudotyped lentiviral vectors for tract-targeting and application for the functional control of selective neural circuits. J Neurosci Methods 344: 108854
14. Watakabe A, Kato S, Kobayashi K et al (2012) Visualization of cortical projection neurons with retrograde TET-off lentiviral vector. PLoS One 7:e46157
15. Hioki H, Kuramoto E, Konno M, Kameda H, Takahashi Y, Nakano T, Nakamura KC, Kaneko T (2009) High-level transgene expression in neurons by lentivirus with Tet-off system. Neurosci Res 63:149–154
16. Watakabe A, Ohtsuka M, Kinoshita M, Takaji M, Isa K, Mizukami H, Ozawa K, Isa T, Yamamori T (2015) Comparative analyses of adeno-associated viral vector serotypes 1, 2, 5, 8 and 9 in marmoset, mouse and macaque cerebral cortex. Neurosci Res 93: 144–157
17. Haery L, Deverman BE, Matho KS et al (2019) Adeno-associated virus technologies and methods for targeted neuronal manipulation. Front Neuroanat 13:93
18. Callaway EM, Luo L (2015) Monosynaptic circuit tracing with glycoprotein-deleted rabies viruses. J Neurosci 35:8979–8985
19. Soudais C, Laplace-Builhe C, Kissa K, Kremer EJ (2001) Preferential transduction of neurons by canine adenovirus vectors and their efficient retrograde transport in vivo. FASEB J 15: 2283–2285
20. Tervo DGR, Hwang B-Y, Viswanathan S et al (2016) A designer AAV variant permits efficient retrograde access to projection neurons. Neuron 92:372–382
21. Watakabe A, Takaji M, Kato S, Kobayashi K, Mizukami H, Ozawa K, Ohsawa S, Matsui R, Watanabe D, Yamamori T (2014) Simultaneous visualization of extrinsic and intrinsic axon collaterals in Golgi-like detail for mouse corticothalamic and corticocortical cells: a double viral infection method. Front Neural Circuits 8:110
22. Kinoshita M, Matsui R, Kato S et al (2012) Genetic dissection of the circuit for hand dexterity in primates. Nature 487:235–238
23. Sadakane O, Watakabe A, Ohtsuka M et al (2015) Vivo two-photon imaging of dendritic spines in marmoset neocortex. eNeuro. https://doi.org/10.1523/ENEURO.0019-15.2015
24. Bortoloci JGT, Motta SC (2022) Failure of AAV retrograde tracer transduction in hypothalamic projections to the periaqueductal gray matter. Heliyon 8:e10243
25. Hoogland PV, Wouterlood FG, Welker E, Van der Loos H (1991) Ultrastructure of giant and small thalamic terminals of cortical origin: a study of the projections from the barrel cortex in mice using Phaseolus vulgaris leucoagglutinin (PHA-L). Exp Brain Res 87:159–172
26. Kaneko T, Fujiyama F, Hioki H (2002) Immunohistochemical localization of candidates for vesicular glutamate transporters in the rat brain. J Comp Neurol 444:39–62

Chapter 10

Lentivirus-Mediated Measurement of Transcription Factor Activity in the Brain

Kentaro Abe

Abstract

Transcription factors (TFs) play a pivotal role in the plastic changes of the brain. Measuring TF activity in vivo and uncovering changes under various physiological and pathophysiological conditions provide valuable insights into brain mechanisms. This protocol outlines a detailed procedure for measuring TF activity in the brain by using lentiviral vector (LV)-based TF reporters. The protocol covers the preparation of LV, injection of the vector into the mouse brain, mRNA processing from the tissue, and quantitative PCR-based calculation of TF activity.

Key words Transcription factor, Lentiviral vector, Transcription reporter, RT-PCR, Quantitative PCR

1 Introduction

Transcription factors (TFs) play a central role in both establishing and dynamically modifying a cell's molecular identity through the regulation of its transcriptome [1–3]. It is increasingly clear that TF activity is far more dynamic than previously thought [4, 5]. The activity of TFs is often inferred through transcriptome analysis, where the expression levels of downstream genes regulated by a specific TF are used to estimate the TF activity [6–8]. However, the complexity of gene regulation, where individual loci are often governed by multiple TFs with synergistic and overlapping effects, makes it difficult to measure precisely based solely on gene expression data [9, 10]. Thus, the methodologies that enable the direct and quantitative measurement of individual TF activities are essential. Here, I describe a detailed protocol for measuring TF activity in the brain by using transcription reporters specific to individual TFs [11]. This approach employs lentiviral vectors (LVs) carrying two promoters to account for the variation in transfection efficiency, a common challenge in using viral vectors in vivo

Kazuto Kobayashi et al. (eds.), *Lentiviral Vectors: Applications for Basic and Clinical Neuroscience*, Methods in Molecular Biology, vol. 2974, https://doi.org/10.1007/978-1-0716-4807-0_10,

[12]. Viral vectors can be injected into adult mouse brain, as outlined in this protocol, or into embryonic brains as described before [13]. Adult injection can achieve highly specific expression of particular brain areas. Embryonic injection can achieve highly efficient and broad transduction of constructs. The same basic procedure can be applied to measure the TF activity in other organs in vivo, beyond the brain.

2 Materials

2.1 Preparation of LV

1. Safety cabinet with biosafety level 2 (BSL-2).
2. Cell culture incubator.
3. Centrifuge.
4. HEK293T cells (Lenti-X 293T cell, Takara #632180).
5. Cell culture dish (diameter, 15 cm).
6. Bottle-top filter (pore size, 0.45 μm).
7. Centrifuge tube.
8. Cell culture medium: Dulbecco's Modified Eagle Medium (DMEM) containing 4500 mg/L glucose, 10% fetal bovine serum, 1% penicillin-streptomycin mix.
9. Plasmid mixture. Plasmids of 1 mg/mL each with high purity such as plasmids purified with Maxi-prep column. Following amounts are required for one 15-cm dish: pRSV-Rev, 3.6 μL (Addgene, #12253); pMD2.G, 5.4 μL (Addgene, #12259); pMDLg/pRRE, 10.8 μL (Addgene, #12251); and pLV-Rep/Ref, 18.0 μL (Addgene, #182054).
10. OptiMEM-I (Thermo Fisher).
11. PEI-max solution (1 mg/mL) (Polysciences).
12. Virus concentration solution (4×): 40 mM HEPES, pH 7.4, 400 mM NaCl, 320 mg/mL polyethylene glycol-6000. Sterile with autoclave.
13. Phosphate buffered saline (PBS). Sterile.
14. Sucrose solution: PBS containing 20% sucrose.

2.2 LV Injection into Mouse Brain

1. Stereotaxic apparatus (Narishige, SR-5 M-HT).
2. Micropipette puller (Narishige, PC-100).
3. Micropipette beveler (Narishige, EG-45).
4. Pressure injector.
5. Oil hydraulic micromanipulator (Narishige, MO-10).
6. Anesthetizer for small animals.
7. Dental drill.

8. Surgical scissors.
9. Fine glass capillaries: Outer diameter, 1.0 mm, inner diameter, 0.75 mm, filamented (Harvard Apparatus, #30-0038).
10. Mouse (*Mus musculus*, C57BL/6 J).
11. Isoflurane.
12. Carprofen solution: 1 mg/mL carprofen in PBS.

2.3 RNA Collection from the Brain Tissue

1. Bench-top centrifuge (Eppendorf).
2. Bead mill shaker (Bio-Medical Science).
3. Stainless beads (diameter, 5 mm).
4. Microtubes (1.5 mL, RNase-free).
5. RNA extraction solution: 1.63 M guanidinium thiocyanate, 11.4 mM sodium citrate, 0.23% N-lauroylsarcosine sodium salt, 0.0033% 2-mercaptoethanol, 6.25 mM sodium acetate, 50% water-saturated phenol. Other acid-guanidinium-phenol-based reagent such as TRIzol (Thermo Fisher, 15596018) or TRI reagent (SIGMA, #T9424) can be used instead.
6. Chloroform.
7. RNase-free water.
8. 2-Propanol.
9. Ethanol solution: 75% ethanol in RNase-free water.

2.4 Quantitative Reverse Transcription PCR (RT-qPCR) to Calculate the TF Activity

1. PCR machine (Bio-Rad Laboratories, T-100).
2. Quantitative PCR system (Bio-Rad Laboratories, CFX384 touch).
3. Eight-string tubes.
4. Micropipette with RNase-free tips.
5. PCR plate of 384 well.
6. Adhesive plastic PCR film.
7. PCR-plate centrifuge.
8. ReverTra Ace qPCR RT Master Mix with gDNA remover (TOYOBO).
9. Genome removal reaction mix: Add 1.33 μL of gDNA Remover (4×) to 0.67 μL of RNase-free water. gDNA Remover (4×) is a component of ReverTra Ace qPCR RT Master Mix with gDNA Remover (TOYOBO).
10. RT reaction mix: RT Master Mix II (5×) containing 0.15 μM gene specific primers. RT Master Mix II (5×) is a component of ReverTra Ace qPCR RT Master Mix with gDNA Remover (TOYOBO).

11. Specific primers for Reporter and Reference genes. Such as aactgcttcatccacaactcca and acgggcttctcggtcttgt for *Reporter-1*, or tactgcccaagaagaccgaga and ccacatagcgtaaaaggagcaac for *Reference-1* gene (*see* **Note 1**).
12. Quantitative PCR reagent: GoTaq qPCR master mix (Promega).
13. Standard plasmids (Maxi-prep purity, 1 mg/mL).

3 Methods

This protocol is divided into four parts: Preparation of LV, injection of the vector into the mouse brain, mRNA processing from the tissue, and quantification of TF activity. TF activity measurement relies on the expression of two transgenes, each driven by separate promoters within the constructs (Fig. 1). One gene, referred to as the *Reference* gene, is driven by the human phosphoglycerate kinase 1 (Pgk1) promoter and serves as a transfection control to account for variation in transduction efficiency in vivo. The second gene, referred to as the *Reporter* gene, is regulated by a TF-dependent promoter composed of a minimal promoter with transcription factor-binding sequences (TFBSs). The specific TF being measured is determined by the TFBSs included in the reporter construct. Therefore, before experimentation, it is needed to prepare the reporter constructs containing an appropriate TFBS for the TF of interest. Experiments involving the use of LV must be conducted in a BSL-2 laboratory and require approval from relevant institutional authorities under the regulation of local government. All materials

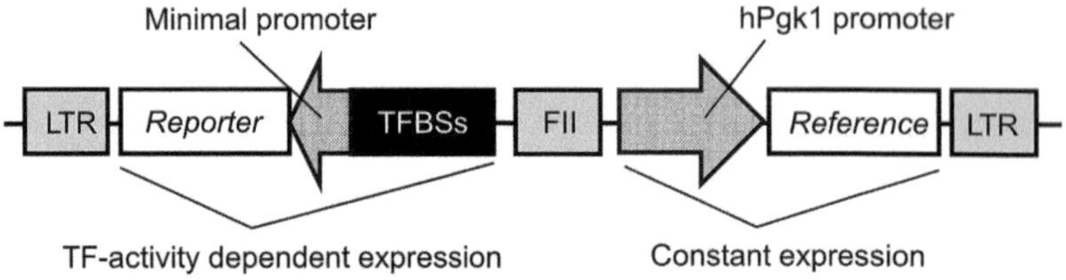

Fig. 1 Structure of the transcription factor (TF) activity reporters. The TF activity reporter construct contains two promoters separated by an insulator element (FII) from the chiken β globin gene locus. The transcription factor-binding sites (TFBSs) are located directly upstream of a synthetic minimal promoter, which drives the expression of *Reporter* mRNA in response to the targeted TF. The expression of *Reference* gene is driven by human Phosphoglycerate Kinase 1 promoter (hPgk1), known for its relatively constant activity. TF activity is calculated as the ratio of the measured amounts of *Reporter* to *Reference* gene expression. TFBSs can be substituted with specific sequence for each TF being measured. To measure multiple TF activities from a single specimen, distinct *Reporter* and *Reference* genes are used for each TF. LTR, long terminal repeats

used during the experiment must be properly disinfected before disposal to ensure safety and compliance with biosafety regulations.

3.1 Preparation of LV-Based TF Reporter for In Vivo Injection

1. All experiments must be conducted inside a BSL-2 safety cabinet.
2. Grow HEK293T cells in cell culture medium at 37 °C under 5% CO_2 in an incubator.
3. Transfer one 15-cm dish of HEK293T cells at ~90% confluency into two 15-cm dishes.
4. Approximately 12 h after transfer, transfect each dish of HEK293T cells with the transfection solution containing plasmid mixture (*see* **Note 2**). For transfection, prepare the plasmid mixture in 500 μL of Opti-MEM I and thoroughly mix it with PEI-max solution in 1000 μL of Opti-MEM I. Incubate the mixture for 15 min at room temperature, then apply it to a dish of HEK293T cells. Gently tilt the dish, mix thoroughly, and then return to the incubator.
5. Within 24 h after the transfection, replace the culture medium with fresh medium prewarmed to 37 °C (*see* **Note 3**).
6. After 48–72 h of transfection, collect the cell culture medium.
7. Filter the collected medium with a bottle-top filter with a pore size of 0.45 μm.
8. Mix the filtered medium with virus concentration solution to achieve a final concentration of 1×. Store the mixture at 4 °C for at least 12 h.
9. Centrifuge at 4000×*g* for 15 min at 4 °C.
10. Remove the supernatant and dissolve the pellet in 1 mL of PBS.
11. Layer the solution on 250 μL of sucrose solution.
12. Centrifuge at 20,000×*g* for 15 min at 4 °C.
13. Remove the supernatant and dissolve the pellet with 50 μL of PBS.
14. Aliquot the solution into small volumes (~4 μL) in microtubes and store at −80 °C (*see* **Notes 4** and **5**).

3.2 Injection of LV-TF-Reporter into the Mouse Brain

1. Before the procedure, determine the stereotaxic coordinates of the desired brain locus. Thaw and dilute the LV solution and keep it on ice before loading into the microinjection needle.
2. Prepare the microinjection needle by pulling the fine glass capillaries using a micropipette puller. If necessary, adjust the tip with a micropipette beveler.
3. Load the LV solution into the microinjection needle, ensuring no air bubbles are present. Attach the needle to micromanipulator on the injection apparatus and confirm the proper flow of

the viral solution by applying pressure. Adjust the pressure based on the required volume.

4. Anesthetize the mouse with isoflurane. Induce anesthetization with ~3% isoflurane and maintain it at ~1.5% throughout the procedure.
5. Confirm the surgical depth of anesthesia by verifying the absence of a reflex response to paw pinching with forceps.
6. Remove hair from the top of the head using a shaver.
7. Place the mouse on a heating pad to maintain body temperature during the procedure.
8. Secure the mouse in the stereotaxic frame by positioning the front teeth on the tooth clamp and fixing the head using the ear bars.
9. Make an incision along the midline of the scalp to expose the skull. Clean the skull with a sterile scrab and disinfect it.
10. Position the tip of microinjection needle at the bregma point on the skull.
11. Identify the stereotaxic coordinates of the desired injection site, and mark the target location on the skull with a marker.
12. Use a dental drill to carefully create a small hole in the skull at the injection point.
13. Gradually lower the microinjection needle to the specified depth for viral delivery.
14. Inject the viral solution slowly, and monitor the injection under a microscope to observe the changes at the surface of viral solution inside the microinjection needle.
15. After completing the injection, wait an additional 5 min before withdrawing the microinjection needle.
16. Slowly retract the needle from the brain.
17. Close the wound by suturing the incision with surgical thread.
18. Remove the ear bars and carefully take the mouse out of the stereotaxic apparatus.
19. Inject 100 μL of carprofen solution to reduce pain.
20. Keep the mouse on heating pad and monitor it closely until it regains consciousness.
21. Return the mouse to the home cage. Check the health condition for 2–3 days.

3.3 Collection of RNA from the Brain Tissue

1. Prepare the equipment and solutions necessary for RNA collection. Keep the solution at room temperature.
2. Perform adequate behavioral experiments with the LV-injected mice for experimental purpose (*see* **Note 6**).

3. Decapitate the mouse and remove the brain, immediately placing it into PBS on ice (*see* **Note 7**).
4. Dissect the brain region of interest under a microscope and collect the tissue into ice-cold PBS in a microtube.
5. Remove PBS and add 1 mL of RNA extraction solution kept at room temperature. Conduct further procedure at room temperature.
6. To homogenize the tissue, add one stainless bead and shake the tube vigorously using a bead mill shaker for 30 s at maximus speed (3800 rpm).
7. If necessary, freeze and store the solution at −80 °C (*see* **Note 8**).
8. Add 300 μL of chloroform to the homogenized tissue solution and mix with a bead mill shaker for 30 s at 3800 rpm.
9. Incubate the samples at room temperature for 3 min.
10. Centrifuge at 15,000×*g* for 15 min at 20 °C. After centrifugation, a clear supernatant layer will appear at the top.
11. Carefully collect around 500 μL of the supernatant into a new tube, avoid touching the intermediate or chloroform layers to maintain RNA purity.
12. Add 300 μL of chloroform to the collected supernatant and mix with a bead mill homogenizer for 30 s at 3800 rpm.
13. Centrifuge at 20,000×*g* for 15 min at 20 °C. After the centrifugation, clear supernatant layer will appear at the top.
14. Collect around 300 μL of the supernatant into a new tube.
15. Add 300 μL of 2-propanol and mix by voltexing.
16. Centrifuge at 20,000×*g* for 15 min at 4 °C. After the centrifugation, a white pellet will appear at the bottom of the tube.
17. Wash the pellet with ethanol solution and centrifuge at 20,000×*g* for 5 min at 4 °C.
18. Completely remove the supernatant and dry the RNA pellet.
19. Dissolve the pellet with 150 μL of RNase-free water at 55 °C, voltexing several times.
20. Store the RNA sample at −80 °C (*see* **Note 9**).

3.4 Performing Quantitative RT-PCR to Calculate the TF Activity

1. Thaw stored RNA sample (*see* **Note 10**).
2. Mix 3.3 μL of RNA samples with 2.0 μL of Genome removal reaction mix in an 8-string tube. Warm the tube at 37 °C for 5 min. Stop the reaction by putting the tube on ice.
3. Perform reverse transcription on the processed RNA samples. Open the lid of the 8-string tube and add 2.0 μL of RT reaction mix to each tube. Mix well, briefly centrifuge and set the tube to PCR machine. Heat the tube for 37 °C for 15 min, 42 °C for

5 min, 50 °C for 5 min, and finally 95 °C for 5 min. Hold the tube at 4 °C.

4. Dilute the resultant cDNA five times with H_2O and keep the sample at −80 °C.
5. Perform quantitative PCR (qPCR) to detect Reporter and Reference gene expression. Mix 2 μL of cDNA solution with 8 μL of qPCR Master mix in a 384-PCR plate (*see* **Note 11**). Prepare standard samples containing the target gene at known concentrations for the qPCR reaction (*see* **Note 12**). Typically, the plasmid used to generate the LV-reporter (LV-Rep/Ref) is used for this purpose. Seal the PCR plate with adhesive plastic film, and briefly spin down the solution with plate centrifuge. Mix gently and spin down the plate again. Load the plate into qPCR machine and run the reaction as follows: Heat activation at 95 °C for 2 min, followed by 60 cycles of 95 °C for 10 s, 64 °C for 55 s. After the cycles, perform a melt-curve analysis to verify that the PCR reaction occurred without undesired cross amplification.
6. Using the expression amount of each set of *Reporter* and *Reference* genes, calculate the TF activity using the following formula (*see* **Notes 13** and **14**):

$$\text{TF activity} = \frac{\text{Value of } \textit{Reporter} \text{ gene}}{\text{Value of } \textit{Reference} \text{ gene}}$$

4 Notes

1. This method can accommodate unlimited numbers of *Reporter* and *Reference* gene pairs. Examples of *Reporter* and *Reference* genes previously used are shown in prior studies [11]. Additional Reporter and Reference can be designed and applied; however, it is essential to ensure that primers for these genes do not interfere with the PCR amplification of all other *Reporter* and *Reference* genes.
2. Achieving maximum transfection efficiency (around 90–100%) is critical for producing LV with a sufficiently high titer for in vivo use. The ratio of DNA to PEI-max should be optimized for each batch of PEI-max. A DNA: PEI-max ratio of 1:4 to 1:10 is typically appropriate.
3. All materials used in the production of LVs must be properly disinfected and discarded according to biosafety protocols.
4. Avoid repeated freeze-thaw cycles of the LV solution. It is recommended to thaw only the required volume of LV from the aliquoted tubes. Before conducting in vivo experiments,

estimate the viral titer using qPCR methods. To quantify the titer, transfect 0.1 μL of prepared LV into HEK293T cells, and after 48 h, harvest the cells. Extract the genomic DNA and perform qPCR to quantify the transgene integrated into the host cell genome.

5. The LVs prepared with this protocol may not achieve high purity. However, these vectors are suitable for measuring TF activity in vivo, provided a sufficient waiting period is allowed after surgery. For applications requiring early quantification immediately after the surgery, it is recommended to use LVs with higher purity to circumvent unwanted changes in TF activity driven by inflammatory responses. This procedure is versatile and can be applied to various experimental species beyond mice.
6. The procedure for behavioral experiment and the timing of sample preparation varies depending on the type of TF activity the experimenter intends to access. If the goal is to measure the basal activity of TFs, animals must be kept calm, undisturbed in their home cages prior to sacrifice. However, if the experiment aims to evaluate change in TF activity in response to specific stimuli or induced behaviors, it is crucial to standardize the timing of behavioral experiments or manipulation for each animal and allow sufficient time for TF activity change to be reflected in *Reporter* gene expression. Each TF has a distinct activation profile, which may be influenced by factors such as activation timing, signal transduction speed, and other cellular processes. Therefore, it is essential to determine the appropriate post-stimulus time window for each TF and strictly adhere to this timing during behavioral experiments and sample collection.
7. It is essential to decapitate and homogenize the brain as quickly as possible, as the activity of some TFs may change during these procedures.
8. Promptly perform the **steps 1**–7 to prevent RNA degradation and unwanted change of TF activity during processing. It is recommended to process only two mice at a time (one control and one experimental), to ensure accuracy and efficiency. After homogenization, store the solution at −80 °C before proceeding with the next set of samples. The homogenized solution can be kept at −80 °C for years. Once a sufficient number of samples have been collected, proceed with the purification process.
9. Purified RNAs can be stored for years.

10. Handle RNA solution with care as RNAs are fragile and tends to degrade by RNase contamination. Use only RNase-free verified materials in this step.
11. The qPCR reaction to quantify each of the pair of *Reference* and *Reporter* gene is preferably performed simultaneously on a single PCR plate. If multiple TF activity reporters are expressed within the same sample, it is crucial to ensure that the primer set specifically amplifies the target sequence without nonspecific or cross amplification. To verify specificity, conduct PCR on negative control samples containing all plasmids with various *Reporter* and *Reference* genes, except for the target gene.
12. Absolute quantification is essential to accurately determine the exact amount of cDNA for the target molecules. Typically, we routinely detect 1000–10,000 molecules of *Reference* genes from 2 μL of cDNA synthesized from the brain samples. Insufficient levels of *Reference* gene expression can distort the calculation of TF activity, as *Reference* gene serves as a critical normalization control.
13. TF activity of multiple TFs can be measured with the same sample. To do this, prepare LV reporter harboring distinct TFBS, *Reporter*, *Reference*, and the specific primer pair for quantification of their expression. It is important that the *Reporter* and *Reference* genes should be independently amplified by qPCR primers without cross amplification of other *Reporter* and *Reference* genes.
14. The TF activity measured using this method should be interpreted as a relative value. Therefore, appropriate controls are necessary to accurately interpret the results. The choice of control will depend on the specific experimental objectives. Examples of suitable controls include home-caged animals, sham operations, or vehicle injections. Comparing TF activity across different TF-reporter is not meaningful, as the activation level depends on the structure of the reporter construct, which is influenced by various features specific to the targeted TFs, such as the affinity of the TFBSs and their arrangements.

Acknowledgments

This work was supported by MEXT/JSPS KAKENHI (JP24H01218, JP22H05482, JP24H02146, JP23K18252), and Tohoku University Research Program “Frontier Research in Duo” grant number 2101.

Conflict of Interest Statement The authors have no conflicts of interest to declare that are relevant to the content of this chapter.

Funding

This study was funded by JSPS/MEXT KAKENHI (JP25H01037, JP24H01218), Takeda Science Foundation, Suzuken Memorial Foundation, and Tohoku University Research Program "Frontier Research in Duo" (Grant No. 2101).

References

1. Vaquerizas JM et al (2009) A census of human transcription factors: function, expression and evolution. Nat Rev Genet 10(4):252–263
2. Spitz F, Furlong EEM (2012) Transcription factors: from enhancer binding to developmental control. Nat Rev Genet 13(9):613–626
3. Lambert SA et al (2018) The human transcription factors. Cell 172(4):650–665
4. Behar M et al (2013) The dynamics of signaling as a pharmacological target. Cell 155(2): 448–461
5. Purvis JE, Lahav G (2013) Encoding and decoding cellular information through signaling dynamics. Cell 152(5):945–956
6. Liao JC et al (2003) Network component analysis: reconstruction of regulatory signals in biological systems. Proc Natl Acad Sci 100(26):15522–15527
7. MacQuarrie KL et al (2011) Genome-wide transcription factor binding: beyond direct target regulation. Trends Genet 27(4):141–148
8. Aibar S et al (2017) SCENIC: single-cell regulatory network inference and clustering. Nat Methods 14(11):1083–1086
9. Georgakopoulos-Soares I et al (2023) Transcription factor binding site orientation and order are major drivers of gene regulatory activity. Nat Commun 14(1):2333
10. Abe K (2025) Dynamic activity changes in transcription factors: unlocking the mechanisms regulating physiological changes in the brain. Neurosci Res 214:16–22
11. Abe H, Abe K (2022) PCR-based profiling of transcription factor activity in vivo by a virus-based reporter battery. iScience 25(3):103927
12. Abe K, Matsui S, Watanabe D (2015) Transgenic songbirds with suppressed or enhanced activity of CREB transcription factor. Proc Natl Acad Sci 112(24):7599–7604
13. Yamamoto H, Abe K (2022) Protocol for viral vector-mediated measurement of transcription factor activity of mouse brain. STAR Protoc 3(3):101633

Chapter 11

Birth Date-Dependent Tracing of Neuronal Cell Lineage

Tatsumi Hirata

Abstract

Neuronal phenotypes are determined by their cell lineage and birth timing. The "neurogenic tagging" method controls these two key factors to classify distinct neuron subsets. The classified neuron subsets can then be used for various experimental manipulations. To spatially constrain the tagged neurons further, we combined this method with focal lentivirus injection. This approach allowed us to identify convergent axon projections of a specific subset of olfactory bulb neurons in the adult mice. This chapter describes the preparation of neurogenic-tagged mice for lentivirus injection.

Key words Neurogenic tagging, Mouse, Neuron, Birth timing, Lineage

1 Introduction

1.1 Principle of Neurogenic Tagging

Neurons are terminally differentiated cells. They no longer divide and are generated solely by dividing neural progenitors one by one during development (Fig. 1a). From a developmental viewpoint, two key factors permanently determine neuronal phenotypes, also known as neuron types; one is cell lineage (where the neurons are born), and the other is birth timing (when the neurons are born) [2].

Based on this principle, we developed the "neurogenic tagging" method, a genetic technique using transgenic mice [1, 3]. This method uses the tamoxifen (TM)-dependent CreER, which is transiently expressed only within a short time window immediately after neuronal fate committed (Fig. 1b). CreER acts as an enzyme that catalyzes the recombination of loxP sequences of the genome in the cell nucleus only in the presence of TM [4]. Administration of TM during the neurodevelopmental time window triggers loxP recombination in the neurons shortly after neuronal commitment (typically 6–12 h after the final mitosis), but not in those born earlier or later (Fig. 1b). Because genetic recom-

Kazuto Kobayashi et al. (eds.), *Lentiviral Vectors: Applications for Basic and Clinical Neuroscience*, Methods in Molecular Biology, vol. 2974, https://doi.org/10.1007/978-1-0716-4807-0_11,

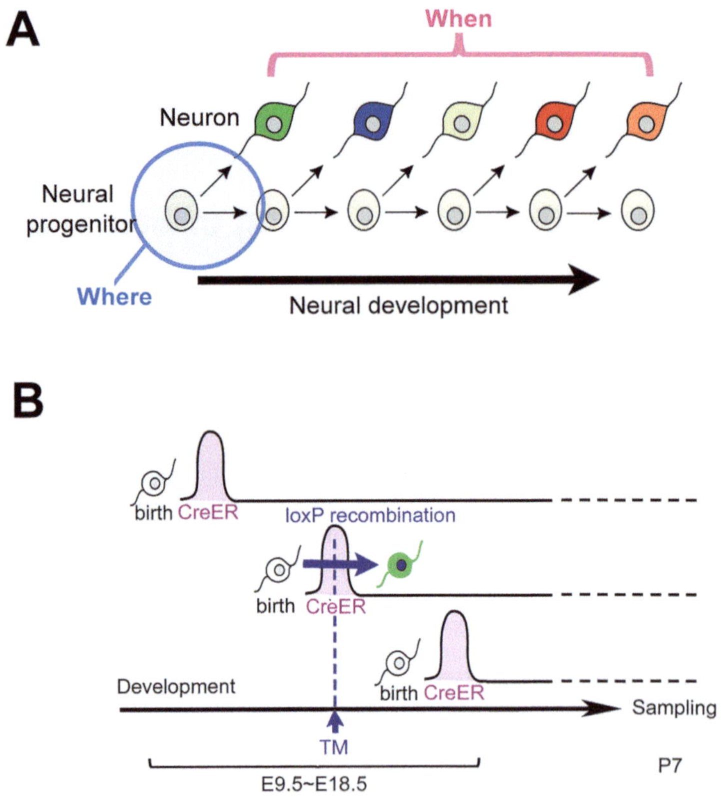

Fig. 1 Neurogenic tagging based on the neurodevelopmental principle. (**a**) Neurons are generated from neural progenitors during development. Neuronal phenotypes are determined by two key factors: cell lineage (where) and birth timing (when). (**b**) In a neurogenic tagging mouse driver, CreER is transiently expressed within a short neurogenic time window in various neurons. Tamoxifen (TM) injection on a single day during E9.5–E18.5 induces *loxP* recombination only in neurons expressing CreER. (Reproduced from ref. [1] under CC-BY license)

bination is an irreversible reaction, the neurogenic tag can be used at a much later time to explore the classified neuronal subsets in various ways. This offers a significant advantage over traditional retrospective histological techniques that use nucleotide analogues to determine neuronal birth dates in sacrificed animals.

The neurogenic tagging method uses driver mouse lines, in which CreER is driven by enhancers of neuronal differentiation genes, ensuring its transient expression within a short time window. We have established four such driver lines using the genes, *neurog1*, *neurog2*, *neurod1*, and *neurod4* (Fig. 2). These genes encode basic helix-loop-helix transcription factors that are transiently expressed during the neuron commitment phase. However, the brain regions expressing CreER differ significantly between driver lines, reflecting the endogenous expression pattern of the driving gene. Thus,

Line Name	Code	Enhancer	Notable regions tagged
Neurog2^{CreER}	G2A	neurogenin2	olfactory bulb, piriform cortex, neocortex, amygdala, hippocampus, mammillary body, inferior colliculus, cerebellum, olive, spinal cord
Neurog1^{CreER}	G1C	neurogenin1	superior colliculus, inferior colliculus, pontine, cochlear nucleus
Neurod1^{CreER}	D1B	neuroD1	subplate neuron, medial cortex, amygdala, hippocampus, retina, habenula, inferior colliculus, cochlear nucleus, sensory ganglia
Neurod4^{CreER}	D4A	neuroD4	olfactory bulb, hippocampus, retina, habenula, supramammilary nucleus, inferior colliculus, cochlear nucleus, olive, spinal cord, peripheral ganglia

Fig. 2 Neurogenic tagging CreER driver mouse lines. Four neurogenic tagging mouse lines have been established to tag different brain regions. The mouse lines are available from RIKEN (http:/www2.clst.riken.jp/arg/TG%20mutant%20mice%20list.html). (Reproduced from ref. [1] under CC-BY license)

each driver line can be used to tag neurons belonging to different cell lineages, following the "where" principle in neuronal development (Fig. 1a). Within each line, injection of TM at different developmental stages (TM stages) tags separate neuron subsets born at different developmental times, reflecting the "when" principle (Fig. 1a).

To facilitate the use of neurogenic tagging resources, we have developed the Brain Atlas of Neurogenic Tagging Mouse Lines (NeuroGT database: https://ssbd.riken.jp/neurogt/), effectively expanding the user base [5–7]. This database provides access to coronal section and whole mount images of neurogenic-tagged neurons throughout the entire brain in postnatal day 7 mice. TM is administrated on a single day during embryonic days 9.5–18.5 (TM stage). The position of neurogenic-tagged cell bodies can be seen by immunostaining for the nucleus-localized beta-Gal reporter, while axons of the tagged neurons can be seen by immunostaining for the membrane localized fluorescent reporter protein (GFP) reporter using a specific reporter line [8]. These sections, tagged at different TM stages by different driver lines, are z-stacked separately in the database. The close spacing of the sections allows researchers to examine even small brain regions by their eyes to identify appropriate mouse lines for their research objectives.

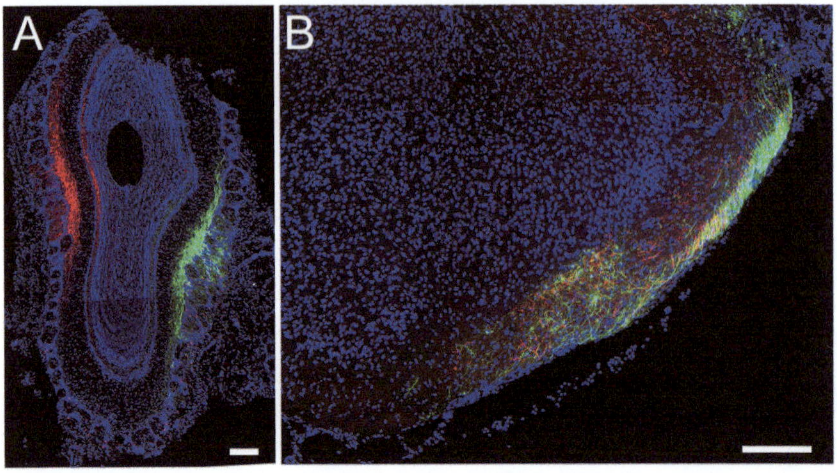

Fig. 3 Tracing of external tufted cells in the medial and lateral olfactory bulb by combining neurogenic tagging with viral injections. Coronal sections prepared from a Neurog2^{CreER}(G2A); Cdhr1^{tTA} mouse that was given TM15.5 and then injected with VSVG-TRE^{GFP} and AAV2-TRE^{TurboFP635} viruses (see Fig. 4a). (**a**) In the olfactory bulb, external tufted cells on the medial and lateral sides of the olfactory bulb are specifically labeled with green and red reporter proteins, respectively. (**b**) The axons make convergent projection onto the anterolateral isolation of the olfactory tubercle. Scale bar = 200 μm. (Reproduced from ref. [3] under CC-BY license)

1.2 Tracing of Neurogenic Tagged Neurons in a Spatially Restricted Region

One effective approach to label a neuronal subset within a specific brain region is to combine neurogenic tagging with focal lentivirus injection. Indeed, this strategy has successfully revealed convergent projections of late-born projection neurons from the medial and lateral olfactory bulbs [3] (Fig. 3). The timing of virus injection is critical. If lentiviruses are injected after the nervous system is fully developed, it may be necessary to take advantage of prolonged expression of a transcription factor until then. For this purpose, we used the Tet-OFF system (*see* **Note 1**) in the previous study (Fig. 4a). Specifically, neurogenic tagging was performed at an embryonic stage (TM15.5), resulting in the production of the tetracycline transactivator (tTA) until adulthood under an olfactory bulb-specific promoter. Subsequently, lentiviruses were focally injected into a restricted portion of the adult olfactory bulb to achieve expression of a fluorescent reporter protein (GFP) under the tetracycline response element (TRE). This method allowed the visualization of neurons born at a specific birth timing only within the restricted region with a fluorescent reporter protein [3] (Fig. 3). Although an olfactory bulb-specific promoter was used in this study, it is probably not essential. If the lentivirus injection is sufficiently restricted, any ubiquitous promoter can be substituted.

Another potential approach is to express a reporter regulated by a combination of two recombinases (Fig. 4b). Reporter mouse lines have been developed specifically for this purpose [9]. In these mice, the expression of a fluorescent reporter protein typically requires the excision of double STOP cassettes by both Cre and

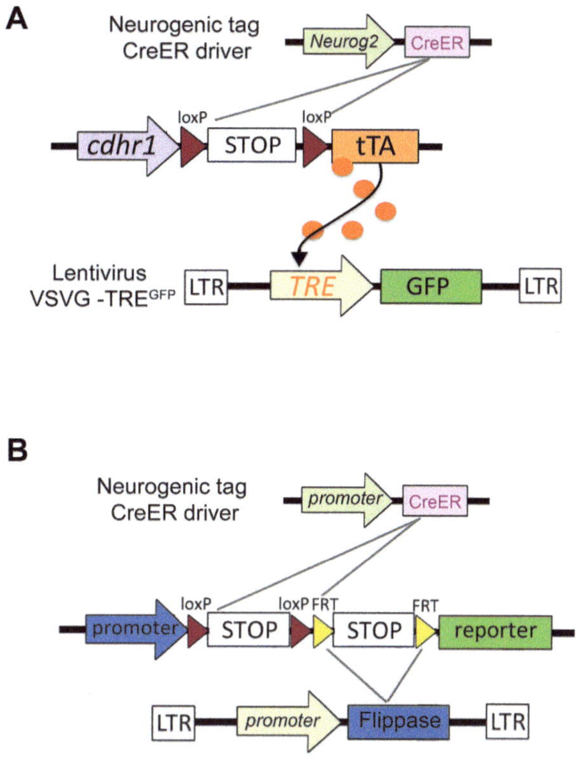

Fig. 4 Gene combination for spatially restricted neurogenic tagging. (**a**) Combination of neurogenic tagging and lentivirus injection used to label olfactory bulb axons [3] (Fig. 3). After neurogenic tagging, tetracycline transactivator (tTA) is continuously expressed until adulthood, and its expression is reported by TRE-driven fluorescent reporter protein (GFP) introduced as a lentivirus. (**b**) A reporter protein is expressed when the stop cassette flanked by loxP sequences is excised by neurogenic tagging, and another stop cassette flanked by FRT sequences is excised by Flippase encoded by a lentivirus

Flippase recombinases. In our own experience with such mice, we have successfully visualized axon projections by combining embryonic neurogenic tagging with adult injection of a Flippase-encoding virus (Iwai et al., unpublished). Because of the potential problems with TRE activity in transgenes [10] (*see* **Note 1**), this experimental design may be more effective under many experimental conditions.

2 Materials

2.1 Mice

Neurogenic tagging mice can be obtained from the RIKEN Center for Biosystems Dynamics Research (BDR) (http://www2.clst.riken.jp/arg/TG%20mutant%20mice%20list.html).

For information on other mouse strains, visit the websites of Mouse Genome Informatics (MGI) (https://www.informatics.jax.org/) and RIKEN BioResource Research Center (BRC) (https://web.brc.riken.jp/en/).

Upon receiving mice, follow appropriate quarantine procedures and health screening to introduce them to your animal facility. Maintain the mice according to established protocols and institutional guidelines. Genotyping can be performed using polymerase chain reaction (PCR) on tail tip genomic DNA (3–5 mm) with specific primers.

2.2 Lentivirus

Lentiviral vectors are prepared as described in Chapter 2.

2.3 Solutions

1. Tamoxifen Stock Solution (0.1 M): Dissolve 37 mg of tamoxifen (CAS_10540-29-1) in 1 mL of ethanol and store at −20 ° C.
2. Progesterone Stock Solution (0.075 M): Dissolve 23.5 mg of progesterone (CAS_57-83-0) in 1 mL of ethanol and store at −20 °C (*see* **Note 2**).
3. TM Solution (for Injection): Mix 1.265 mL of corn oil (CAS_8001-30-7, Sigma C8267) is mixed with 135 μL of Tamoxifen Stock Solution and 100 μL of Progesterone Stock Solution. (final concentrations: 9 mM tamoxifen and 5 mM progesterone). Store at 4 °C and use within 10 days.
4. Phosphate-Buffered Saline (PBS;10×): 1.3 M NaCl and 100 mM Na_2HPO_4, adjusted to pH 7.4.

2.4 Equipment

1. Stereotactic Apparatus: EDMS11-054 SR-5M-S, Narishige. Japan.
2. Glass Pipette: 3B100-75-10, Sutter Instrument.
3. Pipette Holder: HI-7, Narishige, Japan.
4. Micro Grinder with a Diamond Point: UC210, Urawa, Japan with ED101, Argofile, Japan.
5. Oil Syringe Injector: IM-6, Narishige, Japan.

3 Methods

3.1 Preparation of Neurogenic Tagged Mice

1. Place four to six female mice and one male mouse of appropriate genotypes together in a cage overnight from 16:00 to 20:00.
2. On the following morning, visually inspect the female mice for virginal plugs.
3. Isolate plug-detected female mice. The noon of plug detection is designated as embryonic day 0.5 (E0.5).

4. Intraperitoneally inject 250 μL of TM Solution (for Injection) into a pregnant mouse between E8.5 and E18.5 using a 26-gauge needle (*see* **Note 2**).
5. Due to the reduced natural delivery rates after TM injection (*see* **Note 3**), prepare foster mothers. We typically use wild-type female mice in the ICR background in the gestation period a couple of days longer than that of the neurogenic-tagged mice prepared in **step 4** (*see* **Note 4**).
6. If pups are not delivered naturally by the morning of E19.5, perform a cesarean section to retrieve all embryos. Revive the pups on a hot plate at 38 °C for approximately 30 min until their skin turns pink. Combine the pups with a few original pups from the foster mother and allow the foster mother to raise them.

3.2 Injection of Lentivirus into the Olfactory Bulb of Neurogenic Tagged Mice

1. Anesthetize neurogenic-tagged mice aged 2–5 months and head-fix them in a stereotactic apparatus under a dissection microscope (*see* **Note 5**).
2. Place a pulled borosilicate glass pipette filled with the virus solution in a holder mounted on a manual course manipulator.
3. Open a small area of the bone above the olfactory bulb by using a micro grinder.
4. Referencing the mouse brain atlas [11], insert the glass pipette through the midline of the olfactory bulb, avoiding blood vessels, and penetrate 75–100 μm medially or 125–150 μm laterally at an oblique angle to the surface plane.
5. Slowly inject the virus solution (1 μL) over 10 min using an oil syringe injector.
6. Cover the window over the olfactory bulb with the removed bone and seal it with Vetbond Tissue Adhesive.
7. Three to four weeks after the virus injection, dissect the brains for histochemical examination in 1× PBS.

4 Notes

1. Tet-OFF and Tet-ON systems often cause problems when studying the mouse nervous system. We often encounter the problem that the TRE-reporter protein is not expressed even though tTA is expressed. This problem is particularly evident when a TRE reporter is introduced as a transgene. In fact, TRE-driven expression from any transgene is rarely detected in mouse neocortex neurons. In the olfactory bulb, curiously, the TRE-reporter is expressed by tufted cells but not by mitral

cells, showing a neuron type specificity [3]. We speculate that this phenomenon is caused by the silencing of the repetitive TRE elements in the genome, as reported in the previous study [10]. This problem can be solved by knocking in the TRE reporter flanked by insulators in the ROSA or TIGRE locus [3]. Viral encoded TRE reporters also escape from this silencing problem.

2. To shorten the active time frame of TM, direct injection of 4-hydroxytamoxifen, the active metabolite of TM with a shorter half-life, into the mouse tail vein has been shown to be effective [12].
3. TM injection into pregnant mice between E9.5 and E15.5 often delays delivery, necessitating cesarean section. Conversely, TM injection later than E16.5 frequently results in miscarriages. While progesterone is added to the TM solution to reduce miscarriages, it may not be entirely effective. Progesterone may be omitted from the TM solution without significant effect. Because pups born prematurely after E17.5 can have a high probability of survival, the most difficult TM stage to obtain mice is TM16.5.
4. Almost all the TM-injected dams have difficult deliveries. Therefore, it is recommended to prepare the same number of foster mothers as TM-injected neurogenic tagging mice. To prepare foster mothers, we usually keep ICR wild-type male and female mice together continuously for 3 days starting 2 days before mating of neurogenic tagging mice.
5. Anesthesia of mice is an important issue to consider. Previously, we used sodium pentobarbital [3]. However, because the use of this drug for anesthesia is not recommended in recent guidelines, we switched to the use of isoflurane, an inhalational anesthetic. Although surgical maneuverability is reduced, we are satisfied with the conditions and recovery from isoflurane anesthesia. Now that ventilation equipment for isoflurane anesthesia is readily available, we would recommend its use.

Acknowledgments

This work was supported by MEXT/JSPS KAKENHI Grants (23H02581) and IU-REAL Collaborative Research Grant.

Conflict of Interest Statement The authors have no conflicts of interest to declare that are relevant to the content of this chapter.

Funding

This study was funded by KAKENHI Grant Numbers 23K27272 and 17H05587 from Ministry of Education, Culture, Sports, Science and Technology (MEXT); Japan Society for the Promotion of Science (JSPS).

References

1. Hirata T, Tohsato Y, Itoga H, Shioi G, Kiyonari H, Oka S, Fujimori T, Onami S (2021) NeuroGT: a brain atlas of neurogenic tagging CreER drivers for birthdate-based classification and manipulation of mouse neurons. Cell Rep Methods 1:100012
2. Hirata T, Iwai L (2019) Timing matters: a strategy for neurons to make diverse connections. Neurosci Res 138:79–83
3. Hirata T, Shioi G, Abe T, Kiyonari H, Kato S, Kobayashi K, Mori K, Kawasaki T (2019) A novel birthdate-labeling method reveals segregated parallel projections of mitral and external tufted cells in the main olfactory system. eNeuro 6:ENEURO.0234-19.2019
4. Feil R, Wagner J, Metzger D, Chambon P (1997) Regulation of Cre recombinase activity by mutated estrogen receptor ligand-binding domains. Biochem Biophys Res Commun 237:752–757
5. Zhang J, Tran-Anh K, Hirata T, Sugihara I (2021) Striped distribution pattern of Purkinje cells of different birthdates in the mouse cerebellar cortex studied with the Neurog2-CreER transgenic line. Neuroscience 462:122–140
6. Luo Y, Chao Y, Owusu-Mensah RNA, Zhang J, Hirata T, Sugihara I (2023) Neurogenic timing of the inferior olive subdivisions is related to the olivocerebellar projection topography. Sci Rep 13:7114
7. Tran-Anh K, Zhang J, Nguyen-Minh VT, Fujita H, Hirata T, Sugihara I (2020) Common origin of the cerebellar dual somatotopic areas revealed by tracking embryonic Purkinje cell clusters with birthdate tagging. eNeuro 7: ENEURO.0251-20.2020
8. Hippenmeyer S, Vrieseling E, Sigrist M, Portmann T, Laengle C, Ladle DR, Arber S (2005) A developmental switch in the response of DRG neurons to ETS transcription factor signaling. PLoS Biol 3:e159
9. Imayoshi I, Hirano K, Sakamoto M, Miyoshi G, Imura T, Kitano S, Miyachi H, Kageyama R (2012) A multifunctional teal-fluorescent Rosa26 reporter mouse line for Cre- and Flp-mediated recombination. Neurosci Res 73:85–91
10. Zhu P, Aller MI, Baron U, Cambridge S, Bausen M, Herb J, Sawinski J, Cetin A, Osten P, Nelson ML, Kugler S, Seeburg PH, Sprengel R, Hasan MT (2007) Silencing and un-silencing of tetracycline-controlled genes in neurons. PLoS One 2:e533
11. Paxinos G, Franklin KBJ (2008) The mouse brain in stereotaxic coordinates, 3rd edn. Elsevier Academic Press, Amsterdam/Boston
12. Hatanaka Y, Yamada K, Eritate T, Kawaguchi Y, Hirata T (2024) Cerebral Cortex (34)11: bhae439. https://doi.org/10.1093/cercor/bhae439

Chapter 12

Application of Lentivirus Vectors to Adult Neurogenesis and Manipulation of Neural Stem Cells

Itaru Imayoshi, Mayumi Yamada, Shinji C. Nagasaki, Tomonori D. Fukuda, Ikumi Nagano, and Tomohisa Toda

Abstract

Lentivirus vectors are capable of efficient gene transfer to nondividing cells, in addition to dividing cells. Therefore, lentivirus vectors can also be used for gene transfer into dormant stem cells in the postnatal and adult brains of experimental animals, many of which have ceased cell division. For example, it is known that neural stem cells (NSCs) are present around the lateral ventricle and in the dentate gyrus of the hippocampus in the postnatal and adult brain and continue to give rise to neurons. However, the majority of these NSCs are considered to be quiescent, and lentivirus vectors are an important tool to enable gene transfer to dormant stem cells. Lentivirus vectors are also capable of multiple infection of the same cells, making them an important option for experiments requiring high gene expression levels. By selecting transduced cells after infection with high titer lentivirus vectors and repeating infection, genes that are difficult to express or gene sequences that require high copy number can be introduced into cells with high efficiency. In this chapter, we present examples of gene transfer to NSCs localized at the hippocampal dentate gyrus in mice using lentivirus vectors and establishment of stable transgenic NSC lines by multiple infections.

Key words Neural stem cells, Quiescent cells, Hippocampus, Optogenetics, bHLH transcription factor, Lentivirus vector

1 Introduction

1.1 Application of Lentivirus Vectors to Adult Neurogenesis

It was previously thought that the mammalian central nervous system has poor regenerative capacity, that neural stem cells (NSCs) are virtually absent, and that the neurogenesis or regeneration of neurons is generally difficult to achieve [1]. However, in the 1960s, Dr. Altman and colleagues reported that NSCs were present in rodent brains and that active neurogenesis was occurring [2]. Subsequent studies in birds followed [3], and in the late 1980s it was confirmed that NSC cultures could be established from adult rodent brains, and it was also reported that neurogenesis was occurring in the hippocampus of the human brain [4]. Since then, various methods have enabled gene transfer and genetic

Kazuto Kobayashi et al. (eds.), *Lentiviral Vectors: Applications for Basic and Clinical Neuroscience*, Methods in Molecular Biology, vol. 2974, https://doi.org/10.1007/978-1-0716-4807-0_12,

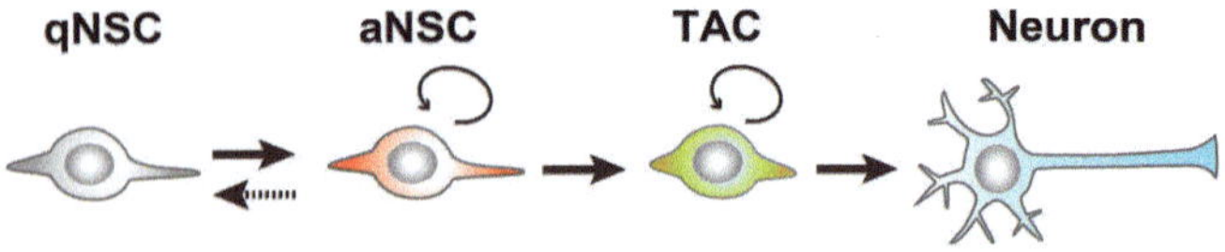

Fig. 1 Neurogenesis in the postnatal and adult brains. In the mammalian brain, neural stem cells (NSCs) change from a quiescent state (quiescent NSC, qNSC) in which they do not proliferate to an active state (active NSC, aNSC) in which they proliferate. Then aNSCs divide to produce transit amplifying cells (TACs). TACs divide a few times and differentiate into neurons

modification of NSCs in the adult rodent brains, and the overall picture of adult brain neurogenesis has become elucidated. Currently, studies on the analysis of the regulatory mechanism of adult NSC differentiation and the functional analysis of newborn neurons are being intensely conducted [5, 6].

In the history of postnatal and adult brain neurogenesis researches, lentivirus vectors have played an important role as essential tools that allow gene transfer to NSCs [7–9]. This is because many populations of NSCs in the postnatal and adult brain are considered to be in the G0 dormant phase, when cell proliferation has ceased, and after being activated by specific stimuli, they are thought to produce neurons after multiple cell divisions (Fig. 1) [1]. Therefore, the required characteristics of viral vectors are the ability to efficiently infect nondividing cells and to continue expression of the transgene without dilution or degradation during cell division of the infected cells, and lentivirus vectors have been used to meet these requirements.

Lentiviruses are a member of the retrovirus family, and the lentivirus life cycle is representative of the retrovirus family [10]. Initially, glycoproteins (such as VSV-G) of the viral envelope bind to specific cellular receptors. The envelope fuses with the host cell membrane, and the viral core is released into the cytoplasm. Immediately after this internalization, single-stranded RNA is reverse transcribed with the help of viral proteins to form the double-stranded genome, which is then incorporated into the host genome. Lentiviruses are also capable of infecting undivided cells via proteins expressed from the *vpr* gene. These properties of lentivirus vectors have made them an essential research tool in the study of neurogenesis in the postnatal and adult brain, starting from neural stem cells, many of which are initially dormant and dividing after activation.

1.2 Application of Lentivirus Vectors to Manipulation of NSCs with Optogenetics

Another advantage of lentivirus vectors is their ability to multiply infected target cells [10]. It is possible to incorporate the transgene into multiple locations of the host genome in a single infection (Fig. 2). Although cell damage and the secondary effects of inserting the transgene into multiple host genome regions must be taken

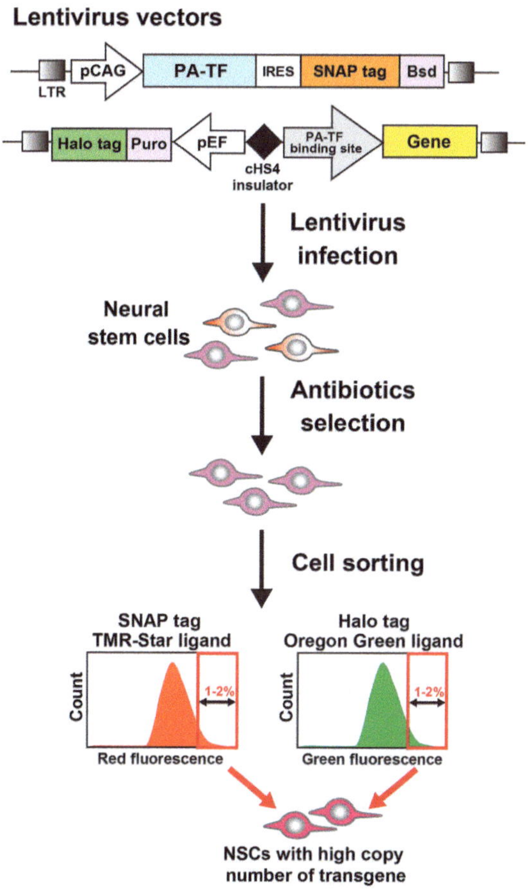

Fig. 2 Flow of lentivirus experiments and cell establishments in cultured NSCs. Schematic illustration of examples of the transgenic constructs (CSII vectors) used for lentivirus production. (upper) The CSII-pCAG-PA-TF-SNAP-Bsd plasmid, which is composed of the CAG promoter, the coding sequence of the photoactivatable transcription factor (PA-TF), SNAP tag, and Bsd gene. (lower) The reporter CSII plasmid is composed of the PA-TF-binding site, coding sequence of the target gene, Halo tag, and Puro gene. Lentiviral particles purified with above CSII plasmid are used to infect NSCs, and antibiotic selection is started after a few days. Once the number of NSCs has increased, substrates for SNAP tag and Halo tag (e.g., SNAP tag TMR-Star and Halo tag Oregon Green ligands) are added and chromogenic, and a cell population that highly expresses both SNAP and Halo tags (top 1–2%) is collected by cell sorter

into consideration, it is possible to select transduced cells with high-titer lentivirus vectors and repeated infection to obtain desired transgenic cells that contains genes or sequences with a high copy number. By selecting transduced cells, genes that are difficult to express or gene sequences requiring high copy number can be introduced into cells with high efficiency. For example, optogenetic tools for gene expression are useful techniques for perturbation experiments of dynamic expression kinetics, such as bHLH-type transcription factors, and for light-mediated manipulation of NSCs [11]. However, efficient light manipulation requires high expression of photo-activatable transcription factors (e.g., hGAVPO or PA-Tet) as well as increasing the number of introduced copies of genetic elements with transcription factor target sequences (e.g., UAS or TRE) [11–15]. It is known that the efficiency of infection varies greatly depending on the host cell type, even when lentivirus vectors of the same titer are used. Cultured NSCs are not inefficiently infected, but often require preparation of high-titer purified viral vectors and multiple infections to achieve incorporation of the transgene into the host genome at high copy number.

2 Materials

2.1 Production and Purification of Lentivirus Vectors

1. Producer cells: HEK293T cells (American Type Culture Collection [ATCC]).
2. DMEM medium: Dulbecco's Modified Eagle Medium (DMEM) supplemented with antibiotics (100 units/mL penicillin and 100 mg/mL of streptomycin) and 10% FBS.
3. Reduced serum medium: Opti-MEM supplemented with antibiotics (100 units/mL penicillin and 100 mg/mL of streptomycin).
4. TE buffer: 10 mM Tris–HCl—1 mM EDTA, pH 8.0. Use to redissolve all plasmids.
5. Envelope plasmid: pMD2.G (addgene, #12259) dissolved at TE buffer.
6. Packaging plasmid: psPAX2 (addgene, #12260) dissolved at TE buffer.
7. Vector plasmid: pCSII [16] (*see* **Notes 1** and **2**) dissolved at TE buffer.
8. Transfection reagent: Polyethylenimine (PEI) or Lipofectamine 3000.
9. Phosphate-buffered saline (PBS), pH 7.4.
10. 0.05% Trypsin/EDTA.
11. 10-cm tissue culture dish.
12. 37 °C humidified incubators, 5% CO_2.
13. 1.5-mL microcentrifuge tubes, sterile, disposable.
14. 15- and 50-mL conical centrifuge tubes, sterile.
15. 50 mL syringes and 0.45-μm pore size PES filters.
16. Filter tips (1000, 200, 20, and 10 μL).
17. 0.2 mL silicone-coated tube (Bio Medical Science Inc., #BC-bmb20200).

18. Centrifuge with an angle rotor.
19. Forskolin.
20. Sodium butyrate.
21. Real-time PCR machine and 96-well Optical Reaction plate and seal (Bio-Rad Laboratories).
22. SYBR Master Mix (TOYOBO, THUNDERBIRD® SYBR qPCR Mix).
23. Primer pair targeting WPRE: forward primer; 5′--GTCCTTTCCATGGCTGCTC-3′, and reverse primer; 5′--CCGAAGGGACGTAGCAGA-3′.
24. WPRE-containing plasmid for standard curve.

2.2 Infection of Lentivirus Vectors into the Adult Mouse Brain

1. Small Animal Stereotaxic Instrument (David Kopf).
2. Micromotor, dental contra-angle.
3. Syringe pump.
4. Hair clipper.
5. Cotton swab.
6. 19G sterile needle.
7. Dental drill.
8. Glass pipette.
9. Suture needle.
10. Parafilm.
11. Vaseline.
12. 70% ethanol.
13. Anti-inflammatory drugs (dexamethasone sodium phosphate).

2.3 Infection of Lentivirus Vectors into Cultured NSCs and Their Selections

1. Infected cells: Cultured NSCs from C57BL6 or ICR mouse embryonic brain.
2. N2 medium: DMEM/F12 Medium supplemented with antibiotics (100 units/mL penicillin and 100 mg/mL of streptomycin,), N-2 MAX supplement, 20 ng/mL Bfgf, and 20 ng/mL EGF.
3. Matrigel.
4. Accutase.
5. 24- and 6-well culture plates.
6. Blasticidin S. solution (10 mg/mL in water).
7. Puromycin solution (10 mg/mL in PBS).
8. HaloTag®, HaloTag® Oregon Green® Ligand (Promega).

9. SNAP-tag® and Bsd fusion protein codon-optimized for human by FASMAC, SNAP-Cell® TMR-Star (New England Biolabs).

3 Methods

3.1 Production and Purification of Lentivirus Vectors

Maintain HEK293T cells in DMEM medium, in 10-cm tissue culture dish in a 37 °C/5% CO_2 incubator, and split them using Trypsin/EDTA.

1. The day before the transfection, seed five dishes at 3.6×10^6 cells per dish (10-cm). Incubate overnight in a 37 °C/5% CO_2 incubator. On the following day, co-transfect the cells according to the following recipes.
2. For one plate of 10-cm, mix in a sterile 15-mL centrifuge tube.

Envelope plasmid	pMD2.G	17.5 μg
Packaging plasmid	psPAX2	32 μg
Vector plasmid	pCSII	37 μg
PEI (1 mg/mL)		86.5 μL
Opti-MEM		0.5 mL

3. Mix well by vortex.
4. Leave the precipitates at room temperature for 10–20 min.
5. Add the 0.5 mL precipitate dropwise to the cells in one culture dish prepared as above. Mix by gentle swirling.
6. Place the dish for 6 h in a 37 °C/5% CO_2 incubator.
7. Aspirate the medium, and gently add 8–10 mL of fresh DMEM medium containing 10 μM forskolin and 1 mM sodium butyrate. Incubate for 24–48 h.
8. Transfer the supernatant from each plate to one 50-mL centrifuge tubes. Close the tubes and centrifuge for 2 min at 1500 rpm at 4 °C, to pellet detached cells and debris.
9. Filter the 40–50 mL of pooled supernatant with 50 mL syringe connected to a 0.45 μm PES filter.
10. For concentration, use 50-mL centrifuge tubes in angle rotor. Centrifuge for overnight (16 h and over) at 6000×g at 4 °C.
11. Decant the medium carefully. Try to leave as little medium (serum) as possible.
12. Quickly add 800 μL PBS on the pellet (not always visible). Do not leave the pellet dry or it may result in significant titer decrease. Incubate for 5 min on ice.

13. Mix by gentle pipetting and transfer the suspension solution into a new 1.5 mL microcentrifuge tube.
14. Quickly add 500 μL PBS in 50 mL centrifuge tube again and transfer the suspension solution into the same 1.5 mL microcentrifuge tube.
15. For concentration, use angle rotor. Centrifuge at 14,000 rpm for 4 h at 4 °C.
16. Remove the supernatant (*see* **Note 3**) and add 50–200 μL cold PBS. Incubate for 5 min on ice.
17. Mix by gentle pipetting until pellet is not visible. Be careful not to foam.
18. Dispense 5–10 μL each into 0.2 mL silicone-coated tubes and freeze at −80 °C in aliquots for long-time storage. Do not freeze and thaw. Note that if you need to run a qPCR analysis on the same sample, you must split virus solution in separate silicone-coated tube.
19. Prepare plasmid stocks of 1×10^4, 1×10^5, 1×10^6, 1×10^7, and 1×10^8 molecules/μL to generate a standard curve. Note that the pCSII plasmid for lentivirus production should be used.
20. The purified lentivirus is diluted with water in the following four patterns; No dilution, 1/10 dilution, 1/100 dilution, and 1/1000 dilution.
21. Prepare a mix containing everything but the sample for number of wells needed for the qPCR analysis, including all lentivirus and standards in duplicates or triplicates, according to the following mixture:

SYBR mix	5 μL
Forward primer (100 μM)	0.1 μL
Reverse primer (100 μM)	0.1 μL
Sample	1 μL
Water	3.8 μL

22. Distribute 9 μL of this mix into the wells of 96-well Optical Reaction plate.
23. Add samples (lentivirus or standards).
24. Close with seal.
25. Centrifuge the plate briefly.
26. Run the following protocol in the qPCR instrument using SYBR detection:

    95 °C, 3 min (**step 1**) /95 °C, 15 s (**step 2**) /60 °C, 1 min (**step 3**); repeat 35 cycles for **steps 2** and **3** for melt curve.

27. Calculate the titer of each lentivirus from the standard curve. Since lentiviruses are single-stranded and plasmids for standard curves are double-stranded, do not forget to multiply the value obtained from the calibration curve by two. The quantification of lentiviral titers by qPCR depends on the target sequence and the primer set. Our lab uses a primer set (Forward: 5′-GTCCT TTCCATGGCTGCTC-3′; Reverse: 5′-CCGAAGGGACGTA GCAGA-3′) targeting the WPRE sequence. When lentiviral vector is produced in 5 10-cm tissue culture dishes, the titer is approximately 10^{10-11} viral genome (VG)/mL.

3.2 Infection of Lentivirus Vectors into the Adult Mouse Brain

1. Anesthetize the adult mouse.
2. Shave hair from the mouse head using a hair clipper.
3. Apply hair removal cream to mouse head. After 2–3 min, remove hairs with cotton swab.
4. Cover the eyes with Vaseline to prevent dryness.
5. Sterilize the mouse head with 70% ethanol.
6. The head of mouse is fixed stably to the stereotaxic frame using mouth and ear bars.
7. Incision in the skin of the mouse head.
8. Put 19G sterile needle on a pipet holder to mark the position of bregma and lambda and level the brain.
9. Once leveled, determine the position of the injection into the dentate gyrus.

   For example, set the following positions from bregma.

   $X = \pm 1.3$ mm, $Y = -1.94$ mm, $Z = -1.82$ mm (Z is the position from the brain surface, not the skull surface)
10. The skull at the injection position is drilled with a dental drill.
11. To prevent the skull from drying out, the area around the drilled skull is covered with saline solution and parafilm is placed over it. This parafilm is also used as a stage for loading the virus solution into the injection pipet.
12. Inject the virus solution into a glass pipette. Using an injection pump, inject the virus into the target site. Inject 1–2 μL lentivirus solution over a 10-min period.
13. Wait 5 min after the injection is finished to prevent the lentivirus solution from leaking out of brain tissue.
14. Slowly pull out the glass pipet from brain tissue.
15. Suture the skin with a suture needle.
16. Administer anti-inflammatory drugs to mouse.
17. Recover from anesthesia on hot plate (*see* **Note 4**).

3.3 Infection of Lentivirus Vectors into Cultured NSCs and Their Selections

1. Culture NSCs in N2 medium in 6-well culture plate in a 37 °C/5% CO_2 incubator, and split them using Accutase.
2. On the day of lentiviral infection, seed 2×10^5 cells per well in a 24-well culture plate. Incubate for 1 h in a 37 °C/5% CO_2 incubator.
3. Reduce N2 medium volume to 0.25 mL.
4. Infect NSCs with lentivirus particles with a multiplicity of infection (MOI) = 100–1000. Incubate overnight in a 37 °C/5% CO_2 incubator.
5. The next day, remove medium and add 0.5 mL of N2 medium.
6. Begin antibiotic selection after the second day of lentivirus infection.

   For example, transduced cells are selected by 2 μg/mL Blasticidin S and 2 μg/mL Puromycin for the lentivirus vectors co-expressing Bsd and Puro.
7. Maintain NSCs in N2 medium, in 6-well culture plate or 10-cm tissue culture dish in a 37 °C/5% CO_2 incubator, and split them using Accutase. Increase NSCs until 1–2 10-cm tissue culture dishes.
8. If the lentivirus vector expressing Halo tag or SNAP tag is used, add a substrate for each tag, and collect the target cell population by fluorescence-activated cell sorting.
9. Cells collected by cell sorting are continued to be cultured in the same manner as above until the number of cells available for experiments is reached.
10. If the expression of the gene introduced with lentivirus vector is low, after the several passages of selected cells, the process of "1" can be repeated for several times to increase the copy number of integrated transgenes (*see* **Notes 5–7**).

4 Notes

1. Although lentivirus vectors can carry transgene sequences larger than 10 kb, it should be noted that the titer of purified viruses decreases drastically when the inserted transgene sequence exceeds about 10-kb. When the purified virus titer is estimated using the genomic copy number of the viral genome as an indicator, it should be noted that the virus may be infected only at a lower efficiency than the estimated copy number when the infection experiment is performed. This is expected to be due to incomplete lentivirus vectors being produced and mixed with purified viral particles.

2. For lentivirus vectors with large transgene sequences, it is difficult to prepare high-titer purified viral particles; therefore it is sometimes necessary to shorten the transgene sequences as much as possible or alternatively to split them into two lentiviral vectors.
3. It is known that the differentiation of NSCs is triggered by serum components in the cell culture medium [11]. Therefore, when purifying viral particles from supernatants of HEK293T packaging cells, it is important to reduce contamination of serum components as much as possible.
4. Neurogenesis may be reduced after injection of purified lentivirus vectors into the hippocampal dentate gyrus. In such cases, lowering the titer or dose of the lentivirus vectors to be injected may resolve the problem.
5. Similarly, cultured NSCs may be damaged during infection with lentivirus vectors and their condition may deteriorate. In such cases, lowering the MOI of the lentivirus vectors to be infected may solve the problem and the condition of the cells may recover with continued cell culture.
6. Several weeks after lentivirus vector infection, expression of the transgene may decrease due to epigenomic modification of chromatin and other factors. Therefore, when conducting long-term experiments, it is required to periodically check the expression and induction of expression of the transgene.
7. When conducting long-term experiments on cells transduced with lentivirus vectors, it is effective to establish single cell-derived clones. The reliability of experiments can be improved by producing single-cell-derived clonal cell lines that can stably express or induce expression of the transgene [12, 14, 15].

Conflict of Interest Statement

The authors have no conflicts of interest to declare that are relevant to the content of this chapter.

Funding

None.

References

1. Mira H, Andreu Z, Suh H, Lie DC, Jessberger S, Consiglio A et al (2010) Signaling through BMPR-IA regulates quiescence and long-term activity of neural stem cells in the adult hippocampus. Cell Stem Cell 7(1): 78–89. https://doi.org/10.1016/j.stem.2010.04.016
2. Altman J, Das GD (1965) Autoradiographic and histological evidence of postnatal hippocampal neurogenesis in rats. J Comp Neurol 124(3):319–335. https://doi.org/10.1002/cne.901240303
3. Nottebohm F (2005) The neural basis of birdsong. PLoS Biol 3(5):e164. https://doi.org/10.1371/journal.pbio.0030164
4. Imayoshi I, Sakamoto M, Ohtsuka T, Takao K, Miyakawa T, Yamaguchi M et al (2008) Roles of continuous neurogenesis in the structural and functional integrity of the adult forebrain. Nat Neurosci 11(10):1153–1161. https://doi.org/10.1038/nn.2185
5. Imayoshi I, Kageyama R (2014) bHLH factors in self-renewal, multipotency, and fate choice of neural progenitor cells. Neuron 82(1):9–23. https://doi.org/10.1016/j.neuron.2014.03.018
6. Imayoshi I, Kageyama R (2014) Oscillatory control of bHLH factors in neural progenitors. Trends Neurosci 37(10):531–538. https://doi.org/10.1016/j.tins.2014.07.006
7. Jessberger S, Clark RE, Broadbent NJ, Clemenson GD, Consiglio A, Lie DC et al (2009) Dentate gyrus-specific knockdown of adult neurogenesis impairs spatial and object recognition memory in adult rats. Learn Mem 16(2): 147–154. https://doi.org/10.1101/lm.1172609
8. Toda T, Bedrosian TA, Schafer ST, Cuoco MS, Linker SB, Ghassemzadeh S et al (2024) Long interspersed nuclear elements safeguard neural progenitors from precocious differentiation. Cell Rep 43(2):113774. https://doi.org/10.1016/j.celrep.2024.113774
9. Toda T, Hsu JY, Linker SB, Hu L, Schafer ST, Mertens J et al (2017) Nup153 interacts with Sox2 to enable bimodal gene regulation and maintenance of neural progenitor cells. Cell Stem Cell 21(5):618–634.e617. https://doi.org/10.1016/j.stem.2017.08.012
10. Warnock JN, Daigre C, Al-Rubeai M (2011) Introduction to viral vectors. Methods Mol Biol 737:1–25. https://doi.org/10.1007/978-1-61779-095-9_1
11. Imayoshi I, Isomura A, Harima Y, Kawaguchi K, Kori H, Miyachi H et al (2013) Oscillatory control of factors determining multipotency and fate in mouse neural progenitors. Science 342(6163):1203–1208. https://doi.org/10.1126/science.1242366
12. Nagasaki SC, Fukuda TD, Yamada M, Suzuki YI, Kakutani R, Guy AT, Imayoshi I (2023) Enhancement of Vivid-based photo-activatable Gal4 transcription factor in mammalian cells. Cell Struct Funct 48(1):31–47. https://doi.org/10.1247/csf.22074
13. Yamada M, Nagasaki SC, Ozawa T, Imayoshi I (2020) Light-mediated control of gene expression in mammalian cells. Neurosci Res 152:66–77. https://doi.org/10.1016/j.neures.2019.12.018
14. Yamada M, Nagasaki SC, Suzuki Y, Hirano Y, Imayoshi I (2020) Optimization of light-inducible Gal4/UAS gene expression system in mammalian cells. iScience 23(9):101506. https://doi.org/10.1016/j.isci.2020.101506
15. Yamada M, Suzuki Y, Nagasaki SC, Okuno H, Imayoshi I (2018) Light control of the Tet gene expression system in mammalian cells. Cell Rep 25(2):487–500.e486. https://doi.org/10.1016/j.celrep.2018.09.026
16. Miyoshi H (2004) Gene delivery to hematopoietic stem cells using lentiviral vectors. Methods Mol Biol 246:429–438. https://doi.org/10.1385/1-59259-650-9:429

Chapter 13

Direct Reprogramming of Microglia into Neurons

Kanae Matsuda-Ito, Eriko Hatai, Takashi Irie, Taito Matsuda, and Kinichi Nakashima

Abstract

In recent years, direct reprogramming methods have attracted attention as they can bypass the immature stem cell state and directly induce the desired cell type by introducing key transcription factors into somatic cells. This technique offers advantages such as shorter induction times compared to inducing the desired cells from induced pluripotent stem cells. Furthermore, this approach enables the provision of the requisite cell types to patients requiring treatment without eliciting an immune rejection response. Microglia, the resident immune cells of the central nervous system, exhibit high proliferative and regenerative capacities, allowing them to repopulate even after almost complete depletion. In response to injury, microglia accumulate in damaged brain regions and become the primary cell type that forms the glial scar. Given these properties, microglia are promising candidates for direct reprogramming into neurons in the injured area without depleting cell sources. This chapter outlines a protocol for directly reprogramming cortical microglia into neurons, utilizing primary culture and lentiviral-mediated gene transfer techniques.

Key words Microglia, Neuron, Direct reprogramming, Lentivirus, Infection, Primary culture

1 Introduction

Microglia, the immune cells residing in the central nervous system, play crucial roles in detecting, phagocytosing, and eliminating pathogens. In recent years, microglia have increasingly been found to be associated with various psychiatric and neurological disorders, including depression and Alzheimer's disease. Consequently, techniques for gene delivery to microglia are indispensable for both functional analysis and therapeutic development. Viral vectors have emerged as prominent methods, with lentivirus and adeno-associated virus (AAV) vectors receiving particular attention. These vectors are favored because they can infect nonproliferating cells such as microglia and ensure stable gene expression with little induction of cytotoxic effects over the long term. Balcaitis et al. were pioneers in the field of gene delivery to microglia using lentiviral vectors [1]. They conducted initial experiments in which

Kazuto Kobayashi et al. (eds.), *Lentiviral Vectors: Applications for Basic and Clinical Neuroscience*, Methods in Molecular Biology, vol. 2974, https://doi.org/10.1007/978-1-0716-4807-0_13,

primary cultured microglia and microglial cell lines N9 and BV2 were infected with lentiviruses expressing EGFP under the control of the murine stem cell virus promoter. This procedure resulted in EGFP expression in approximately 50% of primary cultured microglia and over 95% of cells in the cell lines. Additionally, Åkerblom et al. devised an innovative microglia-specific in vivo gene delivery approach by incorporating microRNA (miRNA)-9 target sequences into self-inactivating lentiviral vectors [2]. miRNA-9 is typically not expressed in microglia but highly expressed in other neuronal cells, and use of this vector enables microglia-specific gene expression. In contrast, AAV vectors have lower efficiency (estimated to be less than 20%) of gene delivery to microglia in vivo [3]. Thus, while AAV vectors are preferred for gene therapy due to their non-genome integrative property, additional modifications are necessary to enhance their efficiency of gene delivery to microglia. Therefore, lentiviral vectors currently serve as the primary tool for gene delivery to microglia owing to their high efficiency. In the present study, we utilized two types of lentiviral vectors for gene transfer into microglia: Tet-inducible lentiviral vectors to express genes in a doxycycline (Dox)-dependent manner and lentiviral vectors to express genes under the control of the promoter for *CD68*, a microglial marker [4]. This chapter provides an overview of the protocol for in vitro and in vivo neuronal conversion, encompassing the isolation and primary culture of microglia derived from the mouse cortex, the infection of cultured microglia for in vitro experiments, and the infection of brain microglia for in vivo experiments.

2 Materials

2.1 In Vitro Experiments

2.1.1 Virus Production and Preparation

1. Human Embryonic Kidney (HEK) 293 T cells.
2. Lentivirus plasmids:

   We use Tet-inducible lentiviruses (Tet-O-FUW-EGFP, Tet-O-FUW-NeuroD1(ND1) and FUW-M2rtta (M2rtta is a modified tetracycline-inducible transactivator); Addgene, 30,130, 30,129, and 20,342) in vitro experiments as well as lentiviruses encoding ND1 under the control of the *CD68* promoter for microglia-specific gene delivery in vivo experiments.
3. Packaging plasmids:

   Gag-Pol expression plasmid, Rev. expression plasmid, VSV-G expression plasmid. We use pCAG-HIVgp and pCMV-VSV-G-RSV-Rev [5].
4. Opti-MEM. Store at 4 °C.
5. Polyethyleneimine “Max”: PEI MAX 1 mg/mL. Store at 4 °C.

6. Virus production medium (N2 medium): 1.27 g DMEM/F-12, $NaHCO_3$, 25 mg insulin, 100 mg apo-transferrin, 16 mg putrescine, 6 μL of 5 mM selenite, 20 μL of 1 mM progesterone. Make up to 1 L with sterile water. Adjust to pH 7.2.
7. Antibiotic-Antimycotic Mixed Stock Solution (100×).
8. Syringe filter (pore size 0.45 μm).

2.1.2 Preparation for Primary Microglia Culture

1. Papain from Papaya latex. Stored at 4 °C.
2. Papain solution: 3000 U DNase 30,000 U, 20 mg L-cysteine, 20 mg BSA, 500 mg glucose. Make up to 60 mL with PBS and filter through a 0.45 μm syringe filter. Stored at −20 °C.
3. HBSS: Hank's Balanced Salts, 0.35 g $NaHCO_3$. Make up to 1 L with sterile water. Adjust to pH 7.2. Store at 4 °C.
4. DNase Solution: 10,000 U DNase, 60 mg $MgSO_4{\cdot}7H_2O$ 60 mg. Make up to 20 mL with HBSS and filter through a 0.45 μm syringe filter. Stored at 4 °C.
5. Tissue culture medium: 1000 mL MEM Alpha, 5% FBS, 6 g glucose.
6. Fraction V.
7. Cell Strainer, 40 μm.
8. Primary Microglia Culture Medium: 500 mL DMEM/Ham's F-12, 20% FBS, 0.5% gentamycin, MEM Non-Essential Amino Acid Solution (100×), 100 mM sodium pyruvate (100×).

2.1.3 Preparation for Inducing Neurons from Primary Cultured Microglia

1. Produced lentivirus.
2. Polybrene.
3. Neuronal Induction Medium: 500 mL Neurobasal Medium, 2% B27 Supplement, 1% GlutaMAX, Antibiotic-Antimycotic Mixed Stock Solution (100×), BDNF (final concentration 10 ng/mL), GDNF (final concentration 10 ng/mL), NT3 (final concentration 10 ng/mL).
4. Dissolve doxycycline in water at 1 mg/mL and filter through a 0.45 μm syringe filter. Store at 4 °C.

2.2 In Vivo Experiments

1. HEK 293 T cells.
2. Lentivirus plasmid.
3. Packaging plasmid.
4. Opti-MEM.
5. Polyethyleneimine "Max": PEI MAX 1 mg/mL.
6. Virus Production Medium.
7. Antibiotic-Antimycotic Mixed Stock Solution (100×).
8. Syringe filter (pore size 0.45 μm).

9. Lenti-X Concentrator.
10. Brain Stereotaxic Apparatus for Mouse.
11. Micro syringe.
12. Electric drill.
13. Mixed Anesthesia: 0.75 mL medetomidine hydrochloride, 2 mL midazolam, 2.5 mL butorphanol tartrate in 19.75 mL Distilled water.

3 Methods

3.1 Production and Purification of Lentivirus Vector

See Fig. 1 for in vitro and in vivo experiments.

1. Transfect the packaging cells. For one 10-cm dish of HEK293T cells, add 10 μg of the desired lentivirus plasmid and 2 μg each of the packaging plasmids to 1 mL of Opti-MEM. Then add 42 μL PEI and mix well. Incubate at room temperature for 15 min. Add 1 mL of this mixture to a dish of cells and culture for 6 h to overnight.
2. Remove the medium and add 5 mL of virus production medium plus 5 μL of Antibiotic-Antimycotic Mixed Stock Solution per dish. Culture at 37 °C and 5% CO_2 for 2 days.
3. Collect the culture supernatant with a syringe, filter it through a syringe filter, and collect the filtrate in a 50-mL tube, and store it at 4 °C until use. The period of usability is approximately 1 week. However, it is preferable to use freshly prepared virus for better infection efficiency (*see* **Note 1**).
4. For in vivo virus infection, perform the following steps: Centrifuge the tube of collected virus particles at 6000×*g* overnight at 4 °C for virus vector concentration (*see* **Note 2**).
5. After removing the supernatant, resuspend the pellet in 500 μL of virus production medium and transfer to a 1.5-mL Eppendorf tube. Add 170 μL of Lenti-X Concentrator (1/3 volume of the suspension) and repeatedly pipette the suspension thoroughly.

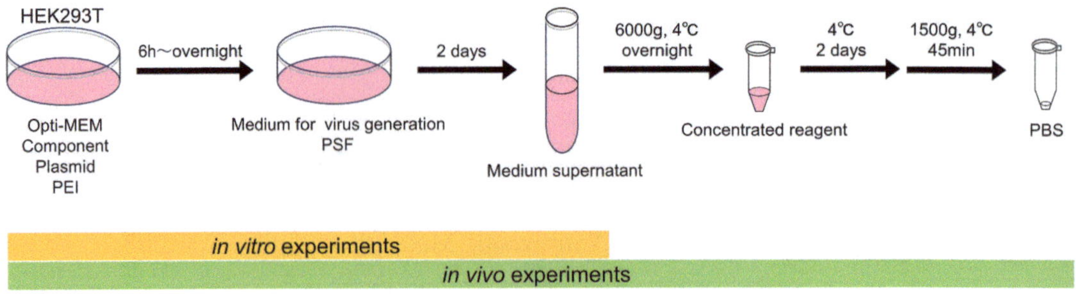

Fig. 1 Outline of lentiviral vector production and purification protocol

6. Incubate at 4 °C for a fixed period (*see* **Note 3**).
7. Centrifuge at 1500×*g* for 45 min at 4 °C.
8. After removing the supernatant, resuspend the pellet in 10 μL of PBS. Store at 4 °C and use within 1 week for in vitro applications. Using freshly prepared virus ensures better infection efficiency.

3.2 Isolation of Primary Cultured Microglia

Microglia for in vitro infection are prepared from postnatal (P)1 mouse cerebral cortex using the shake-off method [6, 7] (*see* Fig. 2).

1. The brain is removed from 4 or 5 P1 mice, the meninges are peeled off and only the cortices are harvested in a tube. The hippocampus still contains many neural progenitor cells at early postnatal stages. Add 3 mL of papain solution and 100 μL of papain, pipette gently, and incubate at 37 °C for 30 min.
2. Remove the supernatant and add 5 mL of tissue culture medium. Once the tissue has precipitated, remove the supernatant, add 1 mL of DNase, and pipet well. Filter it through a cell strainer, add 5 mL of tissue culture medium, and slowly add 600 μL of Fraction V to the bottom of the tube to create two layers. Centrifuge at 1000 rpm for 5 min.
3. Remove only the upper of the two layers, add 5 mL of tissue culture medium, and centrifuge for another 5 min. Remove the supernatant, resuspend the brain cells in 10 mL of primary microglia culture medium, and place them in a T75 flask. Loosen the cap to ensure proper aeration during culture.
4. Culture in the flask for 7 days, changing the entire 10 mL of primary microglia culture medium every 2 days.

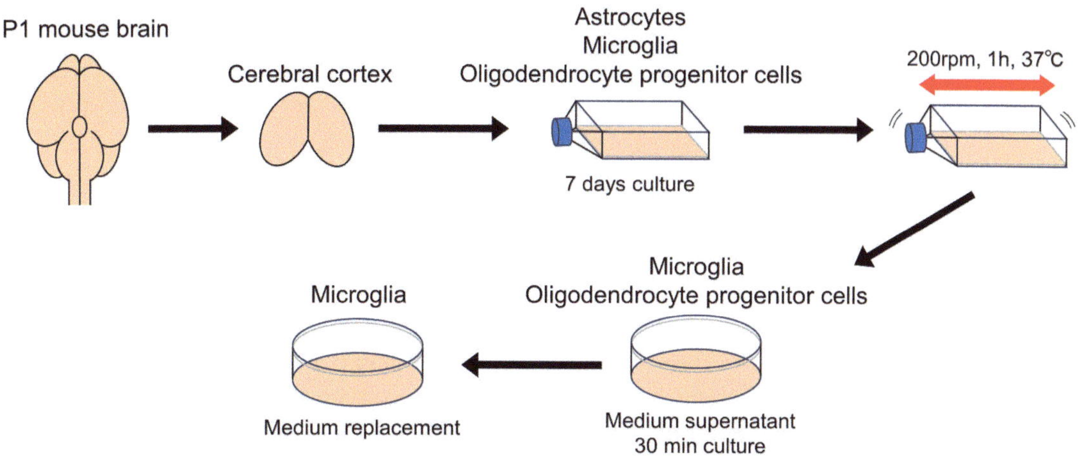

Fig. 2 Outline of microglia isolation by shake-off method in primary culture

5. Shake the flask at 200 rpm for 1 h at 37 °C, ensuring the cap, which was loosened during culture, is securely closed. Astrocytes remain attached to the bottom of the flask, while microglia and oligodendrocyte progenitor cells float.
6. Collect supernatant containing microglia and oligodendrocyte progenitor cells, and seed them directly into dishes of appropriate size.
7. Thirty minutes after seeding, microglia adhere to the bottom of the dish, but oligodendrocyte progenitor cells remain suspended.
8. Change the medium to remove the oligodendrocyte progenitor cells and obtain the dish-attached primary cultured microglia.

3.3 Lentivirus Infection and Reprogramming into Neurons of Primary Cultured Microglia

1. Infection of primary cultured microglia. To increase infection efficiency, add the virus 30 min to 1 h after seeding, once the cells have adhered to the dish (*see* **Note 4**). To neutralize the negative charge of virus particles and facilitate the interaction of the virus envelope with the cell surface, add polybrene to cultured microglia at a final concentration of 8 μg/mL (Polybrene method) (*see* **Note 5**).
2. Add virus solution to cultured cells: Directly add the virus solution to the cultured cells. Use 300 μL of each virus solution (e.g., 600 μL for ND1, M2rtta) for a 3.5-cm dish and 1.2 mL of each solution for a 10-cm dish. Culture for 4–6 h (*see* **Note 6**).
3. Medium change for neuronal reprogramming: Exchange the medium using neuronal induction medium and add Dox at final concentration 1 μg/mL (*see* **Note 7**).
4. After that, perform a half medium change every 2 days for 7 days (*see* **Note 8**).
5. Seven days after induction, reprogrammed neuronal cells can be confirmed by immunostaining using neuronal markers, e.g., tubulin beta III (Tubb3) and microtubule-associated protein 2ba (Map 2ab) (Fig. 3).

3.4 Infection of Microglia in the Brain

Inject lentivirus into the striatum of 8-week-old C57BL/6 mice [8].

1. Administer a mixed anesthetic (0.1 mL per 10 g body weight of the mouse) intraperitoneally.
2. Secure the mouse in a stereotaxic apparatus. Incise the scalp, mark the injection site (0.8 mm rostral and 2.0 mm lateral from bregma), and drill the skull with an electric drill.

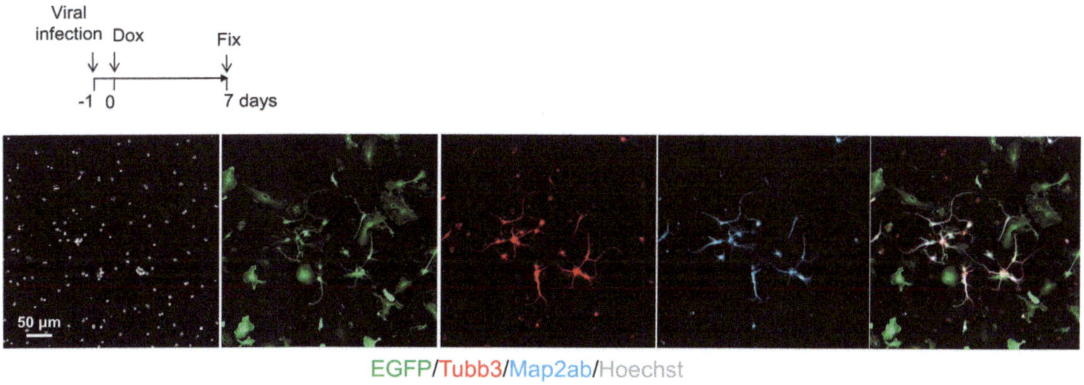

Fig. 3 Reprogramming of microglia into neurons. The primary microglia were infected with a lentivirus encoding *ND1* and *EGFP* genes in a Dox-dependent manner. One day after the virus infection, the medium was replaced with neuronal induction medium containing Dox, and neuronal reprogramming was induced for 7 days. The results of immunofluorescence microscopy revealed the presence of neurons that were positive for EGFP and expressed neuronal markers Tubb3 and Map2ab

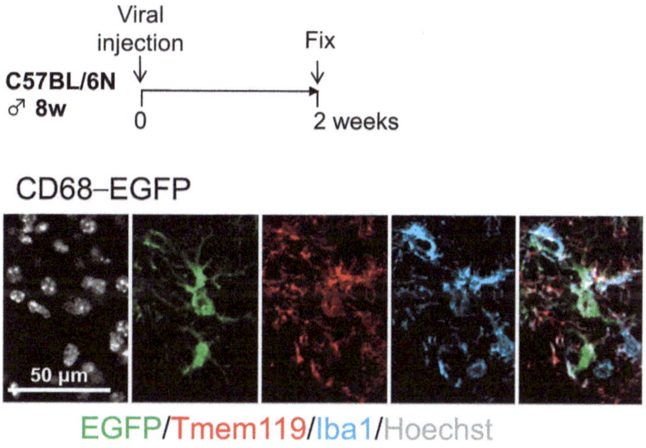

Fig. 4 Examples of gene delivery to microglia using lentivirus in vivo. Injection of a lentivirus encoding EGFP under the control of the CD68 promoter into the striatum of mice, and tissue staining images 2 weeks later. In the striatum, microglia positive for Tmem119 and Iba1 express EGFP

3. Insert the injection needle into the striatum (2.7 mm from the brain surface). Slowly inject 2 μL of virus solution over 5 mins (≤0.4 μL/min). After injection, wait for 4 min to allow the solution to diffuse into the brain parenchyma.
4. Wound closure: Remove the needle and suture the skin.
5. Two weeks after injecting the CD68-EGFP-expressing virus into the striatum, EGFP-positive microglia expressing Tmem119 and Iba1 can be observed within the striatum (Fig. 4).

4 Notes

1. VSVG-pseudotyped viruses can be stored at −80 °C for about 6 months, but their activity decreases by approximately 20–30% with each freeze-thaw cycle. Thaw frozen virus solutions slowly on ice.
2. Since the pellets are difficult to see, the tube is recommended to be marked to indicate the direction of centrifugation.
3. According to the Lenti-X Concentrator protocol, the incubation period can range from 15 min to 1 week; we incubate for 2 days.
4. Prior to infection, verify a sufficient titer (>5 × 10^5 IFU/mL) using easily infected cells such as HEK293T.
5. To avoid toxicity from excessive polybrene exposure (more than 24 h), it is necessary to perform a medium exchange to remove polybrene.
6. Infection of microglia with the virus is most efficient if performed within 1 h after seeding. Additionally, multiple rounds of infection can be used to further enhance the infection efficiency.
7. Dox is added only once to the medium to activate reprogramming factor expression.
8. Perform a half medium change to avoid neuronal cell death, as a whole medium change can induce neuronal death.

Acknowledgments

This work was supported by grants from AMED (JP20gm1310008, JP24wm0625306) and JSPS KAKENHI (JP23H00391) to K.N.; JSPS KAKENHI Grant number (JP23KJ1746) to K.M-I.

Conflict of Interest Statement The authors have no conflicts of interest to declare that are relevant to the content of this chapter.

Funding

This study was funded by AMED (Grant Number JP21bm0404057 to T.M., K.N) and JSPS Research Fellowship for Young Scientists (RPD) (Grant Number JP23KJ1746 to K.M-I).

References

1. Balcaitis S, Weinstein JR, Li S, Chamberlain JS, Möller T (2005) Lentiviral transduction of microglial cells. Glia 50:48–55
2. Åkerblom M, Sachdeva R, Quintino L, Wettergren EE, Chapman KZ, Manfre G, Lindvall O, Lundberg C, Jakobsson J (2013) Visualization and genetic modification of resident brain microglia using lentiviral vectors regulated by microRNA-9. Nat Commun 4:1770
3. Maes ME, Colombo G, Schulz R, Siegert S (2019) Targeting microglia with lentivirus and AAV: recent advances and remaining challenges. Neurosci Lett 707:134310
4. Matsuda T, Irie T, Katsurabayashi S, Hayashi Y, Nagai T, Hamazaki N, Adefuin AMD, Miura F, Ito T, Kimura H, Shirahige K, Takeda T, Iwasaki K, Imamura T, Nakashima K (2019) Pioneer factor NeuroD1 rearranges transcriptional and epigenetic profiles to execute microglia-neuron conversion. Neuron 101: 472–485 e477
5. Miyoshi H, Blömer U, Takahashi M, Gage FH, Verma IM (1998) Development of a self-inactivating lentivirus vector. J Virol 72:8150–8157
6. Matsuda T, Murao N, Katano Y, Juliandi B, Kohyama J, Akira S, Kawai T, Nakashima K (2015) TLR9 signalling in microglia attenuates seizure-induced aberrant neurogenesis in the adult hippocampus. Nat Commun 6:6514
7. Giulian D, Baker TJ (1986) Characterization of ameboid microglia isolated from developing mammalian brain. J Neurosci 6:2163–2178
8. Irie T, Matsuda T, Hayashi Y, Matsuda-Ito K, Kamiya A, Masuda T, Prinz M, Isobe N, Kira JI, Nakashima K (2023) Direct neuronal conversion of microglia/macrophages reinstates neurological function after stroke. Proc Natl Acad Sci USA 42:e2307972120

Chapter 14

Selective Neural Pathway Targeting by Using Lentiviral Vectors

Shigeki Kato

Abstract

Animal behaviors are controlled through a complex neural network in the brain. To explore the nature of brain functions, it is critical to investigate the behavioral and physiological roles of specific neuronal types that comprise this network. Immunotoxin (ITX)-mediated cell targeting is a technology designed to eliminate specific cell types that are genetically engineered to express a foreign receptor, utilizing the cytotoxic activity of a recombinant ITX. Lentiviral vectors with highly efficient retrograde gene transfer (HiRet) and neuron-specific retrograde gene transfer (NeuRet) properties are introduced at neuronal terminal regions and transported retrogradely through axons to the cell body, resulting in the transgene expression. HiRet/NeuRet vectors, combined with ITX-mediated cell targeting technology, enable the selective elimination of specific neural pathways within the network, providing an efficient approach for studying the behavioral and physiological roles of target neuronal pathways. This chapter describes a protocol for selective neural pathway targeting and presents an application of this strategy to eliminate striatal input pathways in the mouse brain.

Key words Lentiviral vector, Highly efficient retrograde gene transfer, Neuron-specific retrograde gene transfer, Selective neural pathway targeting, Recombinant immunotoxin, Striatum, Intralaminar thalamic nucleus, Motor cortex

1 Introduction

Animal behaviors are controlled by transmitting and processing information within complex neural networks in the central nervous system. The neural network consists of multiple types of neurons with distinct structures and electrophysiological properties and capable of forming selective neural connections. To understand how animal behaviors are controlled by the neural network, it is necessary to clarify the behavioral and physiological roles of individual neuronal types within the network and demonstrate the causal relationship between their activity and behavior.

Kazuto Kobayashi et al. (eds.), *Lentiviral Vectors: Applications for Basic and Clinical Neuroscience*, Methods in Molecular Biology, vol. 2974, https://doi.org/10.1007/978-1-0716-4807-0_14,

Immunotoxin (ITX)-mediated cell targeting is a molecular genetic technology designed to remove specific cell types expressing an exogenous receptor from the neural network based on the cytotoxic activity of a recombinant ITX named anti-Tac(Fv)-PE38 [1, 2]. Anti-Tac(Fv)-PE38 is composed of the variable heavy and light regions of an anti-Tac antibody, a monoclonal antibody specific for the human interleukin-2 receptor α-subunit (hIL-2Rα), fused to a truncated form of *Pseudomonas* exotoxin (PE38). In this technology, transgenic animals expressing hIL-2Rα under the control of a cell-type-specific gene promoter are generated and then treated with anti-Tac(Fv)-PE38, which selectively binds to hIL-2Rα on the cell surface and is internalized by endocytosis, leading to apoptosis of the cells by blocking de novo protein synthesis [3]. ITX-mediated cell targeting has been used to study the behavioral and physiological functions of various neuronal types in the rodent brain [4–9].

In ITX-mediated cell targeting, lentiviral vectors with highly efficient capabilities for retrograde gene transfer into neurons provide a powerful strategy for the selective elimination of specific neural pathways (*see* Fig. 1). After being injected into a brain region of interest, vectors encoding *hIL-2Rα* gene are taken up at the terminal regions, retrogradely transported via axons to their cell bodies, where the *hIL-2Rα* gene is expressed. When ITX is injected into one of the brain regions that contain hIL-2Rα-expressing cells, specific neural pathways that connect the two injected regions are eliminated from the network. Lentiviral vectors showing highly efficient retrograde gene transfer (HiRet) and neuron-specific retrograde gene transfer (NeuRet) are pseudotypes with fusion envelope glycoprotein that consists of different segments of rabies virus glycoprotein and vesicula stomatitis virus glycoprotein ([10–14],

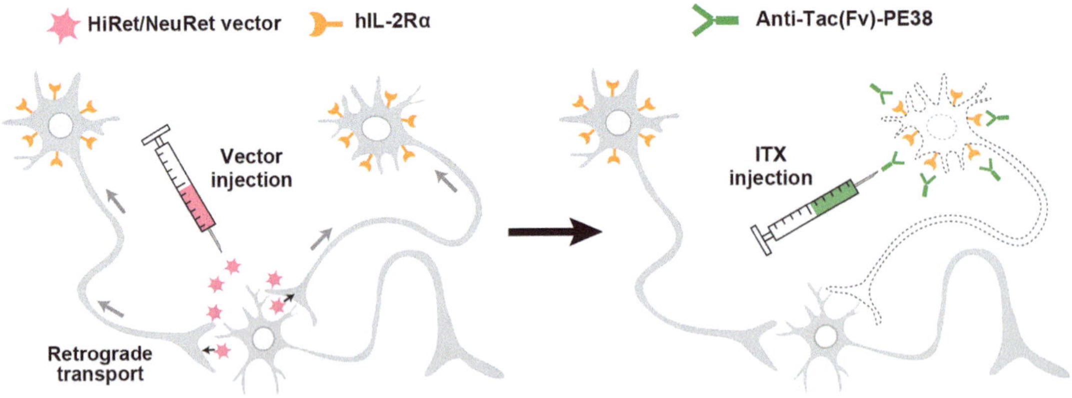

Fig. 1 Strategy for ITX-mediated targeting of selective neural pathways. HiRet/NeuRet vectors encoding the *hIL-2Rα* gene are injected into the target brain region, where they are taken up by neuronal terminals and retrogradely transported to their cell bodies via axon. Injection of anti-Tac(Fv)-PE38 induces the elimination of specific neuronal pathways carrying hIL-2Rα

and *see* [15–17] for reviews). HiRet/NeuRet vectors are combined with ITX cell targeting for selective targeting of neural pathways in the brains of mice [11, 18, 19] and nonhuman primates [20, 21].

In this chapter, the protocol for the selective neural pathway targeting can be divided into two parts: (1) intracranial injection of viral vectors and ITX and (2) immunohistological detection of cells expressing the transgene. The strategy is applied for elimination of striatal input pathways from the cerebral cortex (primary motor cortex or M1) and intralaminar thalamic nucleus (parafascicular thalamic nucleus or PF) in mice. The application of this strategy will make it possible to remove various neural pathways in the brain network and contribute to the elucidation of the behavioral and physiological functions.

2 Materials

2.1 Intracranial Injection of Lentiviral Vectors and ITX

1. Wild-type (C57BL/6) mice.
2. Surgical tools.
3. Stereotaxic apparatus.
4. Infusion pump.
5. Mineral oil or Fluorinert.
6. Microsyringe.
7. Glass capillary.
8. Teflon tube with joint.
9. Stereo microscope for surgery.
10. Lentiviral vectors.

    HiRet/NeuRet vectors encoding *hIL-2Rα* gene fused to green fluorescent protein (GFP) (1.0×10^{12} genomic copies/mL). The procedures for production and purification of these vectors are described in Chapter 2 (*see* **Note 1**).
11. Recombinant immunotoxin (ITX).

    ITX is provided by Dr. Ira Pastan, Laboratory of Molecular Biology, National Cancer Institute, National Institutes of Health; Tel: (240) 760-6470; and E-mail: pastani@mail.nih.gov. ITX is diluted to final concentrations of 40 and 80 ng/μL (for injections into the M1 and PF, respectively) in phosphate buffered saline (PBS) containing 0.1% mouse serum albumin, aliquoted into 10 μL and stored at −80 °C (*see* **Note 2**).

2.2 Immunohistochemical Staining

1. Cryostat.
2. Primary antibody (Nittobo Medical Co., Ltd., anti-GFP antibody, rabbit, #GFP-Rb-Af2020).

3. Secondary antibodies (anti-rabbit IgG antibody conjugated to biotin, #711-065-152, Jackson ImmunoResearch Labs).
4. Avidin-biotin-peroxidase complex kit.
5. Mounting medium.
6. Microscope.

3 Methods

3.1 Intracranial Injection

3.1.1 Injection of Lentiviral Vectors

1. Anesthetize the mice and secure them in a stereotaxic apparatus, and then expose the skull by shaving the head and making an incision.
2. Using the bregma (the intersection of the sagittal and coronal sutures) as the reference point, identify the coordinates of the target brain regions (striatum) according to a mouse brain atlas [22] and mark the corresponding locations on the skull.
3. Drill a hole approximately 5 mm in diameter at the marked location using a dental drill.
4. Tear the dura mater using tweezers to expose the surface of the brain.
5. Connect the infusion pump, microsyringe, teflon tube with joints, and glass capillary. Next, fill the capillary with mineral oil or Fluorinert.
6. Inject the vector solution bilaterally into the target brain region at a rate of 0.1 μL/min (0.5 μL/site, 8 sites) (*see* **Note 3**).

   For injection into the striatum, the anteroposterior, mediolateral, and dorsoventral coordinates (mm) from the bregma and dura are: 0/±2.3/2.3, 0/±2.3/3.0, 1.0/±1.9/2.3, and 1.0/±1.9/3.0 according to the mouse brain atlas.
7. Wait for 5 min after injection before withdrawing the capillary.
8. Stop the bleeding, suture the scalp, and maintain the mice's body temperature until they recover from anesthesia.
9. Wait for 3 weeks after vector injection until sufficient gene transfer has occurred. Perform immunohistochemical staining to detect transgene expression (*see* Subheading 3.2). Administer intracranial ITX injection to eliminate target cells (*see* Subheading 3.1.2).

3.1.2 Injection of ITX

1. Perform **steps 1–4** of Subheading 3.1.1 again on the mice following their intracranial injection of the vector's solution.
2. Inject ITX at a concentration of 40 ng/μL with a volume of 0.5 μL per site into three sites in the M1, at a concentration of 80 ng/μL with a volume of 0.5 μL per site into two sites in the PF or inject PBS as a control. The injection should be

performed unilaterally into the target brain region at a rate of 0.1 μL/min.

3. The anteroposterior, mediolateral, and dorsoventral coordinates (mm) from the bregma and dura are: 1.7/1.7/0.7, 0.7/1.6/0.7, and −0.3/1.0/0.7 for the M1; and −2.1/0.7/3.0 and −2.4/0.7/3.0 for the PF according to the mouse brain atlas.
4. Wait for 5 min after injection before withdrawing the capillary.
5. Stop the bleeding, suture the scalp, and maintain the mice's body temperature until they recover from anesthesia.
6. 5.Wait for 10 days after ITX injection until sufficient cell elimination has occurred.

3.2 Immunohistochemical Staining

1. Anesthetize the mice and perfuse transcardially with PBS.
2. Remove the brain from the skull, postfix for 1–2 days, soak in 10%, 20%, and 30% sucrose solution in PBS at 4 °C, and section into 30-μm thick coronal slices using a cryostat.
3. For transgene detection, prepare brain sections containing the striatum (the vector injection site), M1, the primary somatosensory cortex (S1), PF, the central nucleus and basolateral nucleus of the amygdala (CeA and BLA, respectively), and the substantia nigra pars compacta (SNc) in the ventral midbrain of the mice injected with lentiviral vectors.
4. To evaluate cell elimination, prepare brain sections containing the M1 or PF from the mice injected with ITX into the corresponding brain regions.
5. Incubate the sections with an anti-GFP antibody (1:2000) overnight at 4 °C.
6. Wash the sections three times with PBS.
7. Incubate the sections with a biotin conjugated anti-rabbit IgG antibody (1:500) for 2 h at room temperature.
8. Wash the sections three times with PBS.
9. Treat the sections with an avidin-biotin-peroxidase complex kit. To visualize immune-positive signals, react the sections with 0.05 M Tris–HCl buffer (pH 7.6) containing 0.05% 3,3′--diaminobenzidine tetrahydrochloride and 0.3% H_2O_2 for 5–10 min.
10. Mount the sections on gelatin-coated glass slides, coverslip them, and capture images with a microscope.
11. For transgene detection, observe GFP-positive cells in the M1, S1, PF, CeA, BLA, and SNc of the mice (Fig. 2) (*see* **Note 4**). To evaluate cell elimination, compare the number of immune-positive cells in the M1 or PF between the ITX-injected side and the PBS-injected side (Fig. 3) (*see* **Note 5**).

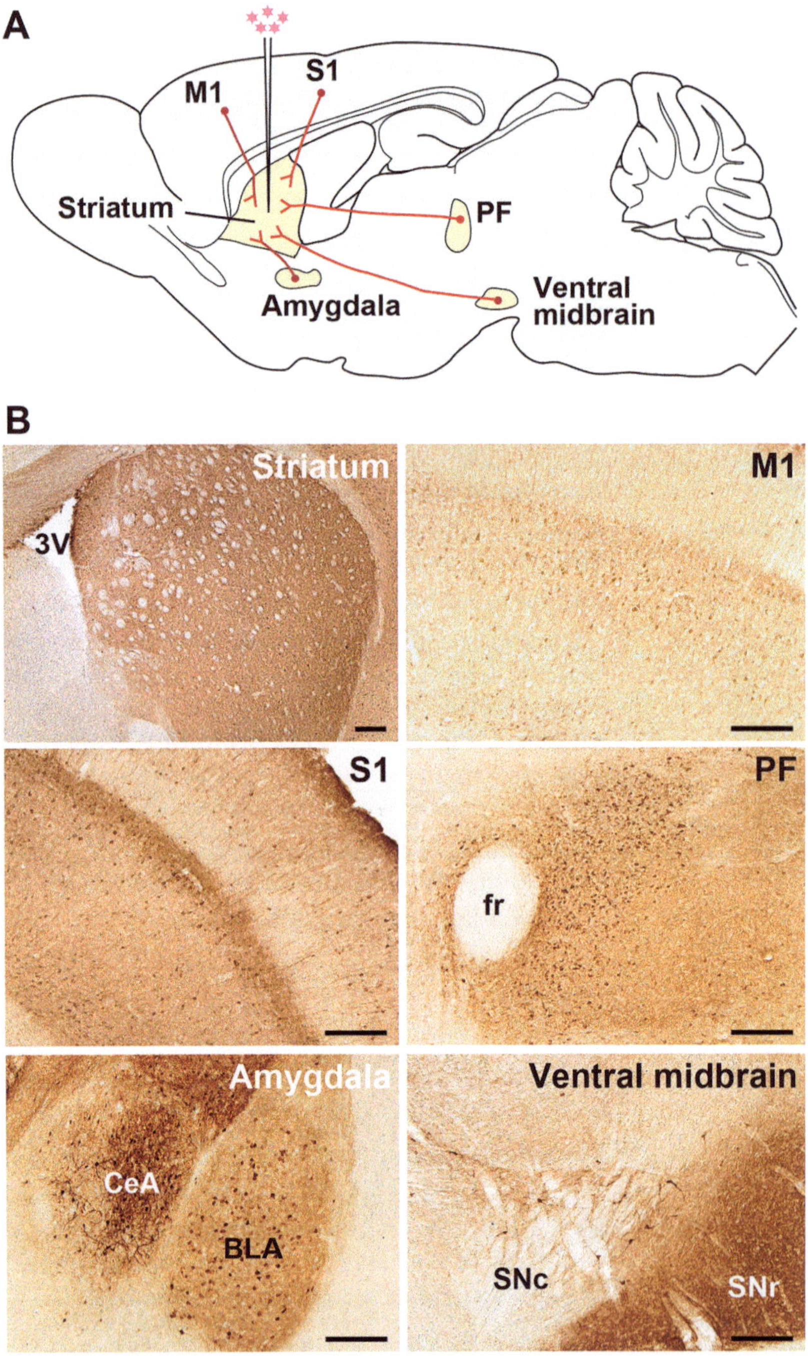

Fig. 2 Retrograde gene transfer into neurons in different brain regions after intracranial vector injection. (**a**) Schematic representation of intracranial injection of lentiviral vectors for retrograde gene transfer. HiRet/NeuRet vectors encoding hIL-2Rα-GFP are injected into the striatum of mice. (**b**) After 3 weeks, brain sections through the striatum, M1, S1, PF, CeA, BLA, and SNc are prepared and stained with an anti-GFP antibody. GFP-positive neurons are visualized in each region. *3V* third ventricle, *fr* fasciculus retroflexus, *SNr* substantia nigra pars reticulata. Scale bar: 200 μm

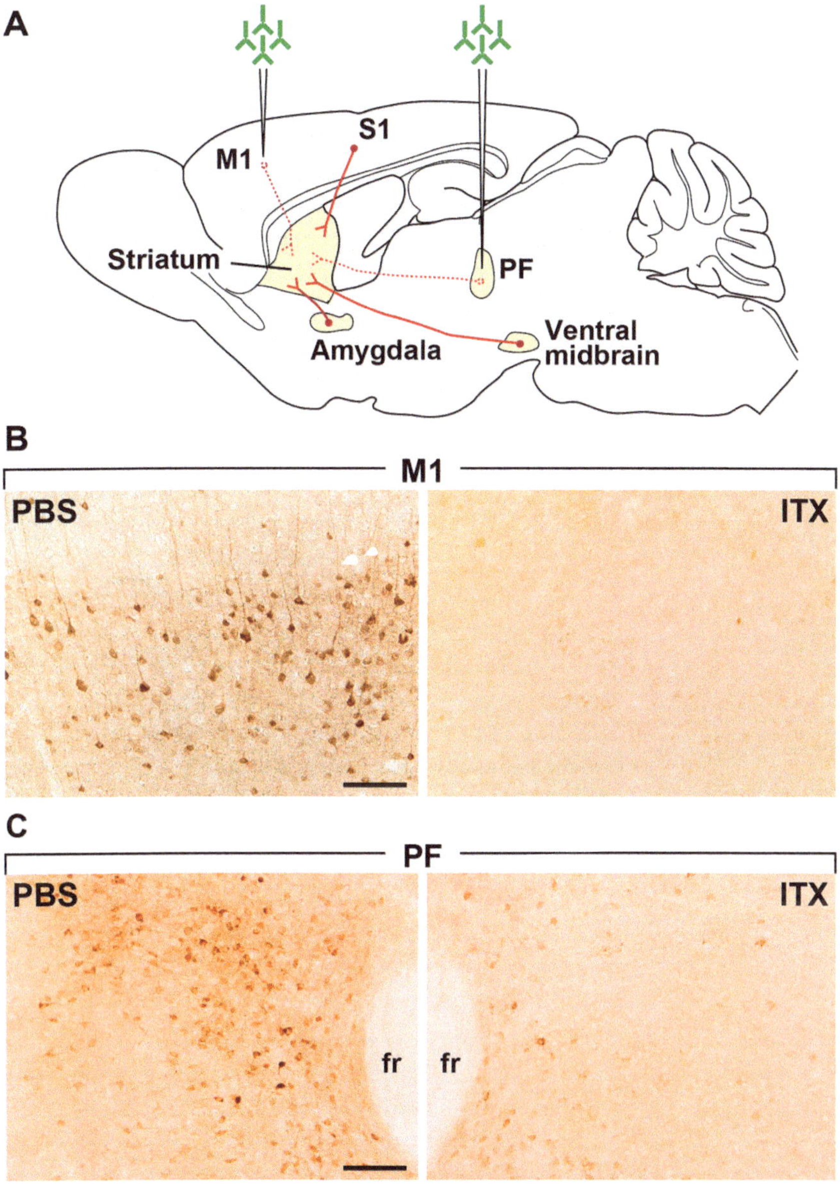

Fig. 3 Targeted cell elimination in the M1 or PF after intracranial ITX injection. (**a**) Schematic representation of intracranial ITX injection for selective pathway elimination. Mice injected with HiRet/NeuRet vectors encoding *hIL-2R*α*-GFP* gene into the striatum are treated with ITX solution or PBS. (**b**, **c**) After 10 days, brain sections through the M1 and PF are prepared and stained with anti-GFP antibody. The number of GFP-positive cells is remarkably reduced on the ITX-injected side. *fr* fasciculus retroflexus. Scale bar: 200 μm

4 Notes

1. The type of fusion envelope glycoprotein used for pseudotyping affects the properties of lentiviral vectors, including host cell specificity, vector entry site, and retrograde gene transfer efficiency. For the selective neural pathway targeting, fusion glycoprotein type E is generally recommended because of its higher efficiency in retrograde gene transfer into neurons and reduced inflammatory reactions around the injection sites in rodent and nonhuman primate brains.
2. Repeated freezing and thawing of ITX solution should be avoided to prevent aggregation of recombinant proteins. Based on our experience, the protein remains functionally stable for at least 3 years under these storage conditions.
3. For HiRet/NeuRet vectors, a 0.5-μL injection typically results in a radius of approximately 0.5 mm.
4. Testing the efficiency of pathway labeling with lentiviral vectors is recommended before conducting experiments. Prepare a solution containing HiRet/NeuRet-hIL-2Rα-GFP vector and cholera toxin subunit B (CTB) conjugated to Alexa Fluor 555 as a retrograde tracer, and inject it into the target brain region as described in Subheading 3.1.1. After 3 weeks, prepare brain slices for double immunohistochemistry for GFP and CTB. Incubate the sections with an anti-GFP antibody, followed by an anti-rabbit IgG antibody conjugated to Alexa Fluor 488. Count the numbers of CTB-positive and doubly positive cells, and estimate the frequency of retrograde gene transfer by calculating the percentage of the number of doubly positive cells relative to the number of CTB-positive cells.
5. The concentrations of ITX required to induce cell removal depend on the expression level of the receptor in the target cell types. In addition, excessive amounts of ITX can cause nonspecific reactions in cells that do not express the receptor. Therefore, preliminary experiments are necessary to determine the appropriate concentrations of ITX. This protocol provides standard concentrations for ITX injections into the mouse M1 and PF regions.

Conflict of Interest Statement

The authors have no conflicts of interest to declare that are relevant to the content of this chapter.

Funding

This study was funded by a Grant-in-Aid for Young Scientists (A) (25702053) from the Japan Society for the Promotion of Science (S.K.).

References

1. Kobayashi K, Morita S, Sawada H, Mizuguchi T, Yamada K, Nagatsu I et al (1995) Immunotoxin-mediated conditional disruption of specific neurons in transgenic mice. Proc Natl Acad Sci USA 92:1132–1136
2. Kobayashi K, Kai N, Kobayashi K, Fujiwara T, Akagawa K, Onda M et al (2008) Transient silencing of synaptic transmitter release from specific neuronal types by recombinant tetanus toxin light chain fused to antibody variable region. J Neurosci Methods 175:125–132
3. Kreitman RJ, Bailon P, Chaudhary VK, Fitz-Gerald DJ, Pastan I (1994) Recombinant immunotoxins containing anti-Tac(Fv) and derivatives of Pseudomonas exotoxin produce complete regression in mice of an interleukin-2 receptor-expressing human carcinoma. Blood 83:426–434
4. Watanabe D, Inokawa H, Hashimoto K, Suzuki N, Kano M, Shigemoto R et al (1998) Ablation of cerebellar Golgi cells disrupts synaptic integration involving GABA inhibition and NMDA receptor activation in motor coordination. Cell 95:17–27
5. Yoshida K, Watanabe D, Ishikane H, Tachibana M, Pastan I, Nakanishi S (1998) A key role of starburst amacrine cells in originating retinal directional selectivity and optokinetic eye movement. Neuron 30:771–780
6. Sano H, Yasoshima Y, Matsushita N, Kaneko T, Kohno K, Pastan I et al (2003) Conditional ablation of striatal neuronal types containing dopamine D2 receptor disturbs coordination of basal ganglia function. J Neurosci 23: 9078–9088
7. Fukabori R, Okada K, Nishizawa K, Kai N, Kobayashi K, Uchigashima M et al (2012) Striatal direct pathway modulates response time in execution of visual discrimination. Eur J Neurosci 35:784–797
8. Nishizawa K, Fukabori R, Okada K, Kai N, Uchigashima M, Watanabe M et al (2012) Striatal indirect pathway contributes to selection accuracy of learned motor actions. J Neurosci 32:13421–13432
9. Kai N, Nishizawa K, Tsutsui Y, Ueda S, Kobayashi K (2015) Differential roles of dopamine D1 and D2 receptor-containing neurons of the nucleus accumbens shell in behavioral sensitization. J Neurochem 135:1232–1241
10. Kato S, Kobayashi K, Inoue K, Kuramochi M, Okada T, Yaginuma H et al (2011) A lentiviral strategy for highly efficient retrograde gene transfer by pseudotyping with fusion envelope glycoprotein. Hum Gene Ther 22:197–206
11. Kato S, Kuramochi M, Kobayashi K, Fukabori R, Okada K, Uchigashima M et al (2011) Selective neural pathway targeting reveals key roles of thalamostriatal projection in the control of visual discrimination. J Neurosci 31:17169–17179
12. Kato S, Kuramochi M, Takasumi K, Kobayashi K, Inoue K, Takahara D et al (2011) Neuron-specific gene transfer through retrograde transport of lentiviral vector pseudotyped with a novel type of fusion envelope glycoprotein. Hum Gene Ther 22:1511–1523
13. Kato S, Kobayashi K, Kobayashi K (2014) Improved transduction efficiency of a lentiviral vector for neuron-specific retrograde gene transfer by optimizing the junction of fusion envelope glycoprotein. J Neurosci Methods 227:151–158
14. Kato S, Sugawara M, Kobayashi K, Kimura K, Inoue K, Takada M et al (2019) Enhancement of the transduction efficiency of a lentiviral vector for neuron-specific retrograde gene delivery through the point mutation of fusion glycoprotein type E. J Neurosci Methods 311: 147–155
15. Kobayashi K, Kato S, Inoue K, Takada M, Kobayashi K (2016) Altering entry site preference of lentiviral vectors into neuronal cells by pseudotyping with envelope glycoproteins. Methods Mol Biol 1382:175–186
16. Kobayashi K, Inoue K, Tanabe S, Kato S, Takada M, Kobayashi K (2017) Pseudotyped lentiviral vectors for retrograde gene delivery into target brain regions. Front Neuroanat 11: 65

17. Kobayashi K, Kato S, Kobayashi K (2018) Genetic manipulation of specific neural circuits by use of a viral vector system. J Neural Transm (Vienna) 125:67–75
18. Kato S, Fukabori R, Nishizawa K, Okada K, Yoshioka N, Sugawara M et al (2018) Action selection and flexible switching controlled by the intralaminar thalamic neurons. Cell Rep 22:2370–2382
19. Sakayori N, Kato S, Sugawara M, Setogawa S, Fukushima H, Ishikawa R et al (2019) Motor skills mediated through cerebellothalamic tracts projecting to the central lateral nucleus. Mol Brain 12:13
20. Inoue K, Koketsu D, Kato S, Kobayashi K, Nambu A, Takada M (2012) Immunotoxin-mediated tract targeting in the primate brain: selective elimination of the cortico-subthalamic “hyperdirect” pathway. PLoS One 7:e39149
21. Takada M, Inoue K, Koketsu D, Kato S, Kobayashi K, Nambu A (2013) Elucidating information processing in primate basal ganglia circuitry: a novel technique for pathway-selective ablation mediated by immunotoxin. Front Neural Circuits 7:140
22. Paxinos G, Franklin KBJ (2001) The mouse brain in stereotaxic coordinates, 2nd edn. Academic, San Diego

Chapter 15

Synaptic Silencing with Transactivation of Tetanus Neurotoxin for Pathway-Selective Manipulation

Tadashi Isa, Kaoru Isa, and Wim Vanduffel

Abstract

To selectively block specific brain circuits using viral vectors, two main approaches have been developed. The first involves injecting a single viral vector into the cell bodies of target neurons, followed by chemical or optogenetic stimulation at their axon terminals. The second method, known as the intersectional double viral vector technique, uses a retrograde vector injected at the axon terminals and an anterograde vector at the cell bodies of target neurons, along with systemic ligand administration. This chapter focuses on the second approach, which allows for the targeted expression of enhanced tetanus neurotoxin in specific neural pathways. This effectively disrupts their transmission by Tet transactivator system, leading to behavioral deficits in macaque monkeys. Additionally, this chemogenetic blocking technique can be combined with the DREADDs (Designer Receptors Exclusively Activated by Designer Drugs) technique to inhibit multiple pathways within a single animal. This approach has the potential to significantly advance systems neuroscience research.

Key words Synaptic silencing, Tetanus neurotoxin, Tet transactivator system, Macaque monkey

1 Introduction

Systems neuroscience studies seek to understand how neural circuits process information to drive complex sensorimotor and cognitive functions. To achieve this, it is crucial to dissect the key components of neural circuits, manipulate their activity through either facilitation or inhibition, and observe the resulting behavioral or cognitive effects. Over the past three decades, advances in cell-type-specific manipulation have been made possible through transgenic mice expressing diphtheria toxin or immunotoxin receptors under cell-type-specific promoters [1, 2], enabling targeted cell ablation. Subsequently, intersectional techniques emerged, combining transgenic Cre expression with adeno-associated viral (AAV) vectors carrying floxed sequences to achieve cell-type-

Kazuto Kobayashi et al. (eds.), *Lentiviral Vectors: Applications for Basic and Clinical Neuroscience*, Methods in Molecular Biology, vol. 2974, https://doi.org/10.1007/978-1-0716-4807-0_15,

specific protein expression. However, while these methods advanced cell-type-specific manipulation, they did not always permit pathway-selective manipulation.

The development of optogenetics revolutionized cell-type-specific manipulation by enabling the expression of depolarizing channelrhodopsins (ChR2) [3, 4] or hyperpolarizing halorhodopsins (NpHR) [5], allowing researchers to control neuronal activity and behavior with light, even in macaques [6]. Researchers soon discovered that these opsins can also be expressed in axon terminals. By illuminating the axon terminals, they were able to induce postsynaptic neuronal responses, making pathway-selective optogenetic facilitation possible [7, 8] (Fig. 1a). For inhibitory pathways, illuminating the presynaptic axon terminals was found to inhibit synaptic transmission, though the effect was only partial [9].

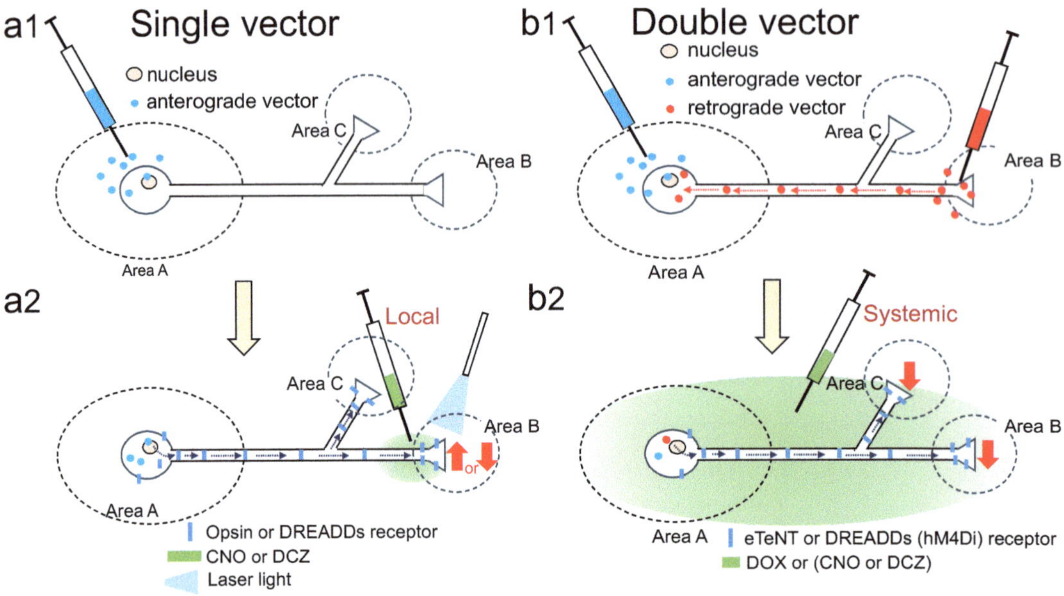

Fig. 1 Comparison of single vector vs. double vector techniques for the pathway-selective manipulation. Neuron in area A project both to area B and area C. In this experiment, the pathway from areas A to B is targeted. (**a**) Single vector technique: An anterograde vector encoding a gene for photo-activated opsins or DREADDs receptors is injected into area A (**a1**). The protein is transported to the axon terminals in both areas B and C. Pathway-selective manipulation is achieved by locally injecting an agonist (e.g., CNO or DCZ) or by illuminating the axon terminals in area B. This method allows selective manipulation of the pathway from areas A to B without affecting the pathway from areas A to C. (**b**) Double-vector technique: A retrograde vector is injected into area B, while an anterograde vector is delivered into area A. This results in double-infected neurons projecting from area A to area B. The synthesized proteins (e.g., eTeNT or DREADD receptors) are transported to axon terminals in both areas B and C. Therefore, systemic administration of drugs (e.g., Dox, CNO, or DCZ) manipulates both the pathways, from areas A to B and areas A to C. In **a2** and **b2**, the upward red arrows indicate increased responses and downward red arrows indicated decreased responses

Meanwhile, chemogenetic manipulation advanced with the development of DREADDs (Designer Receptors Exclusively Activated by Designer Drugs) by Armbruster and colleagues [10], which allowed for cell-type-specific manipulation through the systemic administration of Clozapine N-oxide (CNO). However, pathway-selective chemogenetic manipulation was not possible until 2014, when Mahler et al. demonstrated that DREADD receptors (hM4Di) could be expressed in axon terminals, and that local CNO injections could modulate synaptic transmission in a pathway-selective manner [11]. More recent studies have further refined this approach, showing that expressing hM4Di in target cells and administering a more efficient ligand, DCZ (Deschloroclozapine) [12], to the target areas enabled pathway-selective transmission blockade in the macaque monkey brain [13].

In parallel, Kazuto Kobayashi and his team developed a highly efficient retrograde gene transfer vector that proved effective in the nonhuman primate brain. Initially, a lentiviral vector incorporating sequences from the rabies virus (RV) demonstrated retrograde transport, but its efficiency was somewhat limited [14]. To address this, they engineered a chimeric vector combining critical sequences from RV and vesicular stomatitis virus glycoprotein (VSVG) within a lentiviral backbone. This modification significantly enhanced transcription efficiency following retrograde transport. This vector, named the "highly efficient retrograde gene transfer vector" or HiRet [15], enabled pathway-selective chemogenetic synaptic silencing when used in combination with an anterograde AAV vector (Fig. 1b), as described below [16] (Fig. 1b).

In this approach for synaptic silencing, Dai Watanabe and colleagues optimized the GC content of tetanus neurotoxin (TeNT), a bacterial toxin, increasing it from 30% to 44% without altering its amino acid sequence. This modification enhanced both the expression of TeNT and its ability to cleave the synapse transmission machinery, VAMP2, in mammalian cells (Fig. 2; eTeNT). Additionally, to accelerate the onset of expression using the Tet-activator system [17], the team developed an effective Tet-ON sequence, rtTAV16.

Isa and colleagues demonstrated that by injecting HiRet-TRE-eGFP.eTeNT around the axon terminals of targeted neurons and AAV2-CMV-rtTAV16 around their cell bodies, both rtTAV16 and eTeNT genes could be introduced in the double-infected target neurons (Figs. 1b and 3a). Upon doxycycline (Dox) administration, eTeNT could be expressed, transported to axon terminals, and cleave VAMP2, effectively blocking synaptic transmission. They applied this technique to block the synaptic transmission of a group of spinal cord interneurons, specifically propriospinal neurons, in the macaque spinal cord. This allowed them to demonstrate the role of these neurons in precision grip control (Fig. 3).

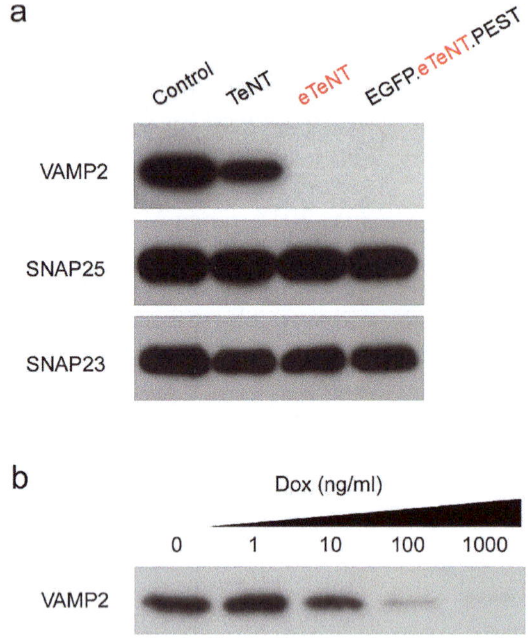

Fig. 2 Macaque VAMP2 cleavage by wild-type and modified TeNTs. (**a**) Codon-optimized or non-codon-optimized TeNT expression plasmids were co-transfected into HEK293 cells with FLAG-tagged SNARE-expression plasmids (pVAMP2, pSNAP25, or pSNAP23). Cell lysates were immunoblotted with an anti-FLAG antibody. Intact VAMP2 was undetectable in the presence of codon-optimized eTeNT or EGFP.eTeNT.PEST, whereas intact VAMP2 remained intact with non-codon-optimized wild-type TeNT. Both wild-type and modified TeNTs did not affect the levels of intact SNAP25 or SNAP23. (**b**) Plasmids for Tet-on expression of eTeNT (pLV-TRE-EGFP.eTeNT.PEST and prtTA2) were co-transfected with pVAMP2. Immunoblot analysis revealed that Dox treatment decreased the amount of intact VAMP2 in a dose-dependent manner. (Adapted from Kinoshita et al. [16])

Around 2010, this intersectional technique combining retrograde and anterograde vectors, which is now widely used, was not yet common. At the time, the mouse neuroscience community favored intersectional transgenic approaches combined with viral vector-based methods. However, Isa and colleagues, working with macaque monkeys, considered transgenic approaches in monkeys to be impractical, and instead pioneered the double viral vector technique for pathway-selective manipulation.

Comparing single and double viral vector transfection techniques reveals distinct advantages and disadvantages (Fig. 1). The single vector technique (Fig. 1a) avoids collateral effects but requires repeated micro-syringe insertions, which can be technically challenging and potentially harmful. This is especially problematic for spinal cord injections in awake animals, as it is impossible to repeat this injection safely. Conversely, double vector techniques

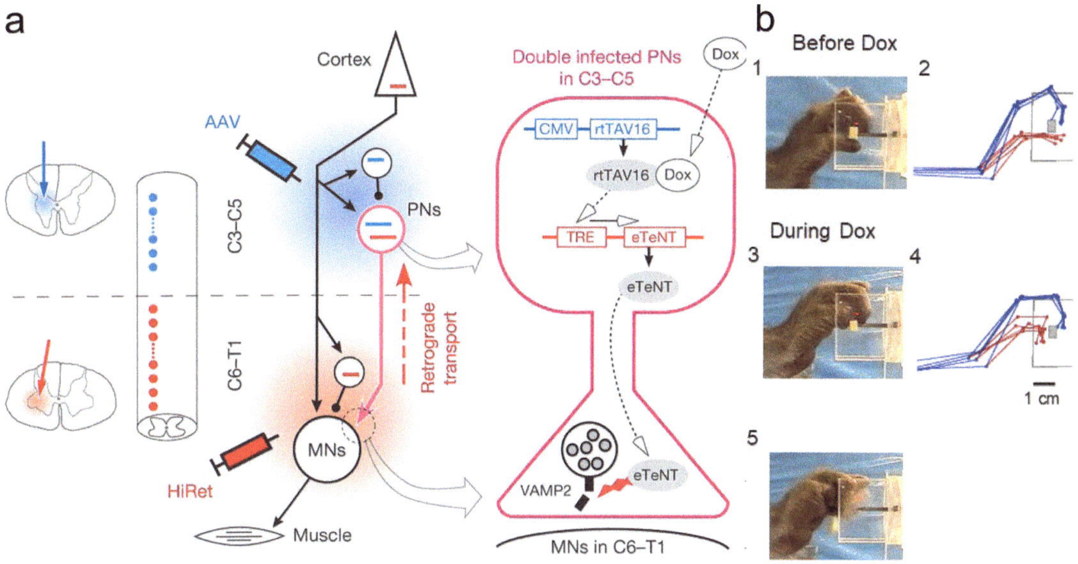

Fig. 3 Role of mid-cervical propriospinal neurons in the control of dexterous finger movements. (**a**) To selectively block the C3-C5 PNs, a retrograde gene transfer vector carrying TRE-EGFP.eTeNT (red) was injected into multiple tracks targeting the motor nuclei at the C6-T1 segments, while an anterograde vector, AAV-CMJ-rtTAV16 (blue), was injected into the intermediate zone at C3–C5 segments, where the cell bodies of the PNs reside. The schematic diagram illustrates how the double vectors interact upon administration of doxycycline (Dox). (**b**) (*1–2*) Photo images (*1*) and superimpositions of stick diagrams showing the precision grip of monkeys (*2*) before Dox administration. Blue sticks indicate the index finger and red sticks the thumb when monkeys start grasping a small piece of sweet potato. (*3–4*) The same arrangement as (*1–2*), but taken on the second day after the start of Dox administration. (*5*) An image captured 300 ms afterward (*3*), showing the dropping of the food. (Adapted from Kinoshita et al. [16])

work well with less invasive systemic drug administration, but they can be more complicated. This is because collaterals to another structure (such as area C in Fig. 1b) may also be affected. As a result, experimenters must be very cautious in inspecting whether axon collaterals of the target neurons are present, which may confound the results (*see* **Note 1**).

As shown below, the double vector approach has been widely adopted in both rodents and nonhuman primate models, exploring normal and injured states. In rodent spinal cord models, Magnuson and colleagues demonstrated that silencing spinal L2 interneurons disrupted hindlimb alteration, shifting locomotion to a hopping pattern [18]. They also showed that silencing long ascending propriospinal neurons disrupted left-right limb coupling [19].

In the context of the neural mechanisms underlying recovery after brain or spinal cord injury, Schwab and colleagues introduced this technique to clarify the contribution of a subset of corticospinal neurons originating from the motor cortex on the intact side in the recovery of skilled forelimb functions. This was demonstrated in a large unilateral stroke rat model, alongside growth-promoting immunotherapy targeting a neurite growth-inhibitory protein [20].

Courtine and colleagues further expanded on this by demonstrating that reorganization of the cortico-reticulo-spinal pathway contributed to recovery of locomotion after severe spinal cord injury caused by contusion. They achieved this by blocking the reticulospinal pathway using the double viral vector technique [21].

In primates, Isa and colleagues explored the superior colliculus to ponto-medullary reticular formation pathway for controlling orienting behavior [22] and the cortico-rubral pathway in stroke recovery models [23].

Isa and colleagues also applied the double viral vector technique to examine the role of mid-cervical propriospinal neurons in recovery after the spinal cord injury [24]. In addition, they investigated the superior colliculus-pulvinar pathway in "blindsight" model monkeys with primary visual cortex lesions [25], as well as pathway connecting the ventral tegmental area with nucleus accumbens in motivation regulation [26]. Moreover, Isoda and colleagues used this technique to study social behavior by blocking the pathway between the ventral premotor cortex and medial prefrontal cortex [27]. Additionally, Sakagami and colleagues demonstrated the role of the prefrontal network in inhibitory control and working memory [28].

Furthermore, by combining the Tet transactivator system and DREADDs, it became possible to manipulate two different pathway in the single animals [29] (*see* **Note 2**).

These studies highlight the versatility and effectiveness of the double viral vector technique in uncovering the complex neural circuits involved in various cognitive and behavioral functions in primates (see also Vanduffel and Isa [30]).

2 Materials

2.1 Vector Injection

1. Lentiviral vector encoding a modified TeNT gene.

   Expression plasmid contains a gene cassette encoding codon-optimized modified TeNT (eTeNT) fused to enhanced green fluorescent protein (EGFP.eTeNT.PEST) under the control of Tet-on inducible promoter with the rtTA2 responsive element (pLV-TRE-EGFP.eTeNT.PEST). Preparation of lentiviral vector is described in Chapter 2 in this book.

   Immunoblots were performed using a mouse anti-Flag M2 monoclonal antibody (F1804, Sigma-Aldrich) and visualized with a horseradish peroxidase-conjugated sheep anti-mouse immunoglobulin-G (IgG) antibody (NA931V, GE Healthcare) (Fig. 2).

2. AAV vector encoding rtTAV16 gene.

   The rtTA variant rtTAV16 contains the V9I, G12S, F67S, F86Y, R171K, and A209T mutations into rtTA2S-M2 of the pTet-On advanced vector (Clontech). Expression plasmid pAAV2-CMV-rtTAV16 is based on pAAV-MCS (Agilent Technologies), constructed by inserting the CMV sequence of the pTet-On advanced vector, rtTAV16, the woodchuck hepatitis virus posttranscriptional regulatory element (WPRE) sequence of pFUGW, and the SV40 polyadenylation signal (SV40pA) of the pCMV-script vector (Agilent Technologies) into the MCS. Preparation of the AAV vector is described by Kaneda et al. [9].

3. Surgical tools.

   General surgical tools for the neurosurgery of monkeys are described below.

   For vector application:

   Stereotaxic apparatus: Special order product by Narishige group, in Japan.

   Infusion pump: Legato® 130 Syringe pump, KdScientific, (https://www.kdscientific.com/legato-130-syringe-pump.html).

   Microsyringe: Hamilton syringe (10 μL, Model 701, Cemented NDL, 26sG, 2 in, point style 2) with special order product was used for deeper area injection.

   Glass capillary: The patch clamp glass, 4 in pich clmp gl 1.65 mm OD (PG10165-4m WPI) was used for deeper area injection. The glass capillary puller to make a fine-tip injection pipette. The diameter of the tube should be 50–100 μm in diameter.

   Teflon tube with joint: PFA tube 0.6 mm diameter outside and 0.4 mm diameter inside, and mixed epoxy resin is used for joint with injection syringe or glass capillary.

   Stereo microscope for surgery: Stereotaxic micromanipulator, SMM-100, One-axis Oil Hydraulic micromanipulator, Narishige.

   To determine the injection sites:

   MR images of the brain were taken with 3T MRI scanner Siemens Prisma Fit.

   Others

   Sliding microtome, RetoratomeREM-710.

   Fluorescent microscopes, Akioplan 2, Zeiss.

   Confocal microscopy A1R, Nikon.

2.2 Doxycycline Administration

Doxycycline (Dox) solution (D9891; Sigma Aldrich, St. Louis, MO)

Concentration: 15 mg/kg/day, total volume: 350 mL, sucrose: 6 g/350 mL (1.71%)

1. Dissolve 30 g of sucrose into 1000 mL of Milli-Q water. It'll stock at 4° in the refrigerator.
2. Dissolve doxycycline sulfate (Sigma) with 200 mL of 3% sucrose solution in a 500 mL light-blocking bottle.
3. Add 150 mL of distilled-water to the bottle and mix well.
4. Stock at 4°.

3 Methods

3.1 Injecting Large Dose of Vector Solution

To inject small doses of vector solution (e.g., 1 μL each time), we use an ordinary glass micropipette. In contrast, to deliver a large dose of viral vector solution, with the aim to spread the vector over a large area of the cortex, we adopt a convection-enhanced delivery technique. This method allows for the controlled injection of a large solution volume at a high rate using a specific injection needle. For instance, the following procedure is used to cover a large area of the premotor cortex in the macaque monkey.

1. Fit a fused silica capillary (450-μm outside diameter) to create a 1-mm "step" from the needle tip, minimizing reflux.
2. Insert the injection needle 1.5 mm below the cortical surface.
3. Inject first 1.25 μL at 0.25 μL/min and then the remaining 4.80 μL at 0.60 μL/min by using a 25-μL Hamilton syringe, totally 6.05 μL with a 32-gauge injection needle.
4. To ensure stability, wait for 5 min before the injection and 10 min afterward for each injection.

3.2 Dox Administration

To prevent adhesion of the AAV vector to the glass micropipettes, 0.001% Pluronic-F68 Solution (Sigma-Aldrich) was added to the vector solution. Because the titer deteriorated by about 10% in 1 month in this stock condition, the vector solution was divided into small aliquots (each contained 10 μL of the vector solution) before being stored at −80 °C to reduce the number of freeze–thaw cycles. The vector solutions were used within 5 months of their preparation. The development of the highly efficient Tet-on transactivator, rtTAV16, enabled the sharp onset of the effect of eTeNT transcription within 2 days, as revealed in this study. The propriospinal (PN) blockade seemed to have occurred earlier than the compensatory process, which enabled us to observe the behavioral effects [16]. We chose Tet-on system over the Tet-off system because the latter, although widely regarded as a highly sensitive inducible system, requires 2 to 3 weeks to take effect (*see* **Note 3**).

Using the Tet-off system could have allowed compensatory processes to occur alongside the gradual initiation of the transmission block, making it impossible to observe the behavioral effects accurately.

Dox solution is mainly administered orally both in monkeys and rodents, or in combination with intraperitoneal injection in rodents.

1. For monkeys, administer orally the solution at a dose of 5–15 mg kg per day every morning.
2. For rodents, daily administration of Dox solution is initiated by a single intraperitoneal injection (10 μg/g body weight) in a 0.9% NaCl solution, followed by continuous oral administration through drinking water (3 mg/mL in a 3% sucrose solution).

3.3 Detection of VAMP-2 Degradation

Immunoblots were performed using a mouse anti-Flag M2 monoclonal antibody (F1804, Sigma-Aldrich) and visualized with a horseradish peroxidase-conjugated sheep anti-mouse immunoglobin-G (IgG) antibody (NA931V, GE Healthcare) (Fig. 2)

3.3.1 Total RNA Extraction from Macaque Brain Tissue

Approximately 2 g of cerebral cortex tissue from Japanese macaque (*Macaca fuscata*) was kindly provided from Primate Research Institute, Kyoto University.

The tissue was homogenized in Sepasol-RNA I (Cat# 09379-55, Nacalai Tesque, Kyoto, Japan) using a mechanical homogenizer. Total RNA was extracted according to the manufacturer's protocol.

3.3.2 RNA Purification and cDNA Synthesis

- DNase Treatment: The extracted RNA was treated with RQ1 RNase-Free DNase (Cat# M6101, Promega, USA) to remove contaminating genomic DNA.
- mRNA Purification: mRNA was isolated using Oligotex-dT30 Super (Cat# 917-50301, Takara Bio Inc., Shiga, Japan).
- cDNA Synthesis: First-strand cDNA was synthesized using SuperScript™ III Reverse Transcriptase (Cat# 18080085, Thermo Fisher Scientific, USA), following the manufacturer's instructions.

3.3.3 Cloning of Macaque VAMP

The macaque VAMP gene was amplified via PCR from synthesized cDNA and cloned into a mammalian expression vector (e.g., pcDNA3.1, Thermo Fisher Scientific) using standard molecular cloning techniques.

3.3.4 Transfection of HEK293 Cells

HEK293 cells were seeded in 24-well plates and transfected with the following plasmids:

- pVAMP2, pSNAP25, or pSNAP23 (substrate plasmids)
- TeNT, codon-optimized TeNT, or control plasmids (neurotoxins)

1. Cell Seeding

   HEK293 cells were seeded in 24-well plates at a density of 5×10^4 to 1×10^5 cells/well and incubated overnight in a CO_2 incubator at 37 °C to reach approximately 50–60% confluence.

2. Transfection Preparation (Day 2)

   On the following day, the cell density was approximately 2×10^5 cells/well. The plate was allowed to sit at room temperature for 15 min prior to transfection.

3. Preparation of Transfection Mixture

   The following components were mixed in 30 μL of serum-free medium per well:

   - Plasmid 1 (substrate): 0.5 μg
     - Examples: VAMP2, SNAP25, SNAP23
   - Plasmid 2 (neurotoxin): 0.5 μg
     - Variants: ctrl, TeNT, eTeNT, eTeNT-EGFP-PEST
   - HilyMax Transfection Reagent (Dojindo, Kyoto, Japan): 3 μL

   The DNA/HilyMax mixture was gently pipetted onto each well and incubated at 37 °C in a CO_2 incubator.

4. Medium Change

   After 2.5 h, the transfection medium was removed and replaced with 0.5 mL of DMEM supplemented with 10% FBS per well. Cells were returned to the CO_2 incubator for continued culture.

5. Cell Harvesting

   After 48 h of transfection, cells were harvested and lysed in 0.5 mL of SDS lysis buffer per well. Lysates were stored at −80 °C until analysis.

   SDS Lysis Buffer Composition:

   - 2% SDS
   - 10% Glycerol
   - 50 mM Tris-HCl (pH 8.0)

3.3.5 SDS-PAGE Sample Preparation Protocol

1. Sample Mixing

   Prepare the protein loading sample by mixing the following components in a microcentrifuge tube:

   - Lysate: 1 μL
   - Sample Buffer: 3 μL

- NuPAGE™ Sample Reducing Agent (10×) (Thermo Fisher, Cat. No. NP0009): 1.2 μL
- Distilled Water (DW): 6.8 μL.

  Total volume: 12 μL.

2. Initial Denaturation

   Incubate the sample at 98 °C for 2 min.
3. Vortexing

   Vortex the sample for 2 min to ensure thorough mixing.
4. Final Denaturation

   Incubate the sample at 70 °C for 10 min.
5. Optional

   Briefly spin down the tube to collect condensation and ensure all liquid is at the bottom. The sample is now ready to be loaded onto the SDS-PAGE gel.

3.3.6 SDS-Page

1. Prepare the 12% resolving gel and cast it into the gel cassette. Overlay with isopropanol to ensure a flat surface and prevent air bubbles. After polymerization, remove the isopropanol and gently rinse with distilled water.
2. Prepare the stacking gel, pour it over the resolving gel, and insert a comb to form wells. Allow the stacking gel to polymerize completely.
3. Assemble the gel into the electrophoresis tank and fill the inner and outer chambers with running buffer.
4. Prepare the samples by mixing with 2× or 4× SDS sample buffer and boiling at 95 °C for 5 min.
5. Load 8 μL of each sample into the wells of the 12% polyacrylamide gel. Include a protein ladder as a molecular weight marker.
6. Run the gel at 200 V for 30–60 min, or until the dye front reaches the bottom of the gel.
7. Proceed to protein transfer or staining after electrophoresis.

NuPAGE™ Bis-Tris Mini Protein Gels, 12%, 1.0 mm, NP0342BOX

(https://www.thermofisher.com/order/catalog/product/jp/ja/NP0342BOX)

NuPAGE™ MOPS SDS Running Buffer (20×), cat. No. NP0001

(https://www.biochemisolutions.com/product/nupage-mops-sds-running-buffer-20x/)

Novex™ Sharp Pre-stained Protein Standard

(https://www.fishersci.com/shop/products/novex-sharp-pre-stained-protein-standard/ LC 5800)

MagicMark™ XP Western Protein Standard

(https://www.thermofisher.com/order/catalog/product/jp/ja/LC5602)

3.3.7 Blotting

1. Dip PVDF membrane in 100% methanol for 10–15 s to activate it.
2. Rinse the membrane briefly with distilled water.
3. Immerse the PVDF membrane, filter papers, and gel in cold transfer buffer for at least 10 min.
4. Assemble the transfer sandwich in the following order (from bottom to top):
   - Sponge → Filter Paper → Gel → PVDF Membrane → Filter Paper → Sponge.
   - Make sure to remove any air bubbles between the layers using a roller or pipette.
5. Place the sandwich into the transfer cassette and insert it into the blotting tank filled with cold transfer buffer.
6. Perform protein transfer at 30 V for 1 h at 4 °C.
7. After the transfer, proceed to the blocking step or store the membrane in TBS or PBS at 4 °C.

NuPAGE™ Transfer Buffer (20X) NP0006; Fisherscience

https://www.fishersci.ca/shop/products/novex-nupage-transfer-buffer-20x-2/np0006

PVDF/Filter Paper Sandwich; Therumofisher, LC2002

https://www.thermofisher.com/order/catalog/product/jp/en/LC2002

3.3.8 Western Blot: Immunodetection Protocol

1. Wash the membrane with PBST (0.05% Tween-20 in PBS) for 5 min at room temperature to remove residual transfer buffer.
2. Block the membrane in 5% skim milk prepared in PBST for 1 h at room temperature on a shaker.
3. Wash the membrane with PBST as follows:
   - 2 min
   - 2 min
   - 5 min
4. Incubate the membrane with the primary antibody diluted in PBST:
   - Anti-Flag M2 monoclonal antibody (mouse, F1804, Merck,) at 1:2000 dilution.
   - Incubate for 1 h at room temperature.

   F1804 https://www.sigmaaldrich.com/JP/ja/search/f1804?focus=products&page=1&perpage=30&sort=relevance&term=f1804

5. Wash the membrane with PBST as follows:
   - 2 min
   - 2 min
   - 10 min
   - 5 min
6. Incubate the membrane with the secondary antibody diluted in PBST:
   - Anti-Mouse IgG, HRP-linked whole antibody (sheep, NA931) at 1:20,000 dilution.
   - Incubate for 1 h at room temperature.
7. Incubate the membrane with ECL substrate for 5 min according to the manufacturer's instructions.
8. Detect the signal using X-ray film or a chemiluminescence imaging system.

4 Notes

1. *Combination with DREADDs:* To block two different pathways in a single stroke rat model, a Tet transactivator-based blocking technique was combined with the interruption of another pathway using DREADDs (Ishida et al. 2019). Here, the cortico-rubral pathway was blocked with the combination of NeuRet-TRE-EGFP.eTeNT (injected into the red nucleus) and AAV-DJ-CMV-rtTAV16 (injected into the motor cortex). Simultaneously, the cortico-reticular pathway was blocked by combining NeuRet-MSCV-Cre (injected into the medial PMRF) and AAV-DJ-EF1a-DIO-hM4Di-mCherry (injected into the motor cortex). In this configuration, Dox administration enabled the blockade of the pathway from the motor cortex to the red nucleus, while Clozapine N-oxide (CNO) administration allowed us to block the pathway from the motor cortex to the ponto-medullary reticular formation. Hence, this approach enabled the independent manipulation of two separate pathways in the same animal. Moreover, by integrating optogenetics using the Cre-LoxP system to express photoactivatable opsins, [32] it is possible to manipulate multiple pathways at widely different temporal scales. This offers a versatile platform for studying complex neural networks.
2. *Waiting time:* To measure the blood concentration of Dox in monkeys, a blood sample (2 mL) was collected in the evening, immediately before, during, and after doxycycline administration. The concentration of doxycycline was measured with high-performance liquid chromatography (LC-MS/MS system 6400, Agilent Technologies; an analysis service was provided by Towa Environmental Science) as described previously. The results (Fig. 4) showed that blood concentration of

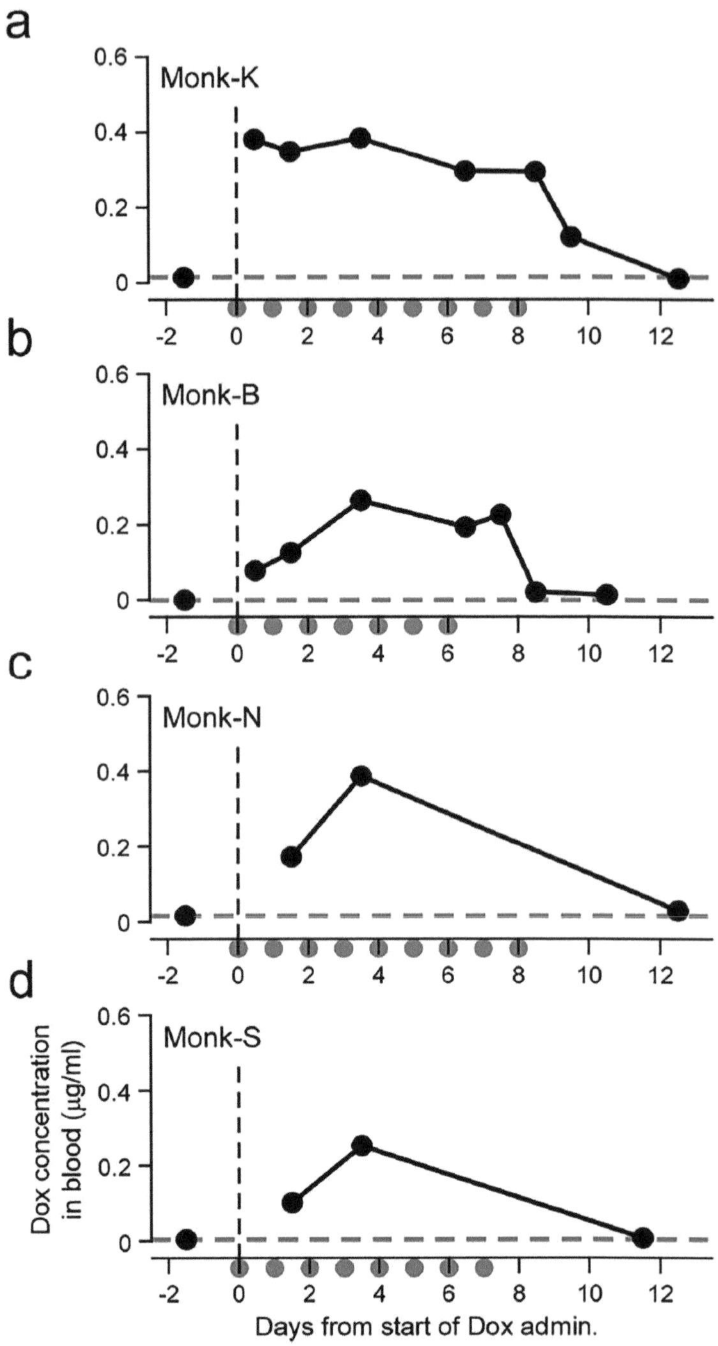

Fig. 4. Dox concentration in the blood before, during, and after Dox administration. Dox concentrations were measured with HPLC in four monkeys (Monk-K, -B, -N, and -S), before, during, and after the Dox administration period, in a, b, c and d, respectively. The days of Dox administration are indicated with gray circles on the horizontal axis. As shown in the figure, Dox concentration in the blood peaked within 3 days after administration began and dropped to virtually zero at 3 days after administration was terminated. (Adapted from Kinoshita et al. [16])

Dox rose immediately, peaked within 3–4 days, remained stable during administration, and completely disappeared within 3 days after final administration. Regarding the interval between repeated Dox administrations, some reporter protein (EGFP)-positive neurons were detectable 14 days but not 28 days after the discontinuing Dox. Consequently, the next round of Dox administration in mice was initiated 28 days after the offset of the preceding Dox administration [22].

3. *Effect of collaterals:* A disadvantage of the double viral vector technique is the need for precise injection of both viral vectors into the intended regions. Furthermore, this method can affect the transmission to other areas if the target neuron has major axon collaterals (Fig. 1b). For instance, mid-cervical propriospinal neurons in macaques (as described above) have collateral projections to the lateral reticular nucleus (LRN) in the brainstem in addition to the spinal motoneurons [16, 31]. As a result, the behavioral deficit of precision grip caused by blocking propriospinal neurons could be attributed either to the disruption of transmission to the motoneurons or to the LRN. To address this, experimenters must first determine the full axonal trajectories of the target neurons post-hoc by immunohistochemical staining with the antibody against the reporter fluorescent protein. If the goal is to clarify the function of "a specific pathway" rather than "a specific neuron type," additional methods may be required. Moreover, since the target pathways often comprise multiple cell types, the double infection technique affects all these cell types and cannot differentiate the functions of individual cell types.

Acknowledgments

We thank Drs. Dai Watanabe, Satoshi Yawata, Ryosuke Matsui, and Kenta Kobayashi for providing crucial information in the process of preparing the chapter.

Conflict of Interest Statement The authors have no conflicts of interest to declare that are relevant to the content of this chapter.

Funding

None.

References

1. Kobayashi K, Morita S, Sawada H, Mizuguchi T, Yamada K, Nagatsu I, Fujita K, Kreitman RJ, Pastan I, Nagatsu T (1995) Immunotoxin-mediated conditional disruption of specific neurons in transgenic mice. Proc Natl Acad Sci USA 92:1132–1136
2. Watanabe D, Inokawa H, Hashimoto K, Suzuki N, Kano M, Shigemoto R, Hirano T, Toyama K, Kaneko S, Yokoi M, Moriyoshi K, Suzuki M, Kobayashi K, Nagatsu T, Kreitman RJ, Pastan I, Nakanishi S (1998) Ablation of cerebellar Golgi cells disrupts synaptic integration involving GABA inhibition and NMDA receptor activation in motor coordination. Cell 95:17–27
3. Boyden ES, Zhang F, Bamberg E, Nagel G, Deisseroth K (2005) Millisecond-timescale, genetically targeted optical control of neural activity. Nat Neurosci 8:1263–1268
4. Nagel G, Szellas T, Kateriya S, Adeishvili N, Hegemann P, Bamberg E (2005) Channelrhodopsins: directly light-gated cation channels. Biochem Soc Trans 33:863–866
5. Zhang F, Wang LP, Brauner M, Liewald JF, Kay K, Watzke N, Wood PG, Bamberg E, Nagel G, Gottschalk A, Deisseroth K (2007) Multimodal fast optical interrogation of neural circuitry. Nature 446:633–639
6. Gerits A, Farivar R, Rosen BR, Wald LL, Boyden ES, Vanduffel W (2012) Optogenetically induced behavioral and functional network changes in primates. Curr Biol 22:1722–1726
7. Petreanu L, Huber D, Sobczyk A, Svoboda K (2007) Channelrhodopsin-2-assisted circuit mapping of long-range callosal projections. Nat Neurosci 10:663–668
8. Cruikshank SJ, Urabe H, Nurmikko AV, Connors BW (2010) Pathway-specific feedforward circuits between thalamus and neocortex revealed by selective optical stimulation of axons. Neuron 65:230–245
9. Kaneda K, Kasahara H, Matsui M, Katoh T, Mizukami H, Ozawa K, Watanabe D, Isa T (2011) Selective optical control of synaptic transmission in the subcortical visual pathway by activation of viral vector-expressed halorhodopsin. PLoS One 6:e18452
10. Armbruster BN, Li X, Pausch MH, Herlitze S, Roth BL (2007) Evolving the lock to fit the key to create a family of G protein-coupled receptors potently activated by an inert ligand. Proc Natl Acad Sci USA 104:5163–5168
11. Mahler SV, Vazey EM, Beckley JT, Keistler CR, McGlinchey EM, Kaufling J, Wilson SP, Deisseroth K, Woodward JJ, Aston-Jones G (2014) Designer receptors show role for ventral pallidum input to ventral tegmental area in cocaine seeking. Nat Neurosci 17:577–585
12. Nagai Y, Miyakawa N, Takuwa H, Hori Y, Oyama K, Ji B, Takahashi M, Huang XP, Slocum ST, DiBerto JF, Xiong Y, Urushihata T, Hirabayashi T, Fujimoto A, Mimura K, English JG, Liu J, Inoue KI, Kumata K, Seki C, Ono M, Shimojo M, Zhang MR, Tomita Y, Nakahara J, Suhara T, Takada M, Higuchi M, Jin J, Roth BL, Minamimoto T (2020) Deschloroclozapine, a potent and selective chemogenetic actuator enables rapid neuronal and behavioral modulations in mice and monkeys. Nat Neurosci 23:1157–1167
13. Oyama K, Hori Y, Nagai Y, Miyakawa N, Mimura K, Hirabayashi T, Inoue KI, Suhara T, Takada M, Higuchi M, Minamimoto T (2021) Chemogenetic dissection of the primate prefronto-subcortical pathways for working memory and decision-making. Sci Adv 7: eabg4246
14. Kato S, Inoue K, Kobayashi K, Yasoshima Y, Miyachi S, Inoue S, Hanawa H, Shimada T, Takada M, Kobayashi K (2007) Efficient gene transfer via retrograde transport in rodent and primate brains using a human immunodeficiency virus type 1-based vector pseudotyped with rabies virus glycoprotein. Hum Gene Ther 18:1141–1151
15. Kato S, Kobayashi K, Inoue K, Kuramochi M, Okada T, Yaginuma H, Morimoto K, Shimada T, Takada M, Kobayashi K (2011) A lentiviral strategy for highly efficient retrograde gene transfer by pseudotyping with fusion envelope glycoprotein. Hum Gene Ther 22: 197–206
16. Kinoshita M, Matsui R, Kato S, Hasegawa T, Kasahara H, Isa K, Watakabe A, Yamamori T, Nishimura Y, Alstermark B, Watanabe D, Kobayashi K, Isa T (2012) Genetic dissection of the circuit for hand dexterity in primates. Nature 487:235–238
17. Yamamoto M, Wada N, Kitabatake Y, Watanabe D, Anzai M, Yokoyama M, Teranishi Y, Nakanishi S (2003) Reversible suppression of glutamatergic neurotransmission of cerebellar granule cells in vivo by genetically manipulated expression of tetanus neurotoxin light chain. J Neurosci 23:6759–6767
18. Pocratsky AM, Burke DA, Morehouse JR, Beare JE, Riegler AS, Tsoulfas P, States GJR, Whittemore SR, Magnuson DSK (2017) Reversible silencing of lumbar spinal interneurons unmasks a task-specific network for

securing hindlimb alternation. Nat Commun 8(1):1963

19. Pocratsky AM, Shepard CT, Morehouse JR, Burke DA, Riegler AS, Hardin JT, Beare JE, Hainline C, States GJ, Brown BL, Whittemore SR, Magnuson DS (2020) Long ascending propriospinal neurons provide flexible, context-specific control of interlimb coordination. elife 9:e53565
20. Wahl AS, Omlor W, Rubio JC, Chen JL, Zheng H, Schröter A, Gullo M, Weinmann O, Kobayashi K, Helmchen F, Ommer B, Schwab ME (2014) Neuronal repair. Asynchronous therapy restores motor control by rewiring of the rat corticospinal tract after stroke. Science 344:1250–1255
21. Asboth L, Friedli L, Beauparlant J, Martinez-Gonzalez C, Anil S, Rey E, Baud L, Pidpruzhnykova G, Anderson MA, Shkorbatova P, Batti L, Pagès S, Kreider J, Schneider BL, Barraud Q, Courtine G (2018) Cortico-reticulo-spinal circuit reorganization enables functional recovery after severe spinal cord contusion. Nat Neurosci 21:576–588
22. Sooksawate T, Isa K, Matsui R, Kato S, Kinoshita M, Kobayashi K, Watanabe D, Kobayashi K, Isa T (2013) Viral vector-mediated selective and reversible blockade of the pathway for visual orienting in mice. Front Neural Circuits 7:162
23. Ishida A, Isa K, Umeda T, Kobayashi K, Kobayashi K, Hida H, Isa T (2016) Causal link between the cortico-rubral pathway and functional recovery through forced impaired limb use in rats with stroke. J Neurosci 36: 455–467
24. Tohyama T, Kinoshita M, Kobayashi K, Isa K, Watanabe D, Kobayashi K, Liu M, Isa T (2017) Contribution of propriospinal neurons to recovery of hand dexterity after corticospinal tract lesions in monkeys. Proc Natl Acad Sci USA 114:604–609
25. Kinoshita M, Kato R, Isa K, Kobayashi K, Kobayashi K, Onoe H, Isa T (2019) Dissecting the circuit for blindsight to reveal the critical role of the pulvinar and superior colliculus. Nat Commun 10(1):135
26. Vancraeyenest P, Arsenault JT, Li X, Zhu Q, Kobayashi K, Isa K, Isa T, Vanduffel W (2020) Selective mesoaccumbal pathway inactivation affects motivation but not reinforcement-based learning in macaques. Neuron 108: 568–581.e6
27. Ninomiya T, Noritake A, Kobayashi K, Isoda M (2020) A causal role for frontal cortico-cortical coordination in social action monitoring. Nat Commun 11(1):5233
28. Oguchi M, Okajima M, Tanaka S, Koizumi M, Kikusui T, Ichihara N, Kato S, Kobayashi K, Sakagami M (2015) Double virus vector infection to the prefrontal network of the macaque brain. PLoS One 10:e0132825
29. Ishida A, Kobayashi K, Ueda Y, Shimizu T, Taijiri N, Isa T, Hida H (2019) Dynamic interaction between conrtico-brainstem pathways during training-induced recovery in stroke model rats. J Neurosci 39:7306–7320
30. Vanduffel W, Isa T (2023) Pathway-selective reversible perturbations using a double-infection technique in the macaque brain. In: Eldridge MAG, Galvan A (eds) Vectorology for optogenetics and chemogenetics. Springer, New York, pp 185–205
31. Isa T, Ohki Y, Seki K, Alstermark B (2006) Properties of propriospinal neurons in the C3-C4 segments mediating disynaptic pyramidal excitation to forelimb motoneurons in the macaque monkey. J Neurophysiol 95:3674–3685
32. Isa K, Sooksawate T, Kobayashi K, Kobayashi K, Redgrave P, Isa T (2020) Dissecting the tectal output channels for orienting and defense responses. eNeuro 7: ENEURO.0271-20.2020

Chapter 16

Optogenetic Approach for In Vivo Manipulation of Neural Circuits Using Lentiviral Vectors

Takashi Matsuda

Abstract

The brain controls various physiological functions through a complex network of neural circuits. To elucidate the function of individual neural circuits, it is necessary to perform both the anatomical mapping and functional manipulation of these circuits in living animals. Our understanding of the neural mechanisms underlying physiological functions has recently been advanced by technological innovations in optical approaches to manipulate neural activity and viral vector systems for in vivo gene delivery. The application of optogenetic tools to neurons in the brain facilitates the selective and reversible manipulation of neuronal activity with millisecond precision using light-sensitive proteins. Regarding gene transfer into specific neural circuits, the lentiviral system for highly efficient retrograde gene transfer (HiRet) offers the advantages of neural circuit selectivity and stable transgene expression. The combined approach of optogenetic tools and HiRet vectors allows the cell type-specific manipulation of neuronal activity and animal behavior with high spatial and temporal precision. I herein describe a protocol for manipulating the activity of neural circuits involved in the control of thirst and salt appetite in freely moving animals as an example of the optogenetic approach using HiRet vectors. This application will be widely utilized to characterize individual neural circuits underlying physiological functions.

Key words Lentiviral vectors, HiRet, Viral injection, Optogenetics, Neural circuits, Thirst, Salt appetite, Behavioral analysis

1 Introduction

The brain supports our daily life by responding to external stimuli and regulating physiological functions throughout the body via neural networks. Billions of neurons in the mammalian brain functionally communicate with each other and organize a complex neural circuit [1]. Neural circuits can be distinguished for each neuronal population responsible for a specific biological function. To elucidate the physiological role of individual neural circuits, it is necessary to examine the structure and function of each circuit in

Kazuto Kobayashi et al. (eds.), *Lentiviral Vectors: Applications for Basic and Clinical Neuroscience*, Methods in Molecular Biology, vol. 2974, https://doi.org/10.1007/978-1-0716-4807-0_16,

living animals. Traditionally, anatomical and functional knowledge of neural circuits in the brain has been obtained using classical technologies, such as neural tracing and electrophysiological analyses [2, 3]. In recent years, breakthroughs in technologies, such as optogenetics, chemogenetics, and calcium imaging, have significantly advanced our understanding of neural circuit functions, ranging from the molecular level to the whole-brain circuit level [3, 4].

Optogenetics is a revolutionary technique in neuroscience that enables the precise control of neural activity using light-sensitive proteins, such as channelrhodopsin-2 (ChR2) and archaerhodopsin-T (ArchT) [5, 6]. When exposed to blue light (~470 nm), ChR2 allows cations, including sodium ions (Na^+) and protons (H^+), to enter the cell, leading to depolarization and neuronal activation [5]. In contrast to ChR2, ArchT pumps H^+ out of the cell in response to yellow-green light (~520 nm), leading to hyperpolarization and neuronal suppression [7]. ArchT offers prolonged suppression under continuous light exposure with minimal phototoxicity. The use of light as a regulator of neural function has many advantages, including no physical contact, no interference with neural activity, and excellent spatial resolution [3]. Combined with optogenetic tools and various viral vectors, the in vivo manipulation of neural activity with millisecond-level temporal precision and high spatial accuracy can be achieved more easily. Neuroscientists recently identified the neural circuits underlying a number of physiological functions using these advanced approaches [8–10].

Traditionally, anterograde and retrograde tracers, such as wheat germ agglutinin and cholera toxin subunit B, were widely used to map neural circuits in anatomical research [2]. However, these tracers lack the ability to select target cell types and carry exogenous genes. In recent years, genetic engineering techniques using viral vectors for in vivo gene delivery have been developed to perform neurological labeling and the gene expression regulation of specific neurons [11]. Several viral vectors, including adeno-associated virus, pseudorabies virus, and lentivirus, are commonly used as important tools in neural circuit studies. Among them, lentiviral vectors have advantages for in vivo gene delivery because they provide stable transgene expression through the permanent integration of transgenes into the host genome of both dividing and nondividing cells, eliminating the need for repeated viral administration [12–15]. Therefore, lentiviral vectors are extensively applicable to gene therapy for neurological disorders as well as for basic neuroscience studies [16–18]. Additionally, a notable benefit of lentiviral vectors over other vector systems is minimal inflammatory responses [19]. The lentiviral system for highly efficient retrograde gene transfer (HiRet) has been developed for the retrograde axonal transport of viral vectors using the pseudotyped glycoprotein constructed from the rabies virus glycoprotein and vesicular stomatitis virus glycoprotein [20, 21]. In analyses of neural networks, this vector offers remarkable advantages, enabling the visualization and manipulation of specific neuronal pathways in combination with modern technologies, such as genetic cell ablation [22], chemogenetics [23], and optogenetics [24].

We previously identified and characterized thirst-driving excitatory neurons (water neurons) and salt appetite-driving excitatory neurons (salt neurons) in the subfornical organ (SFO) by using HiRet vectors encoding optogenetic tools [25]. The HiRet vector encoding ArchT-green fluorescent protein (GFP) was injected into the organum vasculosum of the lamina terminalis (OVLT) and the median preoptic nucleus (MnPO), or the ventral part of the bed nucleus of the stria terminalis (vBNST), and induced the retrograde expression of ArchT-GFP in the cell bodies of neurons in the SFO. Yellow light exposure applied to the SFO resulted in the suppression of the SFO-OVLT/MnPO or SFO-vBNST pathways, which led to a significant reduction in water or salt intake under water- or Na-depleted conditions. In addition, double fluorescent labeling using two patterns of HiRet vectors encoding EGFP or mCherry identified distinct populations of cholecystokinin gene-positive neurons in the dorsal part of the lateral parabrachial nucleus that project to the MnPO or vBNST, but not to the SFO [26, 27]. Moreover, the optical manipulation of OVLT neurons projecting to the paraventricular nucleus (PVN) using HiRet vectors encoding ChR2 and halorhodopsin revealed an essential role of the OVLT-PVN pathway in salt-induced blood pressure elevations [28].

This chapter presents a detail method for the selective manipulation of water neurons and salt neurons in the SFO as an example of optogenetic approaches using HiRet vectors. To improve the success rate of functional studies on neural circuits, it is important to develop experimental skills in the intracranial injection of viral vectors and the implantation of guide cannulas for optical fibers into precise brain regions. In addition, a series of conditional settings in light exposure is known to markedly affect animal behavior. Therefore, the immunohistological detection of the activity status of retrograde-labeled cells from brain sections at the end of the experiment may confirm whether neural activity levels are consistent with the findings of behavioral analyses. This method will be broadly applied to manipulations of the activities of other neural circuits, and advance our understanding of physiological brain mechanisms.

2 Materials

2.1 Stereotaxic Surgeries for Optogenetic Manipulation

1. Wild type (WT) and *Vglut2-Cre* (Jackson Labs, stock no. 016963) mice [29]: Male mice older than 8 weeks are used in experiments. Mice are housed in a room with a temperature range of 23–25 °C under a 12-h light-dark cycle (lights on at 8:00 AM), and have free access to water and food.
2. Anesthetic combination: 0.075 mg/mL medetomidine, 0.4 mg/mL midazolam, and 0.5 mg/mL butorphanol in 0.9% NaCl solution (saline). Store at room temperature for 1 month with shading.
3. An antagonist of the anesthetic: 0.075 mg/mL atipamezole in saline. Store at room temperature for 1 month.
4. Lentiviral vector solutions: The HiRet vectors encoding double-loxP-flanked inverted orientation (DIO)-ChR2 (H134R)-EYFP or ArchT-GFP are prepared. Refer to Chapter 2 for preparation of lentiviral vectors. The H134R variant of ChR2 has a common mutation that increases the photocurrent, making action potentials more efficient than those of native ChR2. HiRet vectors ($>1.0 \times 10^{10}$ genomic copies/ml) are used for in vivo gene transfer. Aliquot the viral vectors into microtubes (10 μL each) and store aliquots in a −80 °C refrigerator (*see* **Note 1**).
5. Stereotactic frame with a micromanipulator and cannula holder (SR-5M-HT, SMM-100, and SHC-1, Narishige).
6. Stereoscopic microscope.
7. Microsyringe pump (Ultra Microliter Syringe Pump, WPI).
8. Gas-tight microsyringe.
9. Joint Teflon tube (0.1 mm in diameter).
10. A pulled-glass pipette: Create a glass pipette with a tip diameter of ~20 μm from a glass capillary using a micropipette puller.
11. Fluorinert (3M Japan).
12. Heating pad.
13. Hair removal cream.
14. Scalpel blade.
15. Forceps.
16. Cotton swabs.
17. 70% ethanol.
18. Dental drill.
19. Surgical silk suture (size 5–0).

20. Stainless-steel guide cannula and dummy cannula (P1 Technologies).
21. Dental acrylic resin.
22. Screw.

2.2 Behavioral Analysis with Optogenetic Manipulation

1. Semi-handmade optical fiber: Separate the connector from the single connector assembly (P1 Technologies) and pass the optical fiber (Lucir) through it. Pass the optical fiber through the stopper separated from the internal cannula (P1 Technologies). Glue the stopper and optic fiber together so that the tip of the optic fiber is exposed approximately 0.1 mm from the guide cannula.
2. Optical swivel (Lucir).
3. Blue light laser (445 nm).
4. Yellow light laser (577 nm).
5. BNC cable.
6. Arbitrary waveform function generators.
7. Optical cable.
8. Automatic measurement system with optical-based counters (BrainScience-Idea).
9. Bottle (Top Corporation).
10. Dropper (TK-U200L, Terumo).
11. Spout (BrainScience-Idea).
12. Test cage (STANK, Instech Laboratories).
13. Diuretic drug: 5 μg/μL Furosemide solution. Prepare the solution beforehand for injection by dissolving 5 mg of furosemide in 100 μL of methanol and heating it at 60 °C with shaking for 3 min. Add 900 μL of saline to complete the preparation.

2.3 Immunohistochemical Detection of Transgene Expression

1. The combined anesthetic: 0.075 mg/mL medetomidine, 0.4 mg/mL midazolam, and 0.5 mg/mL butorphanol in saline.
2. Formalin solution: 20% formalin in PBS.
3. Sucrose solutions: 15% and 30% sucrose in PBS.
4. Optimal Cutting Temperature (OCT) compound.
5. Cryostat.
6. Blocking buffer: 5% Normal donkey serum, 0.1% Triton X-100. Add 2.0 mL normal donkey serum and 40 μL Triton X-100 to 37.96 mL PBS. Store at 4 °C for 3 months.

7. Primary antibody: Rat anti-GFP (1:1000, 04404-84, Nacalai Tesque) and goat anti-Fos (1:500, sc-52G, Santa Cruz Biotechnology).
8. Secondary antibody: Donkey anti-rabbit IgG Alexa Fluor 488 (1:500, A-21206, Thermo Fisher Scientific) and donkey anti-goat IgG Alexa Fluor 555 (1:500, A-32816, Thermo Fisher Scientific).
9. Glass slide.
10. Coverslip.
11. Fluoromount™ (Cosmo Bio).
12. Confocal microscope.

3 Methods

3.1 Stereotaxic Surgeries for Optogenetic Manipulation

Since lentiviral vectors are classified into biosafety level 2 (BSL-2), surgeries that include viral injection and cannulation need to be performed in a biological safety cabinet (class 2) (Fig. 1). In addition, animals must be housed under animal BSL-2 (ABSL-2) conditions during the experimental period. For the optical activation of water neurons, the HiRet vector encoding DIO-ChR2(H134R)-EYFP is injected into the OVLT and MnPO in the *Vglut2-Cre* mouse. Regarding the optical suppression of salt neurons, the HiRet vector encoding ArchT-GFP is injected into the vBNST of the WT mouse.

3.1.1 Intracranial Injection of HiRet Vectors

1. Measure the body weight of a mouse, and intraperitoneally administer the combined anesthetic (250 μL/25 g body weight (gbw)) (*see* **Note 2**).
2. Remove the fur from the top of the scalp using hair removal cream.
3. Mount the anesthetized mouse on the stereotaxic frame to fix the head, and warm the mouse to approximately 37 °C using a heating pad.
4. Disinfect the scalp with cotton swabs soaked in 70% ethanol and incise the scalp along the midline with a scalpel blade to expose the skull. Wipe the connective tissue on the surface of the skull off with the cotton swabs.
5. Adjust the position of the bregma and lambda to the same vertical level.
6. Connect a pulled-glass pipette to the gas-tight microsyringe via a joint tube, and fill with Fluorinert. Immerse the tip of the glass pipette in viral solution, and aspirate the solution into the glass pipette using the microinjection pump (*see* **Note 3**). Attach the glass pipette to the micromanipulator.

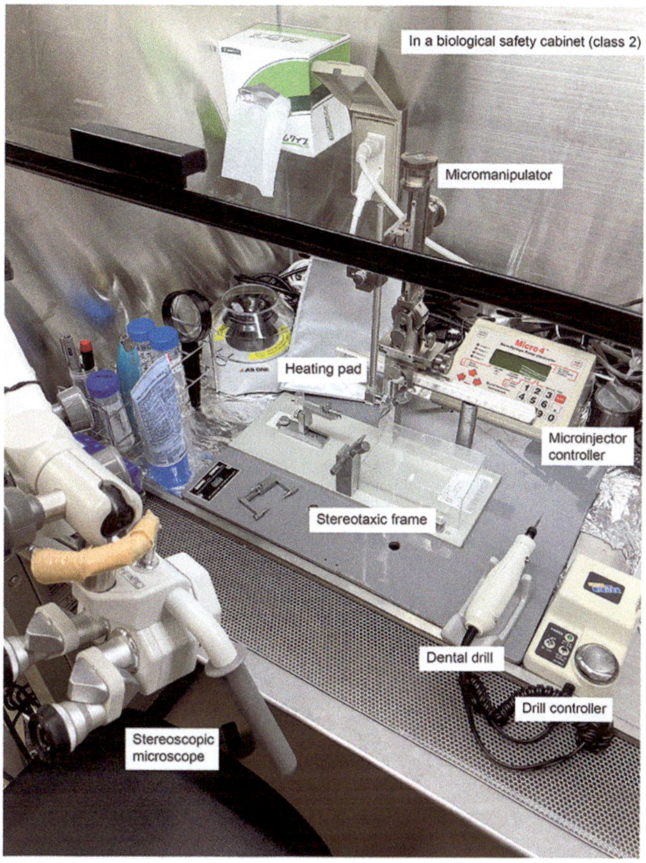

Fig. 1 Surgical instruments for microinjections and cannulation. A stereotaxic system for microinjections includes a stereotaxic frame with a manipulator, a stereoscopic microscope, a microinjector and its controller, a dental drill and its controller, and a heating pad in a biological safety cabinet (class 2)

7. Position the tip of the glass pipette to touch the bregma and set the coordinate to zero.
8. Select the coordinates of the injection sites by referring to the mouse brain atlas. The OVLT are coordinated at Anterior-Posterior (AP), +0.70 mm; Medial-Lateral (ML), 0.00 mm; Dorsal-Ventral (DV), 5.20 mm, the MnPO is coordinated at AP, +0.40 mm; ML, 0.00 mm; DV, 4.50 mm, and the vBNST is coordinated at AP, +0.30 mm; ML, ±1.00 mm; DV, 4.60 mm.
9. Position the tip of the glass pipette at the target AP and ML coordinates, and mark the position of the target site on the surface of the skull with a felt-tip pen.
10. Drill a small hole according to the mark. Puncture the dura mater with a sharp forceps to avoid breaking the tip of the glass pipette.

11. Insert the tip of the glass pipette into the brain through the hole, and gradually reach the DV coordinate (*see* **Note 4**).
12. Start the infusion of viral solution at 0.05–0.10 μL/min. Adjust the final injection volume according to the size of the target brain regions.
13. Wait for 5 min after the infusion, and then withdraw the glass pipette by 0.5 mm every 10 s.
14. Raise the glass pipette from the brain tissue, and close the skin incision with a surgical suture.
15. Administer an anesthetic antagonist intraperitoneally, and place the mouse in a clean cage at 38 °C until it awakens.

3.1.2 Implantation of the Guide Cannula

1. More than 3 weeks after the viral injection, the mouse is again anesthetized and placed in the stereotactic frame.
2. Remove the scalp to expose a large part of the skull (Fig. 2a).
3. Attach the guide cannula to the cannula holder (Fig. 2b, c), and move the tip position to the target AP and ML coordinates of the implant site from the bregma. The coordinate of the SFO for cannulation is at AP, −0.70 mm; ML, 0.00 mm; DV, 2.5 mm (*see* **Note 5**).

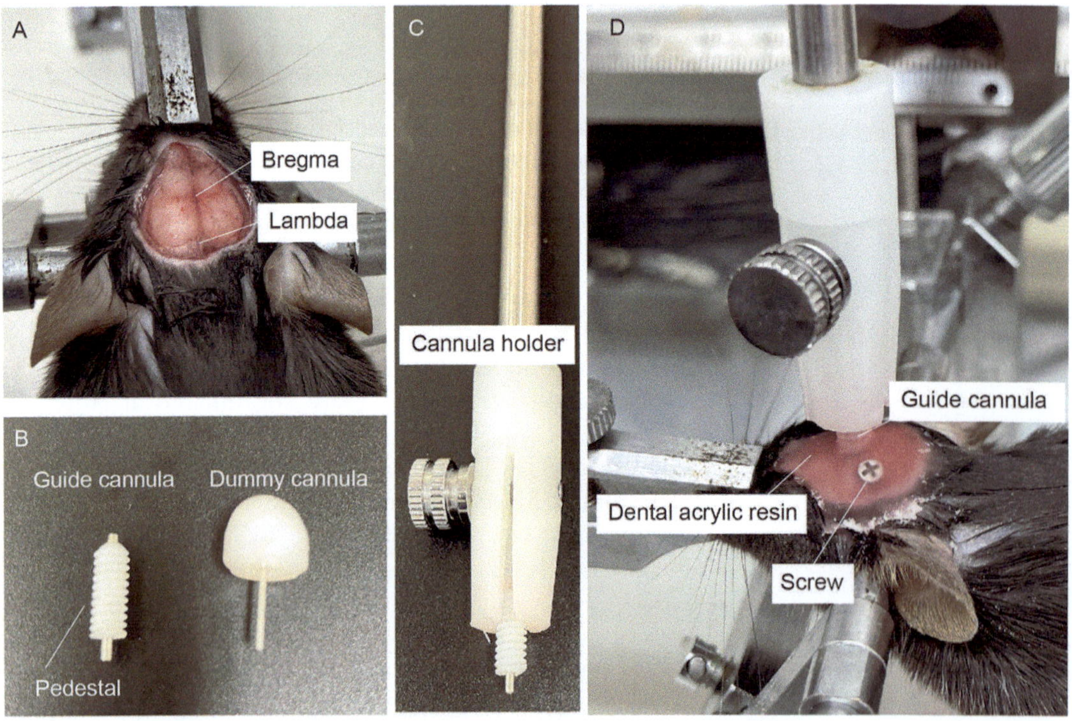

Fig. 2 Implantation of a guide cannula for an optic fiber. (**a**) Mount the mouse on the stereotaxic frame. Cut open the scalp and expose the skull. (**b**) The guide cannula and dummy cannula. (**c**) The guide cannula attached to the cannula holder. (**d**) The guide cannula and screw, fixed to the skull with dental acrylic resin

4. Drill two holes for implantation of the cannula and screw anchored to the skull (*see* **Note 6**).
5. Insert the guide cannula through the hole into the brain to the target DV coordinate (*see* **Note 7**).
6. Attach a screw to the skull and fix the cannula with dental acrylic resin (Fig. 2d).
7. Administer an anesthetic antagonist intraperitoneally, and place the mouse in a clean cage at 38 °C until it awakens.

3.2 Behavioral Analysis with Optogenetic Manipulation

Behavioral experiments are performed more than 1 week after cannulation. In optogenetic experiments, paired behavioral tests are conducted on the same group of mice. Mice under euhydrated conditions are used for the optical activation of water neurons, while mice under Na-depleted conditions are used for the optical suppression of salt neurons. The two-bottle test is used to measure the intake volumes of different solutions simultaneously.

3.2.1 Preparation of the Test Cage for the Two-Bottle Test

1. Set up the automatic measurement system for drinking volumes with optical-based counters that automatically count the number of water drops (0.017 mL/drop) from a dropper (Fig. 3a, b).
2. Mice are acclimated to the test cage with the two spouts providing distilled water or salt solution (0.3 M NaCl) for more than 3 days before the behavioral test (Fig. 3c).

3.2.2 Behavioral Analysis with the Optical Activation of Water Neurons

1. Attach the semi-handmade optic fiber to the guide cannula implanted in the mouse brain (Fig. 4a, b), and habituate the mouse in the test cage for 15 min.
2. Start to measure the intake volume automatically and provide the mouse with distilled water and 0.3 M NaCl through the two spouts without food.
3. Irradiate laser light through the optic fiber with an optical swivel, which is connected to a blue light laser with an optional light on/off frequency (*see* **Note 8**). In the experiment for the optical activation of water neurons, water intake behavior was selectively induced by the optical stimulation (Fig. 5a–d). Additionally, the amount of water ingested by the same individuals was varied by the frequency of the blue light laser (Fig. 5c).

3.2.3 Behavioral Analysis with the Optical Suppression of Salt Neurons

1. Administer furosemide (1 mg/25 gbw) subcutaneously to the mouse and house it with free access to the Na-deficient diet and distilled water for 24 h in order to generate Na-depleted conditions.
2. Attach the semi-handmade optic fiber to the guide cannula implanted in the mouse brain 15 min prior to measurements,

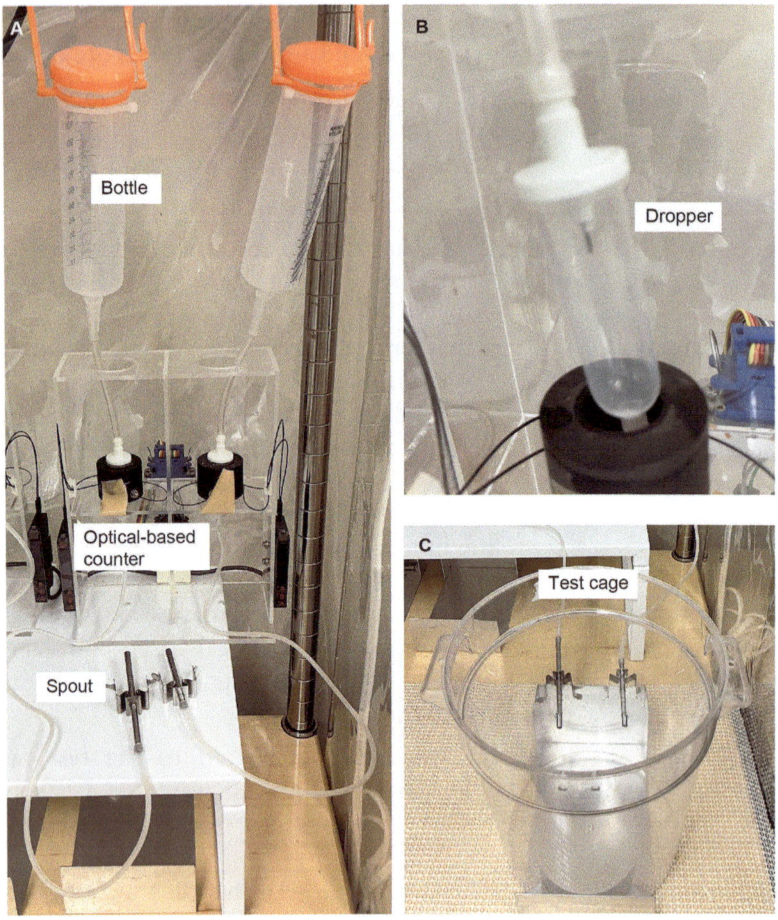

Fig. 3 Instruments for the two-bottle test. (**a**) The automatic drinking volume measuring system connected to optical-based counters combined with bottles, droppers, and spouts. (**b**) A dropper for making water drops (0.017 mL/drop). (**c**) A test cage with two spouts for the behavioral analysis in the two-bottle test

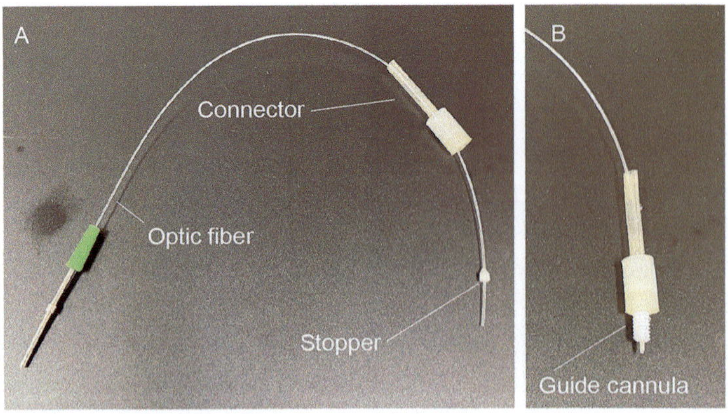

Fig. 4 A semi-handmade optic fiber. (**a**) The optic fiber combined with a connector and stopper. (**b**) The optic fiber is adjusted by the position of the stopper to expose approximately 0.1 mm of the fiber tip from the guide cannula

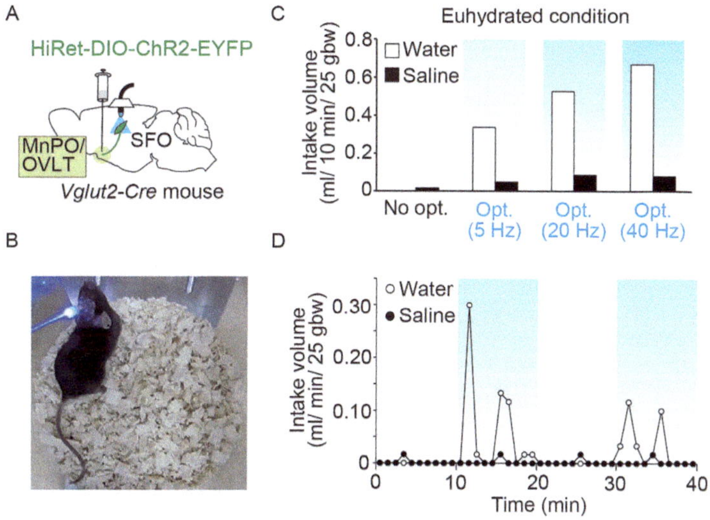

Fig. 5 Optogenetic approach to induce thirst through the activation of water neurons. (**a**) An injection of HiRet-DIO-ChR2(H134R)-EYFP into the OVLT and MnPO of the *Vglut2-Cre* mouse with irradiation of the SFO by blue-light laser. (**b**) Selective water intake of a thirst-induced mouse by the optical activation of water neurons. (**c**) Summary of the intake volumes of a representative mouse in the two-bottle test with an optical stimulation (Opt.) at different frequencies (5, 20, and 40 Hz) for 10 min. Water intake selectively and progressively increased in proportion to the frequencies of light in the same individual. (**d**) Summary of continuous intake volumes with or without the optical stimulation (20 Hz). The optical stimulation for 10 min was started at 10 min and at 30 min

and irradiate laser light through the optic fiber with the optical swivel, which is connected to a yellow light laser, 3 min before the start of the behavioral test (*see* **Note 9**).

3. Start to measure the intake volume automatically and provide the mouse under Na-depleted conditions with distilled water and 0.3 M NaCl without food. In the experiment for the optical suppression of salt neurons, salt intake behavior under Na-depleted conditions was selectively reduced by continuous light exposure (Fig. 6).

3.3 Immunohistochemical Detection of Transgene Expression

All viral injection sites and retrograde-labeled sites are verified by immunohistochemistry after behavioral tests. If low or no gene expression is observed at retrograde-labeled sites, the behavioral data obtained from these animals are excluded from analyses. To observe the ChR2-induced expression of the Fos protein, a marker of neuronal activation, the mouse brain is fixed 2 h after 10 min light exposure for immunohistochemistry.

1. Administer the combined anesthetic to the mouse for deep anesthesia.

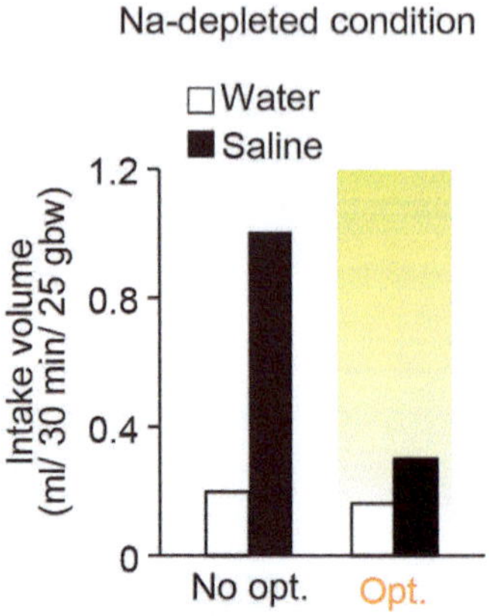

Fig. 6 Optogenetic approach to suppress salt appetite through the inhibition of salt neurons. Summary of the intake volumes of water and saline (0.3 M NaCl) by a representative WT mouse with or without an optical stimulation (Opt.) by the yellow light laser for 30 min. This WT mouse received viral injections of HiRet-ArchT-GFP to the bilateral vBNST. Saline intake was selectively and markedly reduced by the optical stimulation in the same individual under the Na-depleted condition

2. Perform cardiac perfusion with 10 mL of PBS followed by 10 mL of formalin solution in PBS.
3. Remove the guide cannula from the mouse brain.
4. Remove the mouse brain and post-fix it in formalin solution at 4 °C for 2 days.
5. Replace the mouse brain in 15% sucrose solution for 1 day and then 30% sucrose solution for 1 day for cryoprotection.
6. Dissect the mouse brain containing the viral injection sites, embed it in OCT compound, gradually freeze it in liquid nitrogen, and preserve it in a −80 °C deep freezer.
7. Cut 50-μm-thick coronal sections from the mouse brain using a cryostat, and soak the sections in PBS at room temperature.
8. Replace the sections in blocking buffer at room temperature for 1 h, and then incubate them with the primary antibodies at 4 °C for 2 days with gentle shaking.
9. Wash sections twice in PBS at room temperature for 15 min, followed by an incubation with secondary antibodies at 4 °C for 1 day with gentle shaking.
10. Mount brain sections on glass slides, add 3 ~ 4 drops of Fluoromount™, and then slowly lower the coverslip over the slide, taking care to avoid air bubbles.
11. Observe immunostained cells in brain sections under a confocal microscope to confirm the location of the injection site (Fig. 7a, b), cannulation site (Fig. 7c), and retrograde-labeled site (Fig. 7b, d).

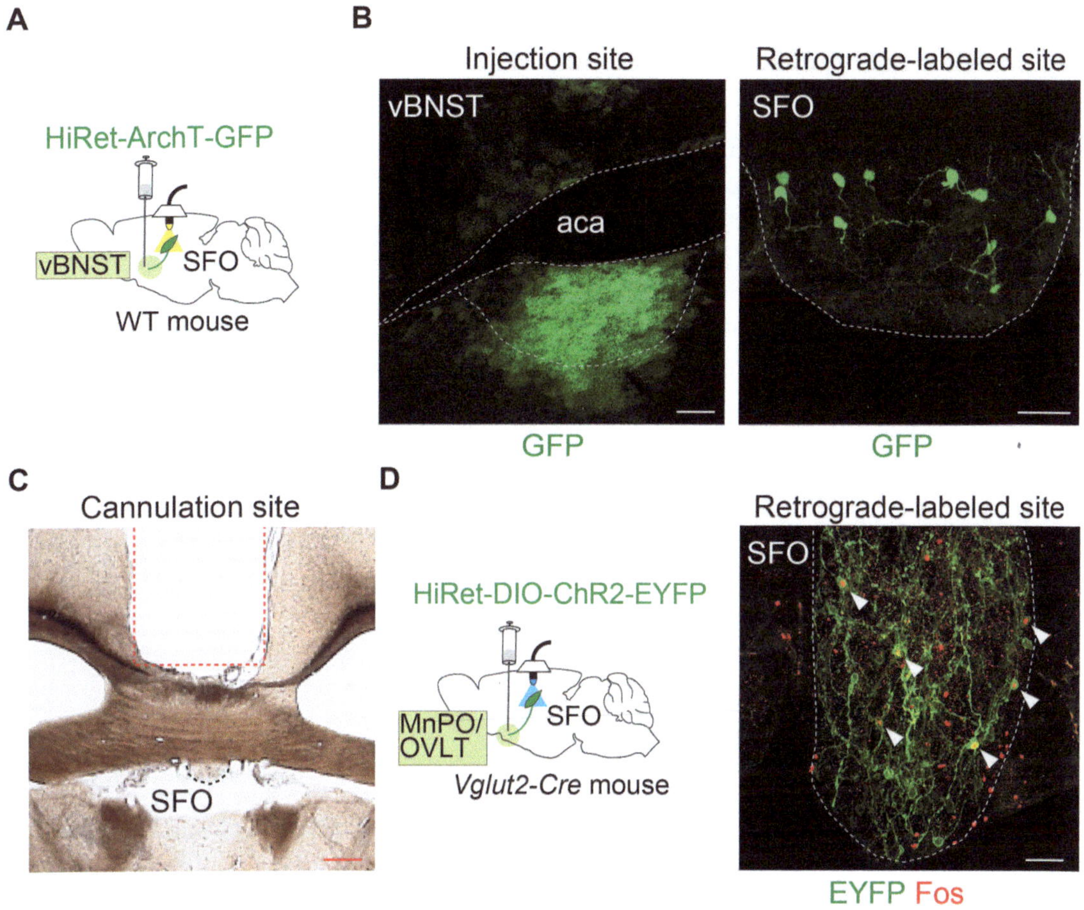

Fig. 7 Fluorescence images of coronal brain sections at injection and retrograde-labeled sites. (**a**) An injection of HiRet-ArchT-green fluorescent protein (GFP) into the ventral part of the bed nucleus of the stria terminalis (vBNST) of a wild type (WT) mouse with irradiation of the subfornical organ (SFO) by yellow-light laser. (**b**) Immunohistochemical detection of GFP in the injection site (left panel, vBNST) and retrograde-labeled site (right panel, SFO). Scale bars, 50 μm. (**c**) Bright-field image of the cannulation site. Red dotted lines show cannulation sites. Red scale bar, 250 μm. (**d**) Immunohistochemical detection of EYFP and Fos after the optical stimulation of the SFO in the *Vglut2-Cre* mouse injected with HiRet-DIO-ChR2(H134R)-EYFP into the MnPO and OVLT. Scale bars, 50 μm

4 Notes

1. To prevent the inactivation of aliquots of lentiviral vectors in a small volume, avoid repeated thawing and freezing. Store thawed aliquots at 4 °C and use on the same day.
2. Check the response of the anesthetized mouse to noxious stimuli (e.g., tail pinch) in order to ensure that the mouse is deeply anesthetized. If not, inject 50 μL/25 gbw more of the

combined anesthetic until the mouse does not respond to noxious stimuli.

3. Fluorinert does not mix with the viral solution, which allows solution boundaries to be observed. To check the dead volume in the syringe, push out approximately 50–100 nL of the viral solution after loading. Note that air bubbles in the glass pipette or tubing will prevent accurate liquid dosing.
4. Taking into account the elasticity of brain tissue, the tip of the glass pipette is advanced another 0.3 mm from the target DV coordinates and then returned to its original position.
5. Select an appropriate guide cannula for the target brain region in order to set the distance from the optical fiber tip to the target tissue at approximately 300 μm. For example, a guide cannula with a guide length of 2.5 mm below the pedestal is better for exposing the SFO to light.
6. Since the screw acts as an anchor between the cannula and the skull, it is better to drill the hole in the area above the temporal lobe, at which the thickness of the skull is greater.
7. If there are large blood vessels on the midline of the brain, avoid piercing them with the guide cannula by using a blunt needle curved to hook the vessels.
8. By connecting the laser and the function generator using a BNC cable, the frequency of the light may be changed as desired. Since the intensity of animal behavior varies with the frequency and intensity of the light, it is necessary to set the appropriate optical conditions for each experimental setting. To achieve the optical activation of the SFO with ChR2, the laser power needs to be set to 5–10 mW/mm^2 measured at the fiber tip with an on/off frequency of 20 Hz.
9. To achieve the optical suppression of the SFO with ArchT, the laser power needs to be set to 7–10 mW/mm^2 measured at the fiber tip with continuous light exposure.

Acknowledgments

I am grateful for the support for research using HiRet vectors from a Cooperative Study Program of the National Institute for Physiological Sciences and JSPS KAKENHI Grant Number JP 22H02732 (AdAMS). This work was also supported by JSPS KAKENHI (grant nos. 14J00037, 17H07331, and 18K14855) and Fusion Oriented Research for Disruptive Science and Technology (FOREST, JST; grant JPMJFR226C).

Conflict of Interest Statement The author has no conflicts of interest to declare that are relevant to the content of this chapter.

Funding

This study was funded by JSPS KAKENHI [Grant Number 14J00037, 17H07331, 18K14855, and 22H02732 (AdAMS)], JST FOREST (Grant Number JPMJFR226C), and JST ACT-X (Grant Number JPMJAX232H). This study was also funded by the Takeda Science Foundation.

References

1. Luo L (2021) Architectures of neuronal circuits. Science 373(6559):eabg7285
2. Lanciego JL, Wouterlood FG (2020) Neuroanatomical tract-tracing techniques that did go viral. Brain Struct Funct 225(4):1193–1224
3. Scanziani M, Hausser M (2009) Electrophysiology in the age of light. Nature 461(7266): 930–939
4. Atasoy D, Sternson SM (2018) Chemogenetic tools for causal cellular and neuronal biology. Physiol Rev 98(1):391–418
5. Boyden ES, Zhang F, Bamberg E, Nagel G, Deisseroth K (2005) Millisecond-timescale, genetically targeted optical control of neural activity. Nat Neurosci 8(9):1263–1268
6. Nagel G, Szellas T, Huhn W, Kateriya S, Adeishvili N, Berthold P et al (2003) Channelrhodopsin-2, a directly light-gated cation-selective membrane channel. Proc Natl Acad Sci USA 100(24):13940–13945
7. Han X, Chow BY, Zhou H, Klapoetke NC, Chuong A, Rajimehr R et al (2011) A high-light sensitivity optical neural silencer: development and application to optogenetic control of non-human primate cortex. Front Syst Neurosci 5:18
8. Adamantidis AR, Zhang F, Aravanis AM, Deisseroth K, de Lecea L (2007) Neural substrates of awakening probed with optogenetic control of hypocretin neurons. Nature 450(7168):420–424
9. Aponte Y, Atasoy D, Sternson SM (2011) AGRP neurons are sufficient to orchestrate feeding behavior rapidly and without training. Nat Neurosci 14(3):351–355
10. Tan CL, Cooke EK, Leib DE, Lin YC, Daly GE, Zimmerman CA et al (2016) Warm-sensitive neurons that control body temperature. Cell 167(1):47–59.e15
11. Wang Y, Hu Z, Ju P, Yin S, Wang F, Pan O et al (2018) Viral vectors as a novel tool for clinical and neuropsychiatric research applications. Gen Psychiatr 31(2):e000015
12. Vodicka MA (2001) Determinants for lentiviral infection of non-dividing cells. Somat Cell Mol Genet 26(1–6):35–49
13. Naldini L, Blomer U, Gallay P, Ory D, Mulligan R, Gage FH et al (1996) In vivo gene delivery and stable transduction of nondividing cells by a lentiviral vector. Science 272(5259):263–267
14. Zufferey R, Dull T, Mandel RJ, Bukovsky A, Quiroz D, Naldini L et al (1998) Self-inactivating lentivirus vector for safe and efficient in vivo gene delivery. J Virol 72(12): 9873–9880
15. Cockrell AS, Kafri T (2007) Gene delivery by lentivirus vectors. Mol Biotechnol 36(3): 184–204
16. Linterman KS, Palmer DN, Kay GW, Barry LA, Mitchell NL, McFarlane RG et al (2011) Lentiviral-mediated gene transfer to the sheep brain: implications for gene therapy in Batten disease. Hum Gene Ther 22(8):1011–1020
17. Azzouz M, Kingsman SM, Mazarakis ND (2004) Lentiviral vectors for treating and modeling human CNS disorders. J Gene Med 6(9): 951–962
18. Kobayashi K, Inoue KI, Tanabe S, Kato S, Takada M, Kobayashi K (2017) Pseudotyped lentiviral vectors for retrograde gene delivery into target brain regions. Front Neuroanat 11: 65
19. Abordo-Adesida E, Follenzi A, Barcia C, Sciascia S, Castro MG, Naldini L et al (2005) Stability of lentiviral vector-mediated transgene expression in the brain in the presence of systemic antivector immune responses. Hum Gene Ther 16(6):741–751
20. Kato S, Kobayashi K, Inoue K, Kuramochi M, Okada T, Yaginuma H et al (2011) A lentiviral strategy for highly efficient retrograde gene transfer by pseudotyping with fusion envelope glycoprotein. Hum Gene Ther 22(2):197–206
21. Kobayashi K, Kato S, Inoue K, Takada M, Kobayashi K (2016) Altering entry site preference of lentiviral vectors into neuronal cells by

pseudotyping with envelope glycoproteins. Methods Mol Biol 1382:175–186

22. Kato S, Fukabori R, Nishizawa K, Okada K, Yoshioka N, Sugawara M et al (2018) Action selection and flexible switching controlled by the intralaminar thalamic neurons. Cell Rep 22(9):2370–2382
23. Ishida A, Kobayashi K, Ueda Y, Shimizu T, Tajiri N, Isa T et al (2019) Dynamic interaction between cortico-brainstem pathways during training-induced recovery in stroke model rats. J Neurosci 39(37):7306–7320
24. Kobayashi K, Kato S, Kobayashi K (2018) Genetic manipulation of specific neural circuits by use of a viral vector system. J Neural Transm (Vienna) 125(1):67–75
25. Matsuda T, Hiyama TY, Niimura F, Matsusaka T, Fukamizu A, Kobayashi K et al (2017) Distinct neural mechanisms for the control of thirst and salt appetite in the subfornical organ. Nat Neurosci 20(2):230–241
26. Matsuda T, Hiyama TY, Kobayashi K, Kobayashi K, Noda M (2020) Distinct CCK-positive SFO neurons are involved in persistent or transient suppression of water intake. Nat Commun 11(1):5692
27. Matsuda T, Kobayashi K, Kobayashi K, Noda M (2024) Two parabrachial Cck neurons involved in the feedback control of thirst or salt appetite. Cell Rep 43(1):113619
28. Nomura K, Hiyama TY, Sakuta H, Matsuda T, Lin CH, Kobayashi K et al (2019) [Na^+] increases in body fluids sensed by central Na_x induce sympathetically mediated blood pressure elevations via H^+-dependent activation of ASIC1a. Neuron 101(1):60–75 e66
29. Vong L, Ye C, Yang Z, Choi B, Chua S Jr, Lowell BB (2011) Leptin action on GABAergic neurons prevents obesity and reduces inhibitory tone to POMC neurons. Neuron 71(1): 142–154

Chapter 17

Selective Chemogenetic Blockade of Cerebello-Rubral Pathway by Double-Virus Infection Method Using Lentivirus and Adeno-Associated Virus in Rats

Hideki Hida and Shinya Ueno

Abstract

Double-virus infection method of lentivirus and adeno-associated virus was used to reveal the contribution of cerebellum in rehabilitative training after intracerebral hemorrhage (ICH) in rats. Chemogenetic blockade of cerebello-rubral pathway was achieved by the injection of a lentiviral vector pseudotyped with fusion glycoprotein type E and carrying Cre recombinase into the red nucleus and an adeno-associated viral vector encoding hM4D(Gi)-mCherry in the double-floxed, inverted orientation into the cerebellar lateral nucleus. This silencing system induced by clozapine-N-oxide significantly reduced the success rate of forelimb reaching test that is improved by forced-limb use after ICH. As a good application example of loss-of-function technique, the protocol of our experiment by double-virus infection method will be explained in detail.

Key words Selective pathway silencing, Double-virus infection method, Motor control system, Cerebello-rubral pathway, Rat, Intracerebral hemorrhage, Rehabilitation

1 Introduction

Lentiviral vectors derived from human immunodeficiency virus (HIV-1) have become a major tool for gene delivery in mammalian cells. Lentiviral vectors are a type of retroviral vector that has the following advantages. One is that the vectors are capable of gene delivery to almost all mammalian cells, including nondividing cells such as hematopoietic stem cells and neurons. The other is that target genes introduced with lentiviral vectors are integrated into the host cell genome, enabling stable gene expression over a long period of time. These are of great advantage, especially in the case of manipulating nonproliferation neurons.

This has made it possible to express inhibitory/excitatory designer receptors that respond to specific compounds in neurons of animal species for which it is difficult to create transgenic

Kazuto Kobayashi et al. (eds.), *Lentiviral Vectors: Applications for Basic and Clinical Neuroscience*, Methods in Molecular Biology, vol. 2974, https://doi.org/10.1007/978-1-0716-4807-0_17,

animals, and to manipulate specific neurons (Designer Receptors Exclusively Activated by Designer Drugs: DREADD) [1–3]. However, simply injecting lentiviral vectors into neurons around the injection site will infect the surrounding neurons to express the target gene, making it difficult to manipulate only the target neurons. To solve this difficult problem, we used a double viral vector infection technique [4–7] in which both retrogradely NeuRet-type lentiviral vector pseudotyped with fusion glycoprotein type E (FuG-E) and adeno-associated virus vector (AAV) that can also transduce into either proliferating or nonproliferating cells was used to manipulate the target neural pathway system. Thus, we have already reported that the recovery of the motor executive system by forced-limb use (FLU: a rehabilitative training like constraint-induced movement therapy) after intracerebral hemorrhage is accompanied by switching from the cortico-spinal tract to the cortico-rubral tract [5].

As our movement system works well because of good coordination between the executive and regulatory systems, it is probable that the switch in motor regulatory system as the switching from the cortico-spinal tract to the cortico-rubral tract is also induced by the rehabilitative training after ICH. In our preliminary data, we can successfully prove the contribution of cerebello-rubral tract in the recovery of forelimb function by FLU after ICH using DREADD system as follows (Fig. 1). That is, a NeuRet vector with FuG-E encoding Cre gene downstream of murine stem cell virus promoter (FuG-E-MSCV-Cre) and an AAV-DJ vector encoding hM4D(Gi)-mCherry gene in the double-floxed, inverted orientation downstream of elongation factor 1a promoter (AAV-DJ EF1a-DIO-hM4D(Gi)-mCherry) were injected into the parvocellular red nucleus and the cerebellar lateral nucleus, respectively. The blockade of cerebello-rubral tract with clozapine-N-oxide (CNO) significantly resulted in the reduction of the success rate.

In this chapter, we show the DREDD system of double-virus vector infection method as one of the good application techniques of loss-of-function experiment.

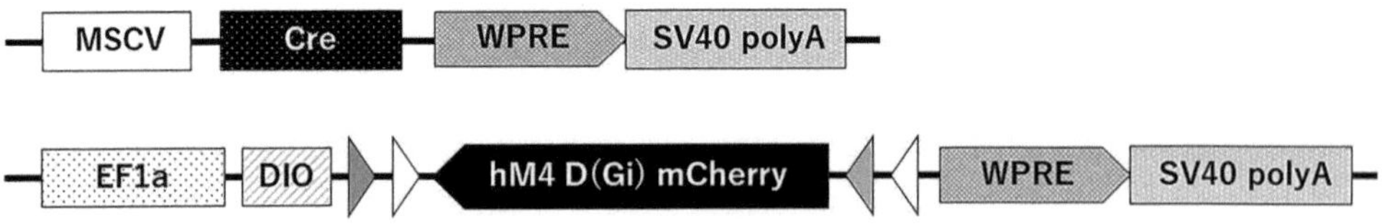

Fig. 1 Transgene constructs in the two viral vector systems. The NeuRet vector pseudotyped with FuG-E encodes Cre gene downstream of the murine stem cell virus (MSCV) promoter (top). AAV-DJ vector encodes hM4D(Gi)-mCherry gene in the double-floxed, inverted orientation (DIO) downstream of the human elongation factor 1a (EF-1a) promoter (bottom). Both constructs have the woodchuck hepatitis virus posttranscriptional regulatory element (WPRE) and SV40 early gene polyadenylation signals (SV40 polyA)

2 Materials

2.1 Animals

1. Male Slc:Wistar/ST rats (250–350 g) are obtained and group-housed under controlled temperature conditions (22–23 °C) and a 12 h light-dark cycle, with chaw and water ad libitum.
2. Experimental procedures must be approved by the Animal Care and Use Committee of each research institution.

2.2 Anesthesia

1. The mixture containing medetomidine hydrochloride (0.1875 mg/mL), midazolam hydrochloride (1 mg/mL), and butorphanol tartrate (1.25 mg/mL) is given for anesthesia.
2. Atropine sulfate (0.1 mg/mL) is given to prevent sputum production and dexamethasone (0.1 mg/mL) is given to reduce brain edema.
3. Atipamezole chloride (0.376 mg/mL) for facilitating awakening from anesthesia is used.
4. For the casting in FLU, isoflurane is used (2% for induction, 1% for maintenance).

2.3 Intracerebral Hemorrhage

1. Type IV collagenase is dissolved in saline at a concentration of 15 units/mL.
2. To get minimum brain damage, a pipette of 60 ~ 70 μm diameter is prepared from a glass micropipette (OD: 1.65 mm, ID 1.20 mm, 100 mm long) using Narishige puller (PP-83), which is connected to a Hamilton syringe of 50 μL by fluorinated ethylene propylene (FEP) tube (ID: 0.25 mm) of 50 cm long and Viton® tubing connector.

2.4 Double-Virus Infection Method

1. For injection of viral vectors, a Nanoliter2020 system (World Precision Instrument) is prepared, with its plunger connected to a WPI No. 1B150-4 (OD: 1.5 mm) glass pipette, suitable for deep nuclei such as the red and cerebellar nuclei.
2. Lentivirus (FuG-E-MSCV-Cre) with a titer of $1.2 \times 10^{10} \sim 2.1 \times 10^{11}$ copies/mL is prepared by Dr. Kenta Kobayashi (the details are shown in Chapter 2), and (AAV-DJ-EF1a-DIO-hM4D(Gi)-mCherry) is used with a titer of 1.1×10^{10} vg/mL [5, 6].
3. To activate DREADD system, 3 mg/kg clozapine-N-oxide (CNO) dissolved in saline is intraperitoneally injected in the rats.

2.5 The Single Pellet Reaching Task Test

1. This test uses a clear acrylic box (13 cm wide × 45 cm long × 40 cm high) with a narrow 1 cm wide longitudinal slit in the front panel.

2. The animal is given 20 trials in which it extended its forelimbs through the slit and attempted to grasp a pellet (Dustless Precision Pellets®, 45 mg, Bio-Serv) on a platform (3 cm wide, 4 cm from the bottom) attached to the outside of the anterior wall.
3. The platform has two depressions in which pellets can be placed, 1.5 cm from the center of the slit.
4. Scores are expressed as the percentage of success in which the rat was able to successfully grasp the pellet and bring it to its mouth.

3 Methods

3.1 Intracerebral Hemorrhage Near the Internal Capsule

1. Rats are anesthetized with an intraperitoneal injection of 2 mL/kg anesthesia mixture containing medetomidine hydrochloride, midazolam hydrochloride, and butorphanol tartrate, followed by injections of 1 mL/kg atropine and 1 mL/kg dexamethasone.
2. Under deep anesthesia waiting for several minutes, rats are set in a stereotaxic apparatus after clipped scalp hair. Thus, stereotaxically fixed brain skull is exposed by midline incision, and the bleeding should be stopped by clean gauze.
3. To create a small hole on the skull, a small mark is made at the inject point (3.8 mm lateral to the midline and 1.8 mm caudal to bregma contralateral to the dominant forelimb) (*see* **Notes 1** and **2**).
4. Type IV collagenase (1.4 μL) is injected into 5 mm below from the brain surface at a rate of 0.2 μL/min over 7 min using a glass micropipette (OD: 60 ~ 70 μm) that is connected to a Hamilton syringe mediated by FEP tube and Viton® tubing connector (*see* **Note 3**).
5. The pipette is left in place for 5 min, and then the pipette is 1.0 mm back keeping further 2 min. Finally, the pipette is back slowly to prevent backflow for more than 1 min.
6. The scalp is sutured, and rats are injected with 1 mL/kg atipamezole chloride, then rats are allowed to recover in their home cage.
7. Twenty-four hours later, the extent of motor dysfunction is assessed with the motor deficit score (MDS) as previously reported [5, 6, 8]. Only animals with MDS more than 8 were recruited and randomized into appropriate groups.

3.2 Forced-Limb Use as a Rehabilitative Training

1. The unimpaired forelimb and upper torso of the rats are restrained by soft felt and plaster cast under isoflurane anesthesia (2% for induction, 1% for maintenance).
2. After the casting, the rats are returned to their home cages and forced limb use (FLU) of the impaired forelimb is given for 7 days from 1 day after ICH as described previously [5, 6, 9].
3. During the 7 days with the cast, water gel and chaws are given in the cage, which facilitated access to food and water.

3.3 Double-Viral Vector Infection Technique

1. Before the start, the cerebello-rubral pathway should be confirmed using neural tracers (*see* **Note 4**).
2. Rats are anesthetized with 2 mL/kg anesthesia mixture as same as in Subheading 3.1.
3. Under deep anesthesia, rats are set in a stereotaxic apparatus as same as in Subheading 3.1.
4. A small hole is made in the skull with a drill at a location corresponding to the ipsilateral red nucleus (AP: −5.2 mm, ML: 1.6 mm) of the dominant forelimb (*see* **Note 1**).
5. Using nanoliter injector, 0.6 μL of FuG-E-MSCV-Cre is injected into the ipsilateral parvocellular part of red nucleus (AP: −5.2 mm, ML: 1.6 mm, DV: −7.5 mm from surface) at a rate of 0.2 μL/min for 3 min with an inward tilt of 5° of the glass pipette (diameter: 50 ~ 60 μm) (*see* **Note 5**).
6. Seven to ten days after lentivirus injection, the skull above the contralateral cerebellar lateral nucleus (AP: −11.0 mm, ML: 3.5 mm and AP: −11.4 mm, ML: 3.6 mm) is exposed by careful craniotomy under anesthesia.
7. Using nanoliter injector, 0.6 μL of AAV-DJ-EF1a-DIO-hM4D (Gi)-mCherry is injected into the contralateral cerebellar lateral nucleus (AP: −11.0 mm, ML: 3.5 mm, DV: −4.0 mm from the surface; AP: −11.4 mm, ML: 3.6 mm, DV: −3.9 mm from the surface) at a rate of 0.2 μL/min for 3 min (*see* **Note 6**).
8. The scalp is sutured after the injection, and rats are intraperitoneally injected with 1.0 mg/kg atipamezole hydrochloride.
9. The rats are allowed to recover in their home cage (*see* **Note 7**).
10. Administration of 3 mg/kg CNO is given to activate DREADD system, silencing the double-infected cerebello-rubral pathway (*see* **Note 8**).

3.4 Behavioral Evaluation by the Single Pellet Reaching Task Test

1. Each rat is placed into a clear acrylic box with a narrow longitudinal slit as the pretraining of the single pellet reaching task test.
2. Dietary restriction is administered during the training period to maintain body weight.

3. The rat is allowed to retrieve the pellets using their dominant limb [5, 6].
4. Each training session consists of 20 trials and is conducted once per day for 2 consecutive weeks.
5. At 2–3 days after double-virus injections (*see* Subheading 3.3), further training session is given to the animals for 1 week.
6. Rats that fail to retrieve more than nine pellets per session in the last two training sessions are excluded from the study.
7. Test performance data from 1–2 days before ICH is used as baseline.
8. A test session consisted of 20 trials and is performed once per day.
9. Dietary restriction is also administered during the test period to maintain body weight.
10. In CNO administration, the test evaluation is performed from 20 min later (*see* **Note 8**).

4 Notes

1. The dominant limb in each rat is determined during training sessions of the single pellet reaching test.
2. In the rat brain map by Paxinos and Watson, the length between bregma and interaural line is 9.1 ± 0.3 mm, but the actual length of each rat is different from the map, so we adjust the AP point using the following function developed from our experience. $Y = (20X - 54)/70$, Y: point on the rat's AP line, X: length between the rat's bregma and interaural line.
3. To minimize the brain damage by the injection, it is strongly recommended to use a glass pipette that has a thinner wall as the use of syringe needle causes in unnecessary tissue damage. This point is very important to make a small hemorrhage near the internal capsule.
4. An anterograde tracer biotin-dextran amine and a retrograde tracer Fluorogold is used in our experiment.
5. To avoid the damage of the blood vessel, nanoliter injector is used with an inward tilt of 5° of the glass pipette.
6. To confirm successful dual viral vector infection, the expression of mCherry in the lateral cerebellar nuclei should be shown.
7. The activation of DREADD system requires 2–3 weeks.
8. Significant effect of 3 mg/kg CNO starts at 20 min after intraperitoneal administration and continued for several hours.

Conflict of Interest Statement

The authors have no conflicts of interest to declare that are relevant to the content of this chapter.

Funding

This study was funded by KAKENHI (Grant Numbers 20H05476 and 22H03456).

References

1. Armbruster BN, Li X, Pausch MH, Herlitze S, Roth BL (2007) Evolving the lock to fit the key to create a family of G protein-coupled receptors potently activated by an inert ligand. Proc Natl Acad Sci USA 104(12):5163–5168. https://doi.org/10.1073/pnas.0700293104
2. Roth BL (2016) DREADDs for neuroscientists. Neuron 89(4):683–694. https://doi.org/10.1016/j.neuron.2016.01.040
3. Nakajima K, Cui X, Li C, Meister J, Cui Y, Fu O, Smith AS, Jain S, Lowell BB, Krashes MJ, Wess J (2016) Gs-coupled GPCR signalling in AgRP neurons triggers sustained increase in food intake. Nat Commun 7:10268. https://doi.org/10.1038/ncomms10268
4. Koshimizu Y, Isa K, Kobayashi K, Isa T (2021) Double viral vector technique for selective manipulation of neural pathways with higher level of efficiency and safety. Gene Ther 28: 339–350. https://doi.org/10.1038/s41434-020-00212-y
5. Ishida A, Isa K, Umeda T, Kobayashi K, Kobayashi K, Hida H, Isa T (2016) Causal link between the cortico-rubral pathway and functional recovery through forced impaired limb use in rats with stroke. J Neurosci 36(2): 455–467. https://doi.org/10.1523/JNEUROSCI.2399-15.2016
6. Ishida A, Kobayashi K, Ueda Y, Shimizu T, Tajiri N, Isa T, Hida H (2019) Dynamic interaction between cortico-brainstem pathways during training-induced recovery in stroke model rats. J Neurosci 39(37):7306–7320. https://doi.org/10.1523/JNEUROSCI.0649-19.2019
7. Kinoshita M, Matsui R, Kato S, Hasegawa T, Kasahara H, Isa K, Watanabe A, Yamamori T, Nishimura Y, Alstermark B, Watanabe D, Kobayashi K, Isa T (2012) Genetic dissection of the circuit for hand dexterity in primates. Nature 487(7406):235–238. https://doi.org/10.1038/nature11206
8. Masuda T, Hida H, Kanda Y, Aihara N, Ohta K, Yamada K, Nishino H (2007) Oral administration of metal chelator ameliorates motor dysfunction after a small hemorrhage near the internal capsule in rat. J Neurosci Res 85(1): 213–222. https://doi.org/10.1002/jnr.21089
9. Ishida A, Misumi S, Ueda Y, Shimizu T, Cha-Gyun J, Tamakoshi K, Ishida K, Hida H (2015) Early constraint-induced movement therapy promotes functional recovery and neuronal plasticity in a subcortical hemorrhage model rat. Behav Brain Res 284:158–166. https://doi.org/10.1016/j.bbr.2015.02.022

Chapter 18

Development of a Gene Therapy with Lentiviral Vectors for Spinocerebellar Ataxias

Tiago Moreira-Gomes and Clévio Nóbrega

Abstract

Among viral vectors, lentiviral vectors allow an efficient delivery of nucleic acid to the target cells due to its ability of gene integration into host genome and broad tropism. In this sense, this tool demonstrated to be very promising in genetic disorders, such as polyglutamine spinocerebellar ataxias. Therefore, we describe a protocol that explains in detail the production, purification, and injection of lentiviral vectors in a spinocerebellar ataxia mouse model.

Key words Lentiviral vectors, HEK 293 T cells, Production of lentiviral vectors, Gene therapy, Spinocerebellar ataxia

1 Introduction

Since 1989, the approved or initiated clinical trials of gene therapy have been increasing worldwide [1]. According to the FDA, gene therapy involves "products that mediate their effects by transcription or translation of transferred genetic material or by specifically altering host (human) genetic sequences" [2]. Gene therapy is generally divided into two groups based on the use of viral or nonviral systems. Nonviral systems take advantage of the safety profile of chemical or physical approaches [3]. However, the main disadvantage of these systems is their low efficiency, which can be overcome by viral systems [4]. In the database "Gene Therapy Clinical Trials Worldwide," viral vectors occupy the leading positions. For gene therapy applications, the most used viral vectors are adenovirus (14.7%), retrovirus (13.8%), and lentivirus (9.6%) [1]. In this chapter, we are going to focus on lentivirus.

Lentiviruses of the *Retroviridae* family most often originated from human immunodeficiency virus (HIV), in which the genome is composed of a 9 kb single-stranded positive RNA. HIV is able to

Kazuto Kobayashi et al. (eds.), *Lentiviral Vectors: Applications for Basic and Clinical Neuroscience*, Methods in Molecular Biology, vol. 2974, https://doi.org/10.1007/978-1-0716-4807-0_18,

infect specific cell populations, including CD4$^+$ T lymphocytes and monocytes/macrophages [5, 6]. Therefore, lentiviral vectors are usually pseudotyped with vesicular somatitis virus G (VSV-G) to expand its natural tropism across multiple cell types [7].

Lentiviral vectors (LVs) are a powerful tool that integrates its RNA into host genome, allowing a long-term expression and a stable transfer of genetic material in several tissues. Other interesting advantages of this viral vector are broad tropism and low immunogenicity. Furthermore, the cloning packaging capacity of LV is almost 8 kb, which is superior in comparison to another viral vector, such as adeno-associated virus (AAV). However, the lentiviral ability to introduce its own RNA into host genome has a possible disadvantage, as insertional mutagenesis might occur [4, 7].

To enhance the biosafety of lentivirus in gene therapy, lentiviral plasmid designs have been modified over time. In the first-generation lentiviral vector, the HIV genome was separated into several components reducing the probability of generating replication-competent lentivirus (RCLs). In the second generation, the main modification was the deletion of accessory genes without changing the number of plasmids decreasing further the probability of RCLs' formation [4]. The third-generation lentiviral vector arises with the elimination of the *tat* gene and the addition of the *rev* gene originating a four plasmids-system: *rev* plasmid, *gag* and *pol* plasmid, *env* plasmid, and the transgene plasmid. *Gag*, *pol*, and *env* plasmids encode crucial proteins for viral integration, transcription, and encapsulation, while *rev* plasmid encodes an RNA-binding protein (RBP) that increases viral gene expression [8].

Gene therapy has demonstrated the potential to induce long-lasting effect in rare disorders [9]. Among them, polyglutamine spinocerebellar ataxias (PolyQ SCAs) are a group of neurodegenerative dominant disorders characterized by progressive ataxia, leading to gradual loss of coordination and voluntary movements control [10]. The main mechanism of this family of disorders is based on the expansion of *CAG* trinucleotide repeats originating a mutant polyglutamine-expanded protein [11]. Despite the extensive research in gene therapies for PolyQ SCAs, their current treatments are still limited to alleviation of symptoms [12]. In preclinical setting, lentiviral vectors have been employed to specifically target therapeutic molecules, such as RBPs, in SCA mouse models, or to induce spinocerebellar ataxia in *wild-type* mice creating models of disease [13]. With the present protocol, we outline the production of lentiviral vectors and their use to transduce neurons in vivo.

2 Materials

2.1 Cell Culture

1. HEK 293 T cells (ATCC, #CRL-3216) (*see* **Note 1**).
2. Dulbecco's Modified Eagle Medium (DMEM).
3. Complete medium: DMEM supplemented with 10% v/v fetal bovine serum (FBS, heat inactivated), 2 mM L-glutamine, and 1% v/v penicillin-streptomycin (Pen-Strep).
4. Trypsin-EDTA solution (0.25%).
5. Phosphate-buffered saline (PBS) solution: 1x concentrated solution obtained from dilution of 10x concentrated solution (PBS powder).

2.2 Transfection of HEK 293T Cells

1. Biosafety hood level 2 (BL-2 compliant).
2. 90 mm cell culture dishes.
3. Mixture of transfection for one 90 mm Petri dish:
   (a) 549.6 ng of pRSV-Rev (Addgene, Plasmid #12253)
   (b) 2381.4 ng of pCMV delta R8.2 (Addgene, Plasmid #12263)
   (c) 687 ng of pMD2.G (Addgene, Plasmid #12259)
   (d) 2381.4 ng of transgene plasmid
   (e) 40 μL of polyethyleneimine (1 mg/mL, PEI Max)
   (f) 50–100 μL of PBS (*see* **Note 2**).

2.3 Collection of Viral Particle-Contained Medium

1. Vacuum filtration system with a 0.22 μm pore size (or Polyethersulfone (PES) membrane syringe filter 0.22 μm).

2.4 Purification of Lentiviral Particles

1. PBS with 1% Albumin Bovine Fraction V (BSA).
2. JS-24.38 Swinging-bucket aluminum rotor and buckets (Beckman Coulter).
3. Beckman Coulter's Avanti JXN-30 high speed centrifuge.
4. Open-top thin wall polypropylene conical tube 31.5 mL (Beckman Coulter).

2.5 Stereotaxic Surgery

1. Anesthesia mixture: Medetomidine (0.5 mg/kg), Midazolam (1.0 mg/kg), and Fentanyl citrate (0.5 mg/kg).
2. Reverser mix: Atipamezole (2.5 mg/kg), Flumazenil (0.5 mg/kg), and Naloxone (1.2 mg/kg) (*see* **Note 3**).
3. Surgical tools (scalpel, surgical scissors, forceps, retractors).
4. Digital mouse stereotaxic frame (Stoelting, Model 51730D).
5. Stereotaxic injector (Stoelting, Model 53,312).

6. Hamilton syringe (volume 10 μL).
7. Hamilton Small Hub Removable Needle (26 G).
8. Skin marker.
9. Nonabsorbable silk suture, size 5/0.
10. Drill.
11. Eye ointment.

3 Methods

The human embryonic kidney cell line (HEK 293 T) is the gold-standard for the lentiviral production due to the presence of the SV40 T-antigen in the producer cells allowing a greater efficiency in comparison to the antigen non-expressing cells [14].

According to NIH Guidelines, the production of lentiviral vectors should be performed at the biosafety level 2 room (BL-2) (*see* **Note 4**).

3.1 Cell Culture

1. Culture HEK 293 T cells in T175-flask with DMEM supplemented with 10% FBS, 2 mM L-glutamine, and 1% Pen-strep.
2. Maintain these cells in a humidified incubator at 37 °C and 5% CO_2 atmosphere.

3.2 Transfection of HEK 293 T Cells

1. Upon reaching 80% confluency, trypsinize HEK 293 T cells with Trypsin-EDTA solution.
2. Plate the cells at 3.8×10^6 density into 90 mm dish in 10 mL of fresh complete medium (*see* **Note 5**).
3. Approximately 24 h after plating, the cell confluency should be around 60–70% (Fig. 1). The transfection is performed using the mix including pRSV-Rev, pCMV delta R8.2, pMD2.G, transgene plasmid, PEI and PBS.
4. Vortex and spin down the mixture.
5. For the formation of the complex between PEI and plasmids DNA, the mix should be steady at room temperature for 10 - 15 min.
6. Add the transfection mixture to the dishes and swirl them gently. Incubate the cells in a humidified incubator at 37 °C for 6 h.
7. After 6 h, replace the medium with fresh one.

3.3 Collection of Viral Particle-Contained Medium

1. During 48 h of incubation, the efficiency of transfection (e.g., through the visualization of GFP fluorescence—Fig. 1) and alteration of pH should be monitored (*see* **Note 6**).

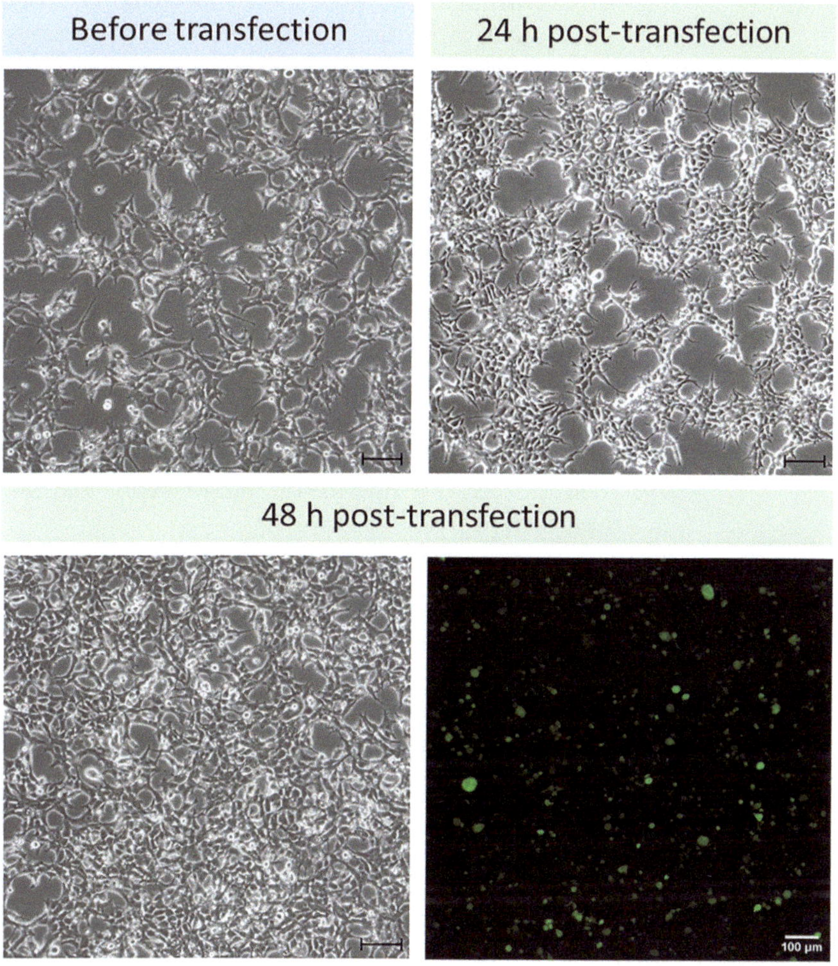

Fig. 1 Photomicrographs of different degrees of cell confluency in three different moments: before transfection, 24- and 48-h post-transfection. Last photo: Fluorescence image of HEK 293 T 24 h post-transfection. Cell type: HEK 293 T (Passage 6). Bar = 100 μm

2. Forty-eight hours after transfection, collect the supernatant from the culture plates containing viral particles and filter it through a vacuum filtration system or syringe with a filter (0.22 μm filter) (*see* **Note 7**).

3.4 Purification of Lentiviral Particles

Before the purification protocol, clean the rotor supports and lids with 70% ethanol, and autoclave all the conical ultracentrifuge tubes. The supernatants must always be handled on ice inside of the hood in BL-2.

1. Transfer the filtered supernatant into an autoclaved conical ultracentrifuge (*see* **Note 8**).
2. Centrifuge the supernatants for 1 h 35 min using the JS-24.38 rotor at 65.000 x *g* and 4 °C.

3. After centrifugation, discard the supernatants, and dry the top of the tubes with paper to remove the supernatant residues.
4. Resuspend the pellets in 400 μl of cold and sterile PBS with 1% BSA.
5. Cover the tubes with parafilm and incubate on ice or at 4 °C for approximately 1 h 30 min.
6. Gently resuspend the pellets and pool the viruses in a single conical ultracentrifuge tube.
7. Fill the tube containing the virus to at least two-thirds of its capacity with cold PBS with 1% BSA to prevent the tube from collapsing.
8. Centrifuge the virus using the JS-24.38 rotor for 1 h 35 min at 65.000 x *g* and 4 °C.
9. Next, discard the supernatant and dry very well the top of the tube with paper.
10. Carefully resuspend the pellet in 45 μl of cold and sterile PBS with 1% BSA without making bubbles.
11. Place parafilm in the tube containing the virus and leave it at 4 °C overnight.
12. On the next day, aliquot the lentiviral suspension in 10 μl per low protein binding tube (0.5 mL) (*see* **Note 9**).
13. For long-term storage, freeze the lentiviral suspension at −80 °C.

3.5 Quantification of Viral Titer

Quantification of viral titer is of outmost importance to determine the transduction efficiency [4]. There are several quantification methods of lentiviral vector, including the quantification of viral genome using RT-qPCR (*see* **Note 10**). With the protocol of lentiviral production, the range of lentiviral particles can go from 1×10^9 to 5×10^9 IU/mL.

The following steps are according to the manufacturer's instructions of qPCR Lentivirus Titer Kit (abm, ct. no LV900).

1. For high concentrated titer viral samples, dilute the virus to 10^6 IU/mL range with 1X PBS or DMEM. For low viral titer samples, collect viral supernatant for direct qPCR step up.
2. Take 2 μl of the sample from step 1 and add it to 18 μl of Virus Lysis Buffer (provided). Incubate at room temperature for 5 min followed by a snap spin (*see* **Note 11**).
3. Perform four tenfold serial dilutions of the standard control DNA for generating the standard curve. In the first dilution, add 5 μl of DNA into 45 μl Nuclease-free H_2O. In the following dilutions, take 5 μl from the previous dilution and add it to 45 μl of nuclease-free H_2O.

Dilution summary:

- 1st dilution: 1/10
- 2nd dilution: 1/100
- 3rd dilution: 1/1,000
- 4th dilution: 1/10,000

4. Make the set up for the qPCR (the volume of the components of the mixture and the cycling conditions are in the kit's protocol).

3.6 Stereotaxic Surgery: Lentiviral Vectors Injection in a Transgenic SCA3 Mice

Depending on the quantification's method and the target organ, the number of lentiviral particles to inject can vary. Previously, our laboratory injected lentivirus particles encoding for G3BP1 or GFP into cerebella of transgenic SCA3 mouse model. The lentiviral concentration determined with ELISA was 800 ng of p24 per μl [15].

1. Before surgery, anesthetize intraperitoneally the animal and confirm if there is an absence of muscle reflex.
2. Shave the scalp of the mouse.
3. Disinfect the shaved area using a suitable disinfectant.
4. Apply eye ointment to the mouse's eyes to prevent corneal drying during surgery.
5. In stereotaxic frame, place gently the front teeth in the incisor bar and the ears in the ear bar (Fig. 2a).

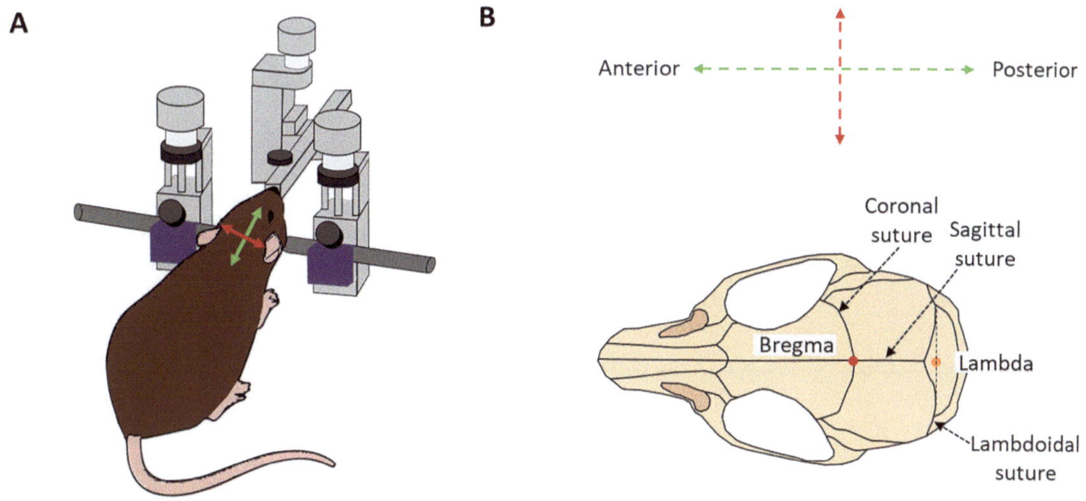

Fig. 2 Anatomical representation of mouse skull and a support of mouse stereotaxic apparatus. (**a**) Representation of the mouse orientation in the stereotaxic apparatus support. (**b**) Schematic representation of anteroposterior axis and dorsal view of the mouse skull, indicating different sutures (coronal, sagittal, and lambdoidal) and anatomical references (bregma and lambda)

6. In the shaved area, make a small incision in the scalp using a sterile scalp.
7. Gently separate the cut tissue and clean the skull's surface.
8. Place the Hamilton syringe with the needle and lentiviral suspension in the syringe holder.
9. Use the **bregma** (the intersection of coronal and sagittal sutures) or **lambda** (the point where the best fit lines passing through the sagittal suture and the left and right segments of the lambdoid suture intersect) as the anatomical reference point, depending on the target region, e.g., for striatum use bregma and for cerebellum use lambda (Fig. 2b). Using the micromanipulator, lower the needle until it touches to the chosen reference point. Record the mediolateral and anteroposterior coordinates and reset the coordinates to zero considering the reference coordinates.
10. Move the needle to the desired target position and mark this location on the skull.
11. Perforate gently a small hole at the marked position on the skull using a drill, exposing dura mater (the membrane covering the brain) (*see* **Note 12**).
12. With a fine needle, cut gently the dura mater exposing the brain surface.
13. Localize the needle to the specific coordinates. For the lentiviral injection into the striatum, the brain coordinates relative to bregma are anteroposterior (+0.6 mm); lateral (±1.8 mm); ventral (−3.3 mm). For the lentiviral injection into the cerebellum, the brain coordinates relative to lambda are rostral (−1.6 mm); midline (0.0 mm); ventral (−1.0 mm). Once the probe reaches the target location, proceed the injection of viral solution at 0.25 μL/min.
14. After the injection, wait 5 min to avoid the injected liquid from being expelled.
15. Five minutes later, perform a subcutaneous administration of reverser mix to reverse the anesthesia's effects and suture the animal's head.

4 Notes

1. HEK 293FT can be used as an alternative. However, the protocol might be adjusted (e.g., cell density), since these cells grow faster than HEK293 T, which can lead to higher decrease of pH, affecting the lentiviral tropism [16].

2. For the mixture of the transfection, DMEM without FBS can also be used, as serum can interfere with the transfection efficiency.
3. The drug concentrations for anesthesia and reverser mix are adjusted to one animal.
4. The lentivirus produced with this protocol present a certain safety level due to the fact of being self-inactivating vector. However, residual transcription activity was verified in SIN vector [17] justifying the need of BL-2 containment.
5. The total Petri dishes should be around 20 to get enough viral titer. The cells should be 70% confluent and distributed uniformly. A suitable confluency is important due to the alterations of pH (over-confluency can lead to a faster decrease of pH).
6. The efficiency of transfection can be determined though the visualization of fluorescence signal or cell selection with an antibiotic, if the plasmid present gene reporter (e.g., GFP) or antibiotic-resistant gene (e.g., puromycin).
7. After the collection of the medium, observe the state of the cells. As they appear healthy and strongly adhere to the flask surface, a second harvest can be performed by adding new medium for additional 24 h [18]. In this sense, the amount of lentiviral titer will be higher in comparison to 48 h regime.
8. During the protocol of purification, ensure the tubes are always balanced.
9. Rahman et al. demonstrated that lentiviral particles underwent a reduction of almost 50% in a period of 3 days when stored at 4 °C. Therefore, lentiviral particles can be stored at 4 °C overnight and − 80 °C for a long-term use [19]. Therefore, we recommend avoiding freeze–thaw cycles and aliquoting the lentiviral vectors for single use.
10. Alternatively, the quantification of p24 antigen, a component of lentiviral vector, can be performed. This approach not only quantifies the protein bound to lentiviral particles, but also free p24 protein and p24 protein associated to nonfunctional vector particles [20]. Typically, the range of p24 quantity varies between 200 and 1000 ng per μL. Both ELISA and qPCR do not evaluate the viral functionality. Thus, there is a more precise technique involving the fluorescence-activated cell sorting (FACS) to analyze the functionality of the lentiviral vectors. In a broad way, serial dilutions of virus expressing green fluorescence protein (GFP) are performed followed by cell transduction. After that, cells are collected and analyzed by flow cytometry. Based on flow cytometry results, it is possible to determine minimal number of vector particles necessary to

infect a cell in a determined volume [21]. This approach will only be feasible if there is GFP tag in the transgene plasmid. It is important to consider that there is a variety of methods that were not mentioned here, but all of them present advantages and disadvantages. Therefore, it is recommended to analyze the titration method suitable for a specific goal.

11. The viral sample has been diluted 1:10, so this dilution factor must be considered when calculating the final titer.
12. Be cautious not to perforate too deeply to avoid damaging the brain.

Acknowledgments

CN's laboratory is funded by the Cure CSB project, by the Viljem Julijan Association for Children with Rare Diseases (Slovenia), and by the Algarve Biomedical Center Research Institute (ABC-Ri). TMG is funded by the Portuguese Science Foundation with the PhD fellowship 2022.11008.BD.

Funding None.

Conflict of Interest Statement The authors have no conflicts of interest to declare that are relevant to the content of this chapter.

References

1. Ginn SL, Mandwie M, Alexander IE, Edelstein M, Abedi MR (2024) Gene therapy clinical trials worldwide to 2023—an update. J Gene Med 26:e3721. https://doi.org/10.1002/jgm.3721
2. FDA (2020) Long term follow-up after Administration of Human Gene Therapy Products; Guidance for Industry
3. Wang C, Pan C, Yong H, Wang F, Bo T, Zhao Y, Ma B, He W, Li M (2023) Emerging non-viral vectors for gene delivery. J Nanobiotechnol 21:272. https://doi.org/10.1186/s12951-023-02044-5
4. Nóbrega C, Mendonça L, Matos CA (2020) A handbook of gene and Cell therapy. Springer, Cham
5. Deeks SG, Overbaugh J, Phillips A, Buchbinder S (2015) HIV infection. Nat Rev Dis Primer 1:1–22. https://doi.org/10.1038/nrdp.2015.35
6. Watts JM, Dang KK, Gorelick RJ, Leonard CW, Bess JW Jr, Swanstrom R, Burch CL, Weeks KM (2009) Architecture and secondary structure of an entire HIV-1 RNA genome. Nature 460:711–716. https://doi.org/10.1038/nature08237
7. Thomas CE, Ehrhardt A, Kay MA (2003) Progress and problems with the use of viral vectors for gene therapy. Nat Rev Genet 4:346–358. https://doi.org/10.1038/nrg1066
8. Warnock JN, Daigre C, Al-Rubeai M (2011) Introduction to viral vectors. In: Merten O-W, Al-Rubeai M (eds) Viral vectors for gene therapy: methods and protocols. Humana Press, Totowa, pp 1–25
9. Fox TA, Booth C (2024) Improving access to gene therapy for rare diseases. Dis Model Mech 17:dmm050623. https://doi.org/10.1242/dmm.050623
10. Klockgether T, Mariotti C, Paulson HL (2019) Spinocerebellar ataxia. Nat Rev Dis Primer 5:1–21. https://doi.org/10.1038/s41572-019-0074-3

11. Stoyas CA, La Spada AR (2018) Chapter 11 – the CAG–polyglutamine repeat diseases: a clinical, molecular, genetic, and pathophysiologic nosology. In: Geschwind DH, Paulson HL, Klein C (eds) Handbook of clinical neurology. Elsevier, pp 143–170
12. Afonso-Reis R, Afonso IT, Nóbrega C (2021) Current status of gene therapy research in Polyglutamine spinocerebellar ataxias. Int J Mol Sci 22:4249. https://doi.org/10.3390/ijms22084249
13. Moreira-Gomes T, Nóbrega C (2024) From the disruption of RNA metabolism to the targeting of RNA-binding proteins: the case of polyglutamine spinocerebellar ataxias. J Neurochem 168:1442–1459. https://doi.org/10.1111/jnc.16010
14. Merten O-W, Hebben M, Bovolenta C (2016) Production of lentiviral vectors. Mol Ther Methods Clin Dev 3:16017. https://doi.org/10.1038/mtm.2016.17
15. Koppenol R, Conceição A, Afonso IT, Afonso-Reis R, Costa RG, Tomé S, Teixeira D, da Silva JP, Côdesso J-M, Brito DVC, Mendonça L, Marcelo A, Pereira de Almeida L, Matos CA, Nóbrega C (2022) The stress granule protein G3BP1 alleviates spinocerebellar ataxia-associated deficits. Brain. J Neurol awac473. https://doi.org/10.1093/brain/awac473
16. Torashima T, Yamada N, Itoh M, Yamamoto A, Hirai H (2006) Exposure of lentiviral vectors to subneutral pH shifts the tropism from Purkinje cell to Bergmann glia. Eur J Neurosci 24:371–380. https://doi.org/10.1111/j.1460-9568.2006.04927.x
17. Xu W, Russ JL, Eiden MV (2012) Evaluation of residual promoter activity in γ-retroviral self-inactivating (SIN) vectors. Mol Ther 20:84–90. https://doi.org/10.1038/mt.2011.204
18. Hirai H (2008) Progress in transduction of cerebellar Purkinje cells in vivo using viral vectors. Cerebellum 7:273–278. https://doi.org/10.1007/s12311-008-0012-5
19. Rahman H, Taylor J, Clack B, Stewart R, Canterberry S (2013) Effects of storage conditions on the morphology and titer of Lentiviral vectors. Fac Publ
20. Geraerts M, Willems S, Baekelandt V, Debyser Z, Gijsbers R (2006) Comparison of lentiviral vector titration methods. BMC Biotechnol 6:34. https://doi.org/10.1186/1472-6750-6-34
21. Ding B, Kilpatrick DL (2013) Lentiviral vector production, titration, and transduction of primary neurons. In: Zhou R, Mei L (eds) Neural development: methods and protocols. Humana Press, Totowa, NJ, pp 119–131

Chapter 19

Lentiviral Vector-Mediated Gene Delivery for iPSC-Related Neuroscience Research

Taro Okunomiya, Keiko Imamura, Taku Kashiyama, Takashi Sakurai, Hiroyuki Hioki, and Haruhisa Inoue

Abstract

The induced pluripotent stem cell (iPSC) technology provides human neural cells unprecedented material for neuroscience research. Lentiviral vectors have the advantage of being able to transfer genes into nondividing cells, which is useful for the identification and purification of specific neurons generated from human iPSCs, as well as for analysis of their function. Furthermore, the lentiviral vector has long-term expression and large gene capacity. This chapter describes protocols for lentiviral generation, labeling of iPSC-derived motor neurons with fluorescent proteins, and imaging of synaptic function.

Key words Lentiviral vector, Gene delivery, Induced pluripotent stem cells, Nondividing cells, Purification, Synapse imaging

1 Introduction

The lentiviral vector is derived from human immunodeficiency virus 1 (HIV-1) [1] and has been successfully used for gene delivery in basic biology and clinical applications [2–4]. Virus tropism can be optimized with modification of envelope glycoprotein. Newly developed lentiviral vectors with a genetically modified envelope glycoprotein derived from rabies virus are powerful methods for retrograde gene transfer from axon terminals [5]. After fusion to cultured cells, the target gene flanked by long terminal repeats is reverse-transcribed in DNA with reverse transcriptase and incorporated into host genomes with DNA integrase. For the transduction of nondividing cells or transient gene expression,

Authors Taro Okunomiya and Keiko Imamura have equally contributed to this chapter.

Kazuto Kobayashi et al. (eds.), *Lentiviral Vectors: Applications for Basic and Clinical Neuroscience*, Methods in Molecular Biology, vol. 2974, https://doi.org/10.1007/978-1-0716-4807-0_19,

Table 1
Characteristics of lentiviral vector and adeno-associated virus vector

	Lentiviral vector	Adeno-associated virus vector
Size	~100 nm	~26 nm
Cell types	Dividing and nondividing cells	Mainly nondividing cells
Genomic integration	Yes (except for nonintegrating vectors)	Very low frequency
Capacity	~9 kb	About 4.7 kb
Clinical applications	Ex vivo administration for hematopoietic cell transplantation	In vivo systemic/local administration

nonintegrating lentiviral vectors have also been developed [6]. Components derived from lentiviral vectors are used for the development of viral-like particles [7].

The induced pluripotent stem cell (iPSC) technology provides human neural cells unprecedented materials for neuroscience research, especially in disease modeling and drug discovery [8]. Identification/purification of target cells in neural cells and organoids, containing heterogeneous types of cells, differentiated from iPSCs, is a crucial step for neuroscience research [9–11]. Precise labeling of specific cell-type markers in living cells is needed for pathophysiological analysis of disease-targeted cells. In this context, among several types of virus vectors, lentiviral vectors have an advantage in gene delivery in nondividing cells, iPSC-derived postmitotic cells (Table 1). Compared to adeno-associated virus (AAV) vector, which is applicable to neural transduction, the lentiviral vector has long-term expression and large gene capacity.

There are applications of gene delivery with lentiviral vectors to neural cells derived from iPSCs for neuroscience research, as listed below. Transducing genes for:

1. Cell-type-specific markers to label target cells [12]
2. Engineered constructs to evaluate cellular functions [13]
3. Disease-related molecules to generate disease models [14]
4. Therapeutic molecules including shRNA or microRNA to conduct rescue experiments using disease models

In this chapter, we describe protocols for the practical application of the first two items. First, we describe the protocol for the production of vesicular stomatitis virus G glycoprotein (VSV-G) pseudotyped lentiviral vector [12, 15–17] and transduction to iPSC-derived motor neurons [12, 18] (Fig. 1). Previously, we characterized motor neurons from amyotrophic lateral sclerosis (ALS) patient iPSCs [12] by labeling with lentivirus expressing green fluorescent protein (GFP) under the control of HB9

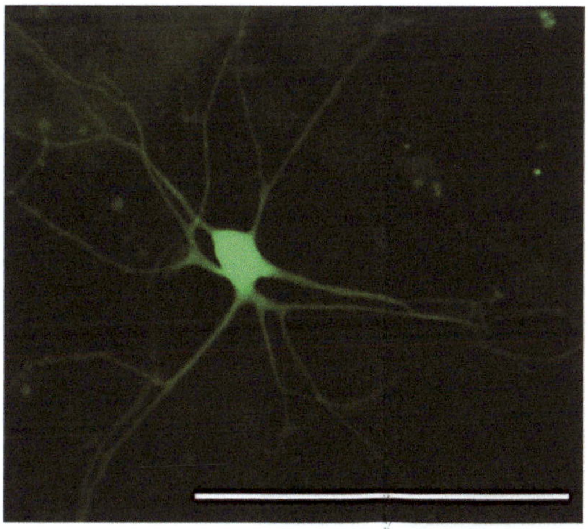

Fig. 1 Induced pluripotent stem cell (iPSC)-derived motor neuron labeled with GFP under the control of HB9 promotor. iPSC-derived motor neurons labeled with GFP under control of HB9 promotor are presented. The figure was acquired under fluorescence microscopy. Scale bar, 100 μm

promoter, a specific marker of spinal motor neurons, thereby avoiding the loss of GFP expression in transiently transfected cells. GFP labeling for motor neurons enabled assessment of neuronal morphology, viability analysis, electrophysiological analysis, and purification by cell sorting for subsequent gene expression analysis [12].

Second, we present the utility of lentiviral vectors for iPSC-related neuroscience research to evaluate neuronal functions. Lentiviral vectors carrying engineered constructs harboring a reporter of synaptic vesicle exocytosis, pHVenus [19] fused to the second intravesicular loop of synaptophysin, were infected with iPSC-derived neurons, and the synaptic function was thereby evaluated (Fig. 2).

2 Materials

2.1 Production of Lentiviral Vector Plasmid (e.g., pLenti6PW-HB9-GFP-BGHpA)

1. pLenti6/ BLOCK-iT™ (Invitrogen).
2. pLenti6PW [16].
3. pENTR1A-SYN-GFP-BGHpA [16].
4. pLP1, pLP2, and pLP/VSV-G (Invitrogen).
5. Endotoxin-free maxi-prep kit.

2.2 Cultivation and Expansion of HEK293FT Cells

1. Complete Medium with or without 500 μg/ml Geneticin®: D-MEM (high glucose), 10% fetal bovine serum (FBS), 0.1 M MEM nonessential Amino Acids (NEAA), 2 mM L-glutamine, 1% Penicillin/Streptomycin.

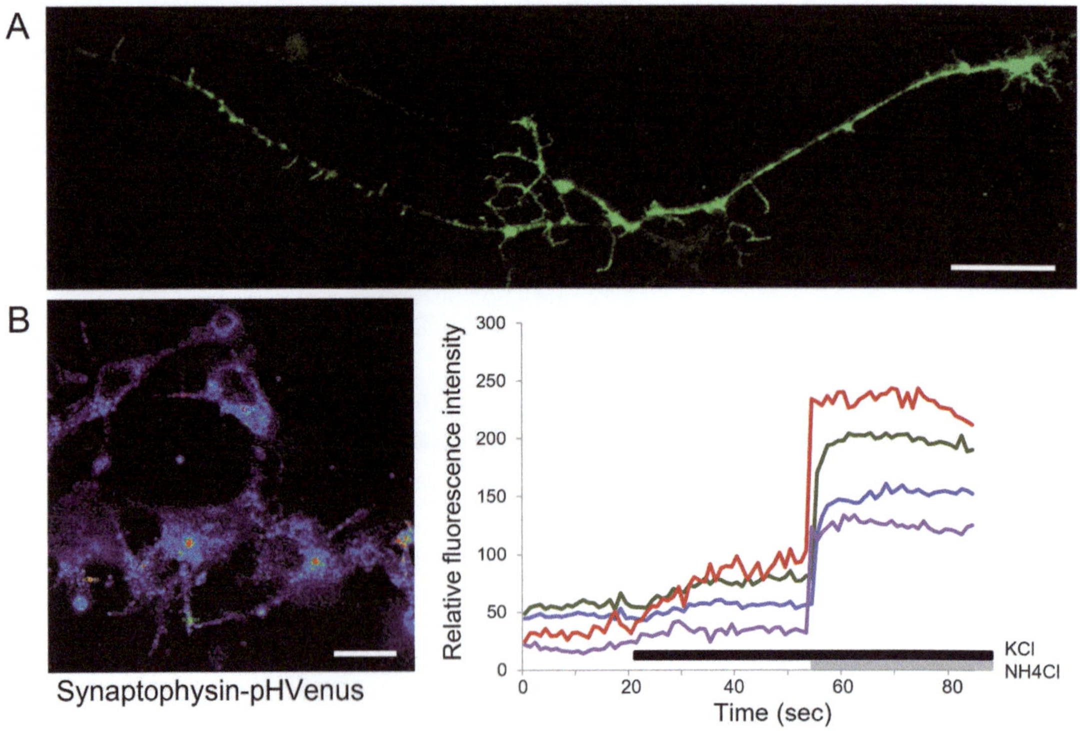

Fig. 2 Application of lentivirus for iPSC research to assess synaptic functions. (**a**) Lentiviral vectors carrying a reporter of synaptic vesicle exocytosis consisting of the synaptic vesicle protein synaptophysin fused to pHVenus, a pH-sensitive form of improved version of yellow fluorescent protein (YFP) (pKa = 7.3), were transfected into iPSC-derived neurons. Similar to sypHy [20], pHVenus is fused to the second intravesicular loop of synaptophysin in the construct. Its fluorescence is quenched in the acidic lumen of synaptic vesicles at rest, but increases upon fusion of synaptic vesicles with the plasma membrane by neuronal activity. Scale bar, 20 μm. (**b**) Activation of neurons by potassium chloride (KCL) caused an increase in the relative fluorescence intensity based on the exposure of pHVenus to the extracellular medium. Areas of high intensity in pHVenus were shown in red (left panel). Scale bar, 10 μm. Total reporter expression levels at synapses were estimated by dequenching of intravesicular pHVenus with NH_4Cl. The change over time of each hot spot is shown by each color (right panel). The presented figures were acquired after the stimulation by confocal microscope with a high-speed imaging system (Leica/Yokogawa)

2. G418 Disulfate Aqueous Solution (Geneticin®).
3. TrypLE Express (Invitrogen 12605010).

2.3 Production of Lentiviral Vector

1. Poly-D-Lysine (PDL)-coated 100 mm dish (*see* **Note 1**).
2. HEK293FT producer cell line.
3. Virus Production Medium (*see* **Note 2**): Serum-free medium (UltraCULTURE; Cambrex 12-725F), 0.1 M Minimum essential medium-nonessential amino acids, 4 mM L-glutamine, 2 mM GlutaMAX, 1 mM sodium pyruvate.
4. 5 ml Polystyrene Round-Bottom Tube (PS tube).

5. Lipofectamine® 2000 (Invitrogen).
6. Opti-MEM®I Reduced Serum Medium (GIBCO 31985-062).
7. Phosphate buffered saline (PBS).
8. Fetal bovine serum (FBS).

2.4 Concentration of Lentiviral Vector Particles

1. Low protein-binding filter with 0.45 μm pore size.
2. Ultrafiltration membranes (Amicon Ultra 100 K).

2.5 Differentiation of Spinal Motor Neurons

1. Human iPSCs.
2. StemFit (Ajinomoto).
3. Low cell adhesion U-shaped 96-well plates.
4. Dulbecco's modified Eagle's medium/Ham's F12.
5. KnockOut Serum Replacement.
6. Minimum essential medium-nonessential amino acids.
7. L-glutamine.
8. 2-mercaptoethanol.
9. Dorsomorphin.
10. SB431542.
11. CHIR99021.
12. Fibroblast growth factor.
13. Retinoic acid.
14. Smoothened ligand.
15. Neurobasal medium (Thermo Fisher Scientific).
16. B27 Supplement (Thermo Fisher Scientific).
17. DAPT.
18. Brain-derived neurotrophic factor.
19. Glial cell line-derived neurotrophic factor.
20. Neurotrophin-3.
21. Accumax (Innovative Cell Technologies).
22. Matrigel (BD Biosciences).

3 Methods

3.1 Production of Viral Plasmid (E.g., pLenti6PW-HB9-GFP-BGHpA)

1. HB9 promoter is amplified by polymerase chain reaction (PCR) with HB9 promoter-Fw (5'-ctcgaggatcctgtctccctggccag-3') and HB9 promotor-Rv (5'-aagcttggggcgcgccggatctgcgg-3').

2. The PCR product is inserted into the entry vector using XhoI and HindIII sites of pENTR1A-SYN-GFP-BGHpA [16] by standard cloning technique, resulting in pENTR1-HB9-GFP-BGHpA [12].
3. pLenti6PW is produced by modifying the destination vector pLenti6/ BLOCK- iT™-DEST (Invitrogen) with insertions of the woodchuck hepatitis virus posttranscriptional regulatory element (WPRE) cassette into XbaI/KpnI sites and central polypurine tract (cPPT) into Eco47III/HpaI sites [16].
4. The insert from the pENTR1-HB9-GFP-BGHpA is transferred to the destination vector pLenti6PW by homologous recombination with LR clonase (Invitrogen), resulting in pLenti6PW-HB9-GFP-BGHpA.
5. Prepare viral plasmids and packaging plasmids (pLP1, pLP2, pLP/VSV-G, and pLenti6PW-HB9-GFP-BGHpA) with endotoxin-free maxi-prep kit according to the manufacturer's instructions (*see* **Note 3**). Final concentration is adjusted to 1 μg/μl with nuclease-free water.

3.2 Cultivation and Expansion of HEK293FT Cells

1. Remove the vial of frozen HEK293FT cells from the freezer and open the cap briefly.
2. Thaw the cells quickly in a water bath at 37 °C.
3. Centrifuge the cells at 400× *g* for 3 min at 4 °C.
4. Decontaminate the outside of the vial with 70% ethanol.
5. Aspirate off the medium and resuspend the cells with 1 ml of complete medium without Geneticin®.
6. Transfer the cells to a 100-mm dish containing 12 ml of complete medium without Geneticin®.
7. Incubate the cells overnight in 5% CO_2 incubator at 37 °C.
8. Discard the medium and replace with complete medium containing 500 μg/ml Geneticin®.
9. Check the cells daily until they are 70% confluent (*see* **Note 4**).
10. Discard the medium and wash the cells once with 5 ml of PBS.
11. Add 2 ml of TrypLE™ Express solution to the monolayer and incubate for 3 min in CO_2 incubator at 37 °C.
12. Briefly pipet the detached cell solution to obtain single cell suspension.
13. Transfer the cells to a 50 ml coning tube and add 2 ml of complete medium containing Geneticin®.
14. Centrifuge the cells at 400× *g* for 3 min at 4 °C.
15. Discard the medium and resuspend the cells with 1 ml of complete medium containing Geneticin®.

16. Transfer 200 μl of the cells to a 100-mm dish containing 12 ml of complete medium with Geneticin®.
17. Incubate the cells and check them daily until they are 70% confluent in 5% CO_2 incubator at 37 °C.

3.3 Production of VSV-G-Pseudotyped Lentiviral Vector

1. The day before plasmid transfection, plate HEK293FT cells in a 100-mm tissue culture plate such that they will be 90–95% confluent on the day of transfection.
2. HEK293FT cells are incubated in 5% CO_2 incubator at 37 °C.
3. The pLenti6PW-HB9-GFP-BGHpA plasmid and a mixture of the packaging plasmids (pLP1, pLP2, and pLP/VSVG) are co-transfected into the HEK293FT producer cell line, using Lipofectamine 2000 as follows:
   (a) In one PS tube, mix 3 μg of pLP1, pLP2, pLP/VSV-G, and pLenti6PW-HB9-GFP-BGHpA with 1.5 mL of Opti-MEM®I medium without serum (*see* **Note 5**). Mix gently and incubate at RT for 5 min.
   (b) In another PS tube, dilute 36 μl of Lipofectamine 2000 with 1.5 mL of Opti-MEM®I medium without serum. Mix gently and incubate at RT for 5 min.
   (c) After the 5 min incubation, mix the diluted DNA with the diluted Lipofectamine 2000 and do pipetting gently.
   (d) Incubate at RT for 20 min to allow the DNA-Lipofectamine complexes to form.
4. While DNA-lipid complexes are forming, prepare HEK293FT cells for transfection.
   (a) Remove the medium and wash the cells once with 5 ml of PBS.
   (b) Add 2 ml of TrypLE™ Express solution to the monolayer and incubate for 3 min in CO_2 incubator at 37 °C.
   (c) Briefly pipet the solution to obtain a single cell suspension (*see* **Note 6**).
   (d) Transfer the cells to a 50 ml coning tube and add 2 ml of Opti-MEM®I medium containing 2% FBS.
   (e) Centrifuge the cells at 800 rpm ($400 \times g$) for 3 min at 4 °C.
   (f) Discard the medium and resuspend the cells with 150 μl of Opti-MEM®I medium containing 2% FBS.
5. Transfer the resuspended cells into 3 ml of DNA-Lipofectamine complexes in PS tube.
6. Incubate the mixture at RT for 5–10 min.
7. Add mixture of the cells and DNA-Lipofectamine complexes to the PDL-coated 100 mm dish containing 7 mL of pre-warmed

Opti-MEM®I medium with 2% FBS. Mix gently by rocking the dish back and forth. Incubate the cells in CO_2 incubator at 37 °C.

8. Eight to twelve hours after transfection, discard the medium containing the DNA-Lipofectamine complexes and replace with the Virus Production Medium.

3.4 Concentration of Lentiviral Vector Particles

1. Sixty hours after transfection, collect the culture supernatant containing the viral particles.
2. Centrifuge the supernatant at 3,000× *g* for 15 min at 4 °C to remove cell debris. Collect the supernatant.
3. Filter the supernatant through the low protein-binding filter using a 20 mL syringe.
4. Apply the viral solution to the ultrafiltration membrane. Centrifuge at 4,000× *g* for 15 min at 20 °C.
5. Discard the flow-through. Add 15 mL of PBS to the remaining solution and centrifuge again at 4,000× *g* for 15 min at 20 °C (*see* **Note 7**).
6. Discard the flow-through and collect the remaining viral solution. The virus solution is stored in 5–30 ul aliquots at −80 °C until use (*see* **Note 8**).

3.5 Generation of Spinal Motor Neurons

1. All of the cultures are maintained in a humidified incubator at 37 °C with 5% CO_2.
2. iPSCs are dissociated to single cells and quickly reaggregated in low cell adhesion U-shaped 96-well plates.
3. Aggregations are cultured in EB medium containing Dulbecco's modified Eagle's medium/Ham's F12, 5% KSR, minimum essential medium-nonessential amino acids, L-glutamine, 2-mercaptoethanol, 2 μM dorsomorphin, 10 μM SB431542, 3 μM CHIR99021, and 12.5 ng/mL fibroblast growth factor in a neural inductive stage for 11 days.
4. 100 nM retinoic acid and 500 nM Smoothened ligand are added on day 4.
5. After patterning with neurobasal medium supplemented with B27 Supplement, 100 nM retinoic acid, 500 nM Smoothened ligand, and 10 μM DAPT, the aggregates are incubated with Accumax for 10 min followed by dissociation into single cells, and seeded to Matrigel-coated dishes on day 16.
6. Cells are cultured in neurobasal medium with 10 ng/ml brain-derived neurotrophic factor, 10 ng/ml glial cell line-derived neurotrophic factor, and 10 ng/ml neurotrophin-3.

7. On day 21, cells are dissociated to single cells using accumax and seeded to iMatrix-coated 24-well plates at 2×10^5 cells/well for viral infection.

3.6 Transduction of hiPSC-Derived Motor Neurons with Lentiviral Vector

1. Thaw an aliquot of viral stock on ice.
2. Mix the viral solution by pipetting repeatedly.
3. The virus solution is diluted with the culture medium, neurobasal medium containing BDNF, GDNF, and NT-3 at 1:500 (*see* **Note 9**), and replace the entire medium of the motor neuron culture with the medium containing the virus.
4. One day after lentivirus transduction, the medium is replaced with fresh medium. Thereafter, replace half of the medium every 2 days.
5. Find GFP expression peaks about 2 weeks after lentivirus infection.
6. Labeled motor neurons can be used for subsequent analysis (Figs. 1 and 2).

4 Notes

1. We use poly-D-Lysine or poly-L-Lysine-coated dishes to promote the adherence of HEK293FT cells.
2. Alternatively, use 10% FBS DMEM: Dulbecco's Modified Eagle's Medium 10% fetal bovine serum, 2 mM L-glutamine, 100 U/ml penicillin, and 100 mg/ml streptomycin sulfate.
3. Plasmids are purified with endotoxin-free maxi-prep kit. Endotoxins can reduce transfection efficiency.
4. When maintaining HEK293FT cells, passage the cells before reaching confluence.
5. We use polystyrene tubes to minimize adhesion of cells and DNA to the wall of the tube.
6. Before plasmid transfection, HEK293FT cells should be dissociated into single cells for improving efficiency of transfection.
7. When serum-free culture medium was not used for viral production, it is recommended to remove the medium component. In that case, ultracentrifuge the filtered solution at $100{,}000\times g$ for 3 h at 4 °C. After ultracentrifugation, remove the supernatant and resuspend the pellet with PBS by incubation at 4 °C while rocking the tubes with shaker for at least 8 h (or overnight, recommended).

8. Minimize the freeze-and-thaw process where possible because lentiviral particles are susceptible to repeated freeze-and-thaw cycles. We generally limit the freeze-and-thaw cycles to two times to avoid dropping of functional titer. The volume of each aliquot depends on the amount of virus particles used in a single experiment.
9. To determine the optimal viral dilution, we recommend performing a multiplicity of infection (MOI) assays.

Acknowledgments

The authors would like to thank Atsushi Miyawaki for providing pHVenous; Fred H. Gage, Maria C.N. Marchetto, and Samuel L. Pfaff for providing HB9 promoter; and Kayoko Tsukita, Naohiro Egawa, Shiho Kitaoka, and Masataka Ishizuka for their technical support and suggestions.

Funding This work was supported in part by AMED under Grant Numbers JP23bm1323001, JP23bm1423014, JP23bm1423012, JP23dk0207066, JP24wm0625501, and JP24wm0625201.

Conflict of Interest Statement The authors have no conflicts of interest to declare that are relevant to the content of this chapter.

References

1. Merten OW, Hebben M, Bovolenta C (2016) Production of lentiviral vectors. Mol Ther Methods Clin Dev 3:16017
2. Sakuma T, Barry Michael A, Ikeda Y (2012) Lentiviral vectors: basic to translational. Biochem J 443:603–618
3. Hughes SM, Parr-Brownlie LC, Bosch-Bouju-C, Schoderboeck L, Sizemore R, Abraham W (2015) Lentiviral vectors as tools to understand central nervous system biology in mammalian model organisms. Front Mol Neurosci 8:14
4. Bulcha JT, Wang Y, Ma H, Tai PWL, Gao G (2021) Viral vector platforms within the gene therapy landscape. Signal Transduct Target Ther 6:53
5. Kato S, Kobayashi K (2020) Pseudotyped lentiviral vectors for tract-targeting and application for the functional control of selective neural circuits. J Neurosci Meth 344:108854
6. Yáñez-Muñoz RJ, Balaggan KS, MacNeil A, Howe SJ, Schmidt M, Smith AJ, Buch P, MacLaren RE, Anderson PN, Barker SE, Duran Y, Bartholomae C, von Kalle C, Heckenlively JR, Kinnon C, Ali RR, Thrasher AJ (2006) Effective gene therapy with nonintegrating lentiviral vectors. Nat Med 12:348–353
7. Gee P, Lung MSY, Okuzaki Y, Sasakawa N, Iguchi T, Makita Y, Hozumi H, Miura Y, Yang LF, Iwasaki M, Wang XH, Waller MA, Shirai N, Abe YO, Fujita Y, Watanabe K, Kagita A, Iwabuchi KA, Yasuda M, Xu H, Noda T, Komano J, Sakurai H, Inukai N, Hotta A (2020) Extracellular nanovesicles for packaging of CRISPR-Cas9 protein and sgRNA to induce therapeutic exon skipping. Nat Commun 11:1334
8. Takahashi K, Tanabe K, Ohnuki M, Narita M, Ichisaka T, Tomoda K, Yamanaka S (2007) Induction of pluripotent stem cells from adult human fibroblasts by defined factors. Cell 131:861–872
9. Inoue H, Nagata N, Kurokawa H, Yamanaka S (2014) iPS cells: a game changer for future medicine. EMBO J 33:409–417

10. Shi Y, Inoue H, Wu JC, Yamanaka S (2017) Induced pluripotent stem cell technology: a decade of progress. Nat Rev Drug Discov 16: 115–130
11. Karagiannis P, Takahashi K, Saito M, Yoshida Y, Okita K, Watanabe A, Inoue H, Yamashita JK, Todani M, Nakagawa M, Osawa M, Yashiro Y, Yamanaka S, Osafune K (2019) Induced pluripotent stem cells and their use in human models of disease and development. Physiol Rev 99:79–114
12. Egawa N, Kitaoka S, Tsukita K, Naitoh M, Takahashi K, Yamamoto T, Adachi F, Kondo T, Okita K, Asaka I, Aoi T, Watanabe A, Yamada Y, Morizane A, Takahashi J, Ayaki T, Ito H, Yoshikawa K, Yamawaki S, Suzuki S, Watanabe D, Hioki H, Kaneko T, Makioka K, Okamoto K, Takuma H, Tamaoka A, Hasegawa K, Nonaka T, Hasegawa M, Kawata A, Yoshida M, Nakahata T, Takahashi R, Marchetto MC, Gage FH, Yamanaka S, Inoue H (2012) Drug screening for ALS using patient-specific induced pluripotent stem cells. Sci Transl Med 4:145ra104
13. Andersen J, Revah O, Miura Y, Thom N, Amin ND, Kelley KW, Singh M, Chen X, Thete MV, Walczak EM, Vogel H, Fan HC, Paşca SP (2020) Generation of functional human 3D Cortico-motor Assembloids. Cell 183:1913–1929.e1926
14. Shimada H, Sato Y, Sasaki T, Shimozawa A, Imaizumi K, Shindo T, Miyao S, Kiyama K, Kondo T, Shibata S, Ishii S, Kuromitsu J, Aoyagi H, Ito D, Okano H (2022) A next-generation iPSC-derived forebrain organoid model of tauopathy with tau fibrils by AAV-mediated gene transfer. Cell Rep Methods 2:100289
15. Hioki H, Kameda H, Nakamura H, Okunomiya T, Ohira K, Nakamura K, Kuroda M, Furuta T, Kaneko T (2007) Efficient gene transduction of neurons by lentivirus with enhanced neuron-specific promoters. Gene Ther 14(11):872–882
16. Hioki H, Kuramoto E, Konno M, Kameda H, Takahashi Y, Nakano T, Nakamura KC, Kaneko T (2009) High-level transgene expression in neurons by lentivirus with Tet-off system. Neurosci Res 63:149–154
17. Sohn J, Takahashi M, Okamoto S, Ishida Y, Furuta T, Hioki H (2017) A single vector platform for high-level gene transduction of central neurons: adeno-associated virus vector equipped with the Tet-off system. PLoS One 12:e0169611
18. Matsuzono K, Imamura K, Murakami N, Tsukita K, Yamamoto T, Izumi Y, Kaji R, Ohta Y, Yamashita T, Abe K, Inoue H (2017) Antisense oligonucleotides reduce RNA foci in spinocerebellar ataxia 36 patient iPSCs. Mol Ther Nucleic Acids 8:211–219
19. Tojima T, Akiyama H, Itofusa R, Li Y, Katayama H, Miyawaki A, Kamiguchi H (2007) Attractive axon guidance involves asymmetric membrane transport and exocytosis in the growth cone. Nat Neurosci 10:58–66
20. Granseth B, Odermatt B, Royle SJ, Lagnado L (2006) Clathrin-mediated endocytosis is the dominant mechanism of vesicle retrieval at hippocampal synapses. Neuron 51:773–786

Chapter 20

Phase 1/2a Clinical Trial Protocol for Lentiviral Vector-Based Retinal Gene Therapy to Slow the Progression of Retinitis Pigmentosa

Takahiro Hisai, Yusuke Murakami, Kohji Kusano, Yuko Kobayakawa, and Yasuhiro Ikeda

Abstract

Retinitis pigmentosa (RP) is a group of inherited retinal dystrophies (IRDs) that cause progressive rod and cone photoreceptor cell death. Currently, there is no treatment for RP, and it is a significant cause of blindness worldwide. Neuroprotective retinal gene therapy is a promising therapeutic strategy applicable to various causative genes. We have developed a third-generation lentiviral vector based on a simian immunodeficiency virus from African green monkeys (SIVagm), which allows efficient gene transfer into the retinal pigment epithelium via subretinal injection. In preclinical studies, we used the human gene encoding pigment epithelium-derived factor (PEDF), a potent neuroprotective molecule, and the SIVagm vector encoding human PEDF (SIV-hPEDF) significantly delayed retinal degeneration in rodent models of RP and achieved stable, safe, and long-term transgene expression over 5 years in nonhuman primates. Based on these findings, we conducted a Phase 1/2a investigator-initiated clinical trial using SIV-hPEDF to slow the disease progression and prevent blindness in RP patients, and we here outline the study protocol.

Key words Retinitis pigmentosa, Retinal gene therapy, Lentiviral vector, Pigment epithelium-derived factor, Neuroprotection

1 Introduction

Retinitis pigmentosa (RP) is a major form of inherited retinal dystrophies (IRDs) that causes degeneration of rod and cone photoreceptor cells [1]. Rod photoreceptors are sensitive to dim light in the dark, while cone photoreceptors perceive high-resolution daylight and color vision. Because the causative genetic mutations in RP are related to rod photoreceptor homeostasis and function, the clinical features of RP start with night blindness; however, as the disease progresses, cone photoreceptors also degenerate,

Authors Takahiro Hisai and Yusuke Murakami have equally contributed to this chapter.

Kazuto Kobayashi et al. (eds.), *Lentiviral Vectors: Applications for Basic and Clinical Neuroscience*, Methods in Molecular Biology, vol. 2974, https://doi.org/10.1007/978-1-0716-4807-0_20,

leading to visual field constriction, central vision loss, and eventually blindness [2, 3]. Approximately 2.5 million people worldwide are affected by RP, and there is no effective treatment to prevent or delay blindness [4]. In Japan, there are about 30,000 patients with RP, and RP is the second most common cause of visual impairment [2].

New treatments for IRDs, such as such as gene therapy [5–8], regenerative medicine [9–11], and neuroprotective and anti-inflammatory therapies [12], have been developed globally. In a breakthrough, voretigene neparvovecryzl (Luxturna®), an adeno-associated virus (AAV)-based vector encoding a normal copy of the *RPE65* gene, received Food and Drug Administration (FDA) approval in 2017 for patients with Leber's congenital amaurosis (LCA) associated with biallelic *RPE65* mutations [5, 6]. Patients with LCA develop severe vision loss and nystagmus from early childhood [13], and without treatment they go blind at a young age. Luxturna has restored the dark vision and improved the quality of life of LCA patients with only a single subretinal surgery, proving the efficacy and usefulness of retinal gene therapy. However, *RPE65*-related IRD is extremely rare, accounting for only 0.23–3% of IRDs [14], and most IRD patients are not eligible for treatment.

In contrast to gene replacement or gene editing strategies, neuroprotective and anti-inflammatory therapies target common biological pathways where heterogeneous genetic variants in RP/IRDs eventually converge. Pigment epithelium-derived factor (PEDF) is a protein isolated from the cultured medium of human retinal pigment epithelium (RPE) cells, and exhibits neuroprotective, antiangiogenic, and antitumor activity [15–19]. The levels of PEDF in aqueous humor are decreased in patients with RP compared with control subjects [20]. Moreover, we have shown that overexpression of PEDF via retinal gene transfer significantly prevents photoreceptor apoptosis and delays retinal degeneration in rodent models of RP [21, 22].

AAV vectors can achieve efficient neuronal transduction and have been widely used in gene therapy for central nervous system (CNS) disorders, including those involving the retina [23, 24]. However, AAV vectors have the disadvantage of relatively low gene transduction efficiency, and therefore require extremely high-titer doses such as from 1E10 vg to 1E12 vg [25]. Indeed, clinical trials have demonstrated that subretinal administration of AAV vectors sometimes causes retinal toxicity and immunogenicity [25–28], including severe panuveitis leading to hypotony. (NCT04418427). In contrast, we are developing a simian immunodeficiency virus from an African green monkey (SIVagm)-based lentiviral vector with the advantage of efficient, long-term, and stable transgene expression [29–33]. Preclinical safety studies have shown that subretinal injection of SIV-hPEDF achieves efficient gene transfer into the retinal pigment epithelium (RPE) without causing retinal toxicity or tumor formation in the eye or systemic organs in rodents and nonhuman primates [30–32, 34]. Based on these results, we conducted a Phase 1/2a clinical trial (jRCT2073180024) to evaluate the safety and efficacy in patients with RP from 2019 to 2023, and we here present the trial protocol.

2 Materials

Details of the manufacture of SIVagm-based vector encoding the *hPEDF* have been described previously [29, 31, 34]. These earlier reports describe the following:

1. A third-generation lentiviral vector with deletion of the U3 region in the 3′ and 5′ long terminal repeats for self-activation.
2. Transfection of human embryonic kidney (HEK) 293 T cells with the packaging vector, the gene transfer vectors encoding *hPEDF* driven by the cytomegalovirus promotor, the Rev. expression vector, and the envelope vector pseudo-typed with vesicular stomatitis virus glycoprotein G.
3. Vector purification and assessment of viral titers by transduction of the HEK 293 T cell line.

3 Methods

3.1 Design

This is a prospective, open-labeled, parallel-group and intra-individual comparison, investigator-initiated clinical trial. The observation period is 12 months. Participants are divided into three groups ($N = 4$ in each): a Low-dose (2.5×10^7 TU/ml), Medium-dose (1.0×10^8 TU/ml), and High-dose (2.5×10^8 TU/ml) group. 100 μl of SIV-hPEDF is administered to the worse-seeing eye. After the safety of the Low-dose group is confirmed at 29 days after administration of SIV-hPEDF in the last patient in the group, administration to the Medium-dose group is initiated. Then, after the safety of the Medium-dose group is confirmed at 29 days after administration of SIV-hPEDF in the last patient in the group, administration to the High-dose group is initiated (Fig. 1).

3.2 Inclusion and Exclusion Criteria

Inclusion Criteria:

1. A diagnosis of typical RP (genetic diagnosis is not necessary).
2. Age of 20–70 years at the time of informed consent.
3. Best-corrected visual acuity worse than 1.0.

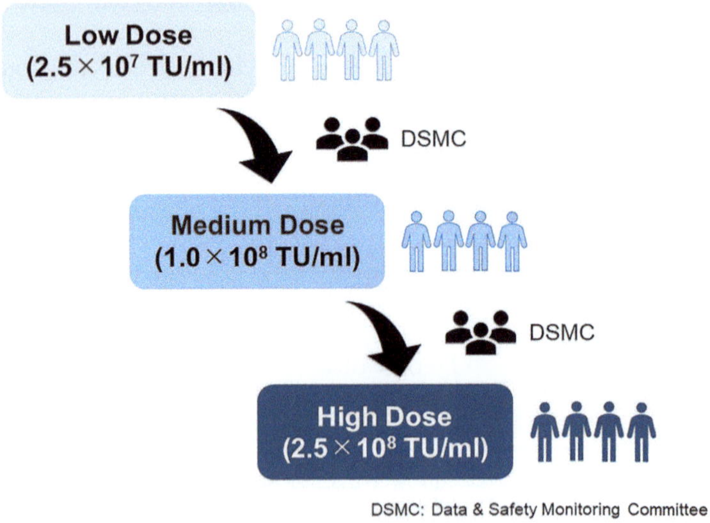

Fig. 1 Study protocol for SIV-hPEDF Phase 1/2a clinical trial. DSMC: Data & Safety Monitoring Committee

4. Mean retinal sensitivity within the central 4 points of below 30 decibels (dB) in the perimetry test of the Humphrey Field Analyzer 10–2 SITA standard program (HFA10-2).
5. Interocular difference in logMAR visual acuity of 0.21 or less.
6. Interocular difference in the visual field area of twice or less (tested with V-4e stimuli in Goldmann perimetry [GP]).
7. Interocular difference in mean deviation (MD) in HFA10-2 of 5 dB or less.
8. Interocular difference in central fovea thickness of 50 μm or less in the optical coherence tomography (OCT) measurement.
9. No difference in the ellipsoid zone between eyes.
10. Changes in MD in HFA10-2 tests and logMAR visual acuity* are less than twice the changes in the other eye#.

    *The difference between the oldest measurement within 12 months from the time of enrollment and the measurements obtained during the screening period (averaged score from the two screening tests) is defined as the amount of change.

    #If the amount of change in logMAR visual acuity of one eye is less than 0.1, the upper limit of the amount of change in the other eye is set at 0.2. If the amount of change in the MD value of one eye was less than 2.0 dB, the upper limit of the amount of change for the other eye is set at 4.0 dB.
11. A difference in mean retinal sensitivity of 3 dB or less in both eyes on two HFA10-2 tests performed within 60 days prior to treatment; if the second test does not meet the criteria, a third measurement should be performed within 7 days of the second

test and the difference between the third measurement and the first or second measurement should be within 3 dB.

12. No severe pigmentary deposition in the fundus.
13. Understanding of the study design and objectives and consent by participants.

Exclusion Criteria:

1. Seropositivity for human immunodeficiency virus antibody.
2. Visual acuities below 0.1 or blindness in either eye.
3. Cone-rod dystrophy.
4. Differences in lens condition between eyes.
5. Glaucoma or ocular hypertension (intraocular pressure: 22 mmHg or above).
6. Uveitis or optic neuritis.
7. Retinal and/or subretinal disorders (e.g., retinal hemorrhage, retinal edema, proliferative tissue) that are not related to RP.
8. V-4e stimuli area in Goldmann perimetry below 10 degrees.
9. Abnormal RPE layer line in the subretinal injection areas in OCT.
10. Severe allergy.
11. Allergy to bovine-derived ingredients.
12. Hemodialysis.
13. Severe cardiac disease.
14. Severe hepatic dysfunction.
15. Systemic inflammatory disease.
16. History of cerebral hemorrhage or stroke within 6 months prior to consent.
17. Hematological disease.
18. Alcoholism, drug addiction, or a psychiatric disorder potentially interfering with the trial.
19. Pregnant, suspected pregnant, or lactating patient.
20. Malignancy, suspected malignancy, or treatment for malignancy within the past 5 years.
21. No contraceptive consent for 1 year after vector administration.
22. Administration of Helenien, Unoprostone, or calcium channel blockers within 30 days prior to consent.
23. Use of an antiviral drug.
24. Plan to undergo other ophthalmic surgery within the trial period.

25. Participation in another clinical trial within 6 months prior to consent.
26. Any other judgment of ineligibility by the investigators.

3.3 Endpoints

The primary endpoint is the ratio of adverse events after subretinal injection of SIV-hPEDF. A secondary endpoint is the efficacy of SIV-hPEDF, and the main secondary endpoint is the rate of change of mean retinal sensitivity in the central 12 points in HFA10-2. Other secondary endpoints include the rate of change of mean retinal sensitivity in the central four points in HFA10-2 and log-MAR visual acuity. The exploratory endpoint is the expression levels of hPEDF protein in the aqueous humor.

3.4 Subretinal Injection

1. Administration of intravenous antibiotics is performed from the day before the surgery to 3 days post-surgery. Topical antibiotics (1.5% levofloxacin) and steroid (0.1% dexamethasone) are administered starting the day before surgery and continued until investigators determine that they are no longer necessary.
2. General anesthesia.
3. Collection of aqueous humor with a 30-gauge needle.
4. Three-port vitrectomy using a CONSTELLATION® Vision System (Alcon, Geneva, Switzerland).
5. Creation of posterior vitreous detachment.
6. Dissolution of SIV-hPEDF to the target concentration with balanced salt solution.
7. Preparation of the MedOne MicroDose™ Injection Kit with a 41G subretinal injection needle, connected to a syringe filled with SIV-hPEDF.
8. Subretinal injection of SIV-hPEDF (total 100 ± 20 μl) in the mid-peripheral, non-degenerated retina using the CONSTELLATION® Viscous Fluid Control function.
9. Irrigation of the vitreous cavity with balanced salt solution.
10. Closure of the three ports.
11. Subconjunctival administration of antibiotics and steroid.

3.5 Data Collection and Registration

The schedule for data collection is listed in Table 1. Postoperatively, in addition to ophthalmologic tests, the presence of virus exclusion is confirmed by SIV-hPEDF nucleic acid sequence detection tests of lacrimal fluid, blood, and urine. The VSV-G antibody titer is also followed. The presence of malignancy is confirmed at the last observation. The data are written on a case report form by the investigators and registered into the electronic data capture (EDC) system (REDCap) by certified clinical research coordinators. The accuracy and reliability of data were reviewed by independent monitors.

			• Outpatient							Inpatient									Outpatient					
			Screening period					Adm		Observation period														
								1D																
		IC	IC to -4D	Registration	-3D	-2D	-1D	Before Adm	After 6h	2D	3D	4D	5D	6D	7D	8D	9D to 15D[2]	29D	2M	3M	Every 1M	6M	Every 2M	12M
IC		○																						
Background	Age, gender, diagnosis, past history, allergy, psychiatric consultation		○																					
Registration				○																				
Pre- & Post-treatment	Transvenous antibiotic								← ● →															
	Antibiotic eyedrop												← ● →											
	Steroid eyedrop														← ● →									
Ophthalmologic examination	Visual acuity		○			○						○			○	○	○	○	○	○	○	○	○	○
	Intraocular pressure		○			○				○	○	○	○	○	○	○	○	○	○	○	○	○	○	○
	Slit-lamp examination		○			○				○	○	○	○	○	○	○	○	○	○	○	○	○	○	○
	Fundus examination		○			○				○	○	○	○	○	○	○	○	○	○	○	○	○	○	○
	Visual field (Goldmann Perimetry [GP])		○															○				○		○
	Visual field (Humphrey Field Analyzer 10-2 [HFA10-2])		○																	○		○		○
	Fundus angiography (FA、IA)		○															○		○		○		○
	Full-field ERG, multifocal ERG		○															○		○		○		○
	Optical coherence tomography (OCT)		○									○				○	○	○	○	○		○		○
	Fundus autofluorescense (FAF)		○																					
Quality of life	Questionnaire on subjective symptoms (VFQ-25)		○															○		○		○		○
Vital signs	Body weight		○			○										○		○						○
	Body temperature		○				○	○	○	○	○	○	○	○	○	○	○	○						○
	Blood pressure and pulse		○				○	○	○	○	○	○	○	○	○	○	○	○	○	○	○	○	○	○
Respiratory function test	Chest X-ray		○																					○
Cardiac functional test	ECG, UCG[1]		○																					○
Renal functional test	BUN, Cr, urine test (PRO, OB)		○			○			○	○		○				○	○	○		○		○		○
Hepatic function test	Alb, Ig (IgG, IgA, IgM, IgE), T-Bil, D-Bil, AST, ALT, ALP, LDH, γ-GTP		○			○			○	○		○				○	○	○		○		○		○
Blood coagulation test	RBC, Hb, Ht, WBC, WBC fraction, plt, PT, APTT, fibrinogen		○			○			○	○		○				○	○	○		○		○		○
Inflammatory marker	CRP		○			○			○	○		○				○	○	○		○		○		○
Diabetes test	FBS, HbA1c		○																					
Serum electrolyte test	Na、K、Cl、Ca		○			○			○	○		○				○	○	○		○		○		○
Cancer screening test	Brain, chest and abdominal CT		○																					○
	Fecal occult blood test (FOBT)		○																					○
	PSA (Male) / Cervical cytology (Female)		○																					
	Rectal examination		○																					
	Esophagogastroduodenoscopy (EGD)		○																					
Pregnancy test	Urine human chorionic gonadotropin (hCG)		○																					
Virus test	HBs-Ag, HCV-Ab, HIV screening		○																					
Serum cytokines	IL-1, IL-4, IL-6, IL-8, TNF-α, IFN-γ					○										○	○	○						
Detection of SIV-hPEDF-derived nucleic acid sequences	In lacrimal fluid, blood and urine with Taqman-PCR					○			○	○						○	○	○		○		○		○
VSV antibody titer	In serum with ELISA					○										○	○	○		○		○		○
Human PEDF protein level	In aqueous humor with ELISA							○[3]										○		○		○		○
Causative gene	Extracted from blood					○																		

1) UCG is done if the patient has cardiac disease.
2) Each test is measured only once between 9D and 15D.
3) Aqueous humor is collected during the surgery (before administration).

Adm: administration; IC: informed consent; FA: fluorescein angiography; IA: indocyanine green angiography; ERG: electroretinography; ECG: electrocardiogram; UCG: ultrasound cardiography; BUN: blood urea nitrogen; Cr: creatinine; PRO: protein. OB: occult blood; Alb: albumin; Ig: immunoglobulin; T-Bil: total-bilirubin; D-Bil: direct-bilirubin; AST: asparate aminotransferase; ALT: alanine aminotransferase; ALP: alkaline phosphatase; LDH: lactate dehydrogenase; γ-GTP: γ-glutamyl transpeptidase; RBC: red blood cell; Hb: hemoglobin; Ht: hematocrit; WBC: white blood cell; plt: platelet; PT: prothrombin time; APTT: activated partial thromboplastin time; CRP: C-reactive protein; FBS: fasting blood sugar; HbA1c: hemoglobin A1c; Na: sodium; K: potassium; Cl: chlorine; Ca: calcium; CT: computed tomography; PSA: prostate specific antigen; HBs-Ag: hepatitis B surface-antigen; HCV-Ab: hepatitis C virus antibody; IL: interleukin, TNF: tumor necrosis factor; IFN: interferon; PCR: polymerase chain reaction; VSV: vesicular stomatitis virus; ELISA: enzyme-linked immunosorbent assay.

Acknowledgments

This clinical trial was supported by a grant from The Japan Agency for Medical Research and Development, Practical Research Project for Rare/Intractable Diseases (#JP 20ek0109343h9903, to YI). Additional funding was provided by ID Pharma Co. (Tsukuba, Japan). The sponsor and funding organization had no role in the design or conduct of this research.

Funding This study was funded by a grant from the Japan Agency for Medical Research and Development, Practical Research Project for Rare/Intractable Diseases (Grant Number #JP20ek0109343h9903). Yasuhiro Ikeda has a received research grant from ID Pharma Co. (Tsukuba, Japan).

Conflict of Interest Statement The authors have no conflicts of interest to declare that are relevant to the content of this chapter.

References

1. Hartong DT, Berson EL, Dryja TP (2006) Retinitis pigmentosa. Lancet 368(9549): 1795–1809. https://doi.org/10.1016/S0140-6736(06)69740-7
2. Matoba R, Morimoto N, Kawasaki R et al (2023) A nationwide survey of newly certified visually impaired individuals in Japan for the fiscal year 2019: impact of the revision of criteria for visual impairment certification. Jpn J Ophthalmol 67(3):346–352. https://doi.org/10.1007/s10384-023-00986-9
3. Mawatari G, Hiwatashi S, Motani T et al (2024) Efficacy of a wearable night-vision aid in patients with concentric peripheral visual field loss: a randomized, crossover trial. Jpn J Ophthalmol 68(4):321–326. https://doi.org/10.1007/s10384-024-01068-0
4. Dias MF, Joo K, Kemp JA et al (2018) Molecular genetics and emerging therapies for retinitis pigmentosa: basic research and clinical perspectives. Prog Retin Eye Res 63:107–131. https://doi.org/10.1016/j.preteyeres.2017.10.004
5. Russell S, Bennett J, Wellman JA et al (2017) Efficacy and safety of voretigene neparvovec (AAV2-hRPE65v2) in patients with RPE65-mediated inherited retinal dystrophy: a randomised, controlled, open-label, phase 3 trial. Lancet 390(10097):849–860. https://doi.org/10.1016/S0140-6736(17)31868-8
6. Maguire AM, Russell S, Chung DC et al (2021) Durability of Voretigene Neparvovec for Biallelic RPE65-mediated inherited retinal disease: phase 3 results at 3 and 4 years. Ophthalmology 128(10):1460–1468. https://doi.org/10.1016/j.ophtha.2021.03.031
7. Lam BL, Pennesi ME, Kay CN et al (2024) Assessment of visual function with Cotoretigene Toliparvovec in X-linked retinitis Pigmentosa in the randomized XIRIUS phase 2/3 study. Ophthalmology 131(9):1083–1093. https://doi.org/10.1016/j.ophtha.2024.02.023
8. Pierce EA, Aleman TS, Jayasundera KT et al (2024) Gene editing for CEP290-associated retinal degeneration. N Engl J Med 390(21): 1972–1984. https://doi.org/10.1056/NEJMoa2309915
9. Schwartz SD, Regillo CD, Lam BL et al (2015) Human embryonic stem cell-derived retinal pigment epithelium in patients with age-related macular degeneration and Stargardt's macular dystrophy: follow-up of two open-label phase 1/2 studies. Lancet 385(9967):509–516. https://doi.org/10.1016/S0140-6736(14)61376-3
10. Liu Y, Chen SJ, Li SY et al (2017) Long-term safety of human retinal progenitor cell transplantation in retinitis pigmentosa patients. Stem Cell Res Ther 8(1):209. https://doi.org/10.1186/s13287-017-0661-8
11. Hirami Y, Mandai M, Sugita S et al (2023) Safety and stable survival of stem-cell-derived retinal organoid for 2 years in patients with

retinitis pigmentosa. Cell Stem Cell 30(12): 1585–1596 e1586. https://doi.org/10.1016/j.stem.2023.11.004
12. Dave AD, Chen KG, Chiang TT et al (2023) Oral minocycline for the treatment of retinitis pigmentosa-associated cystoid macular edema: results of a phase I/II clinical trial. Graefes Arch Clin Exp Ophthalmol 261(8): 2209–2220. https://doi.org/10.1007/s00417-023-05986-6
13. Koenekoop RK (2004) An overview of Leber congenital amaurosis: a model to understand human retinal development. Surv Ophthalmol 49(4):379–398. https://doi.org/10.1016/j.survophthal.2004.04.003
14. Sallum JMF, Kaur VP, Shaikh J et al (2022) Epidemiology of mutations in the 65-kDa retinal pigment epithelium (RPE65) gene-mediated inherited retinal dystrophies: a systematic literature review. Adv Ther 39(3): 1179–1198. https://doi.org/10.1007/s12325-021-02036-7
15. Steele FR, Chader GJ, Johnson LV et al (1993) Pigment epithelium-derived factor: neurotrophic activity and identification as a member of the serine protease inhibitor gene family. Proc Natl Acad Sci USA 90(4):1526–1530. https://doi.org/10.1073/pnas.90.4.1526
16. Taniwaki T, Becerra SP, Chader GJ et al (1995) Pigment epithelium-derived factor is a survival factor for cerebellar granule cells in culture. J Neurochem 64(6):2509–2517. https://doi.org/10.1046/j.1471-4159.1995.64062509.x
17. Tombran-Tink J, Johnson LV (1989) Neuronal differentiation of retinoblastoma cells induced by medium conditioned by human RPE cells. Invest Ophthalmol Vis Sci 30(8): 1700–1707
18. Dawson DW, Volpert OV, Gillis P et al (1999) Pigment epithelium-derived factor: a potent inhibitor of angiogenesis. Science 285(5425): 245–248. https://doi.org/10.1126/science.285.5425.245
19. Becerra SP, Notario V (2013) The effects of PEDF on cancer biology: mechanisms of action and therapeutic potential. Nat Rev Cancer 13(4):258–271. https://doi.org/10.1038/nrc3484
20. Ogata N, Matsuoka M, Imaizumi M et al (2004) Decreased levels of pigment epithelium-derived factor in eyes with neuroretinal dystrophic diseases. Am J Ophthalmol 137(6):1129–1130. https://doi.org/10.1016/j.ajo.2003.11.080
21. Murakami Y, Ikeda Y, Yonemitsu Y et al (2008) Inhibition of nuclear translocation of apoptosis-inducing factor is an essential mechanism of the neuroprotective activity of pigment epithelium-derived factor in a rat model of retinal degeneration. Am J Pathol 173(5): 1326–1338. https://doi.org/10.2353/ajpath.2008.080466
22. Hernandez-Pinto A, Polato F, Subramanian P et al (2019) PEDF peptides promote photoreceptor survival in rd10 retina models. Exp Eye Res 184:24–29. https://doi.org/10.1016/j.exer.2019.04.008
23. Hudry E, Vandenberghe LH (2019) Therapeutic AAV gene transfer to the nervous system: a clinical reality. Neuron 101(5): 839–862. https://doi.org/10.1016/j.neuron.2019.02.017
24. He X, Fu Y, Ma L et al (2023) AAV for gene therapy in ocular diseases: Progress and prospects. Research (Wash D C) 6:0291. https://doi.org/10.34133/research.0291
25. Xiong W, Wu DM, Xue Y et al (2019) AAV cis-regulatory sequences are correlated with ocular toxicity. Proc Natl Acad Sci USA 116(12):5785–5794. https://doi.org/10.1073/pnas.1821000116
26. Khabou H, Cordeau C, Pacot L et al (2018) Dosage thresholds and influence of transgene cassette in adeno-associated virus-related toxicity. Hum Gene Ther 29(11):1235–1241. https://doi.org/10.1089/hum.2018.144
27. Cehajic-Kapetanovic J, Xue K, Martinez-Fernandez de la Camara C et al (2020) Initial results from a first-in-human gene therapy trial on X-linked retinitis pigmentosa caused by mutations in RPGR. Nat Med 26(3): 354–359. https://doi.org/10.1038/s41591-020-0763-1
28. Gange WS, Sisk RA, Besirli CG et al (2022) Perifoveal Chorioretinal atrophy after subretinal Voretigene Neparvovec-rzyl for RPE65-mediated Leber congenital Amaurosis. Ophthalmol Retina 6(1):58–64. https://doi.org/10.1016/j.oret.2021.03.016
29. Nakajima T, Nakamaru K, Ido E et al (2000) Development of novel simian immunodeficiency virus vectors carrying a dual gene expression system. Hum Gene Ther 11(13): 1863–1874. https://doi.org/10.1089/10430340050129486
30. Ikeda Y, Goto Y, Yonemitsu Y et al (2003) Simian immunodeficiency virus-based lentivirus vector for retinal gene transfer: a preclinical safety study in adult rats. Gene Ther 10(14): 1161–1169. https://doi.org/10.1038/sj.gt.3301973
31. Ikeda Y, Yonemitsu Y, Miyazaki M et al (2009) Acute toxicity study of a simian

immunodeficiency virus-based lentiviral vector for retinal gene transfer in nonhuman primates. Hum Gene Ther 20(9):943–954. https://doi.org/10.1089/hum.2009.048

32. Miyazaki M, Ikeda Y, Yonemitsu Y et al (2003) Simian lentiviral vector-mediated retinal gene transfer of pigment epithelium-derived factor protects retinal degeneration and electrical defect in Royal College of surgeons rats. Gene Ther 10(17):1503–1511. https://doi.org/10.1038/sj.gt.3302028

33. Li X, Le Y, Zhang Z et al (2023) Viral vector-based gene therapy. Int J Mol Sci 24(9). https://doi.org/10.3390/ijms24097736

34. Ikeda Y, Yonemitsu Y, Miyazaki M et al (2009) Stable retinal gene expression in nonhuman primates via subretinal injection of SIVagm-based lentiviral vectors. Hum Gene Ther 20(6):573–579. https://doi.org/10.1089/hum.2009.009

Chapter 21

CAR-T Cell Therapy for Central Nervous System Tumors

Masasuke Ohno and Mitsugu Fujita

Abstract

Chimeric antigen receptor (CAR) T cell therapy has emerged as a promising immunotherapy for central nervous system (CNS) tumors, despite the unique biological barriers posed by the blood–brain barrier (BBB) and immunosuppressive tumor microenvironment (TME). This chapter comprehensively reviews the evolution of CAR designs, gene delivery methods using lentiviral vectors (LVVs), and clinical applications for CNS tumors. We discuss current issues in CAR-T cell therapy for CNS tumors, while highlighting the strategies about new CAR designs and novel delivery methods to overcome these issues.

Key words Chimeric antigen receptor T cell therapy, Central nervous system tumors, Glioblastoma, Lentiviral vectors, Tumor microenvironment, Blood–brain barrier

1 Introduction

Chimeric antigen receptor (CAR) T cell therapy is an innovative cell therapy that uses genetic engineering to modify a patient's own T cells. Since the first FDA approval in 2017, six products for CAR-T cell therapy targeting CD19 or B cell maturation antigen have been approved; they have become effective treatment options for relapsed or refractory hematological cancers [1].

In contrast, its application to solid tumors remains challenging due to complex physical and immunological barriers [2]. Central nervous system (CNS) tumors are a particularly challenging area, resistant even to immune checkpoint inhibitors that have dramatically improved outcomes in both hematological and solid tumors [3]. This therapeutic resistance has reinforced the notion that the CNS is an immunoprivileged organ, making researchers cautious in their approaches to immunotherapy. Recently, however, the CNS has been shown to have its own unique immune system [4]. This distinct immune milieu, initially perceived as an obstacle, has become the basis for innovative therapeutic strategies, such as CAR-T cell therapy for CNS tumors.

Kazuto Kobayashi et al. (eds.), *Lentiviral Vectors: Applications for Basic and Clinical Neuroscience*, Methods in Molecular Biology, vol. 2974, https://doi.org/10.1007/978-1-0716-4807-0_21,

This chapter provides a comprehensive overview of the basic structure and therapeutic mechanisms of CAR-T cell therapy. It then reviews the current status and issues regarding its clinical applications in the solid tumors. In addition, distinctive strategies regarding the CAR-T cell therapy for CNS tumors will be discussed.

2 CAR-T Cell Therapy

2.1 Overview of CAR-T Cell Therapy

CAR-T cell therapy is a cancer immunotherapy that integrates cellular and genetic engineering technologies. The therapy involves harvesting a patient's own T cells, genetically modifying the T cells to express CARs, and reinfusing them back into the patient [5]. CARs are synthetic receptor-like molecules that enable T cells to specifically recognize cancer cells expressing a target antigen (Fig. 1) [6, 7]. Conventional T cells recognize peptide antigens presented by MHC molecules through their T cell receptor (TCR) complex after intracellular processing of tumor molecules (Fig. 2a). In contrast CAR-expressing T cells recognize cell surface molecules of tumor cells in an MHC-independent manner to kill the tumor cells (Fig. 2b). The uniqueness of this therapy is that it integrates the innate cytotoxic capacity of T cells with the specificity of antibodies to produce a potent antitumor response.

CAR-T cell therapy was first clinically validated in hematological malignancies and demonstrated outstanding efficacy [1]. After the remarkable success in this area, the next challenge was to expand this therapy to solid tumors [2]. These challenges include the immunosuppressive environment of solid tumor tissues, the

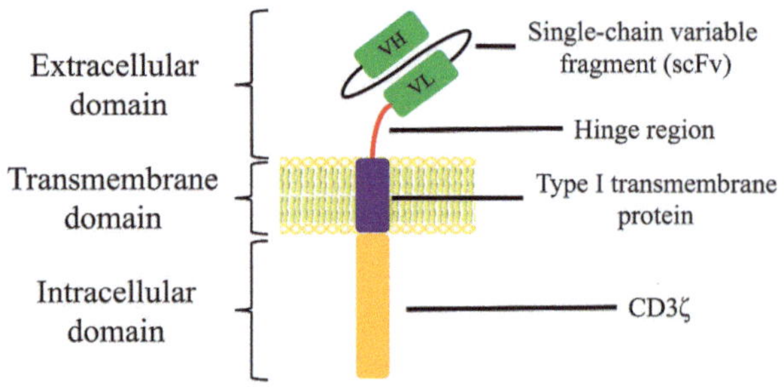

Fig. 1 Basic structure of a chimeric antigen receptor (CAR). CAR consists of three fundamental domains: the extracellular domain, transmembrane domain, and intracellular domain. The extracellular domain contains a single-chain variable fragment (scFv) composed of variable heavy (VH) and variable light (VL) chain regions, and a hinge region. The transmembrane domain is derived from type I transmembrane proteins and anchors the CAR in the cell membrane. The intracellular domain contains CD3ζ signaling components that initiate T cell activation

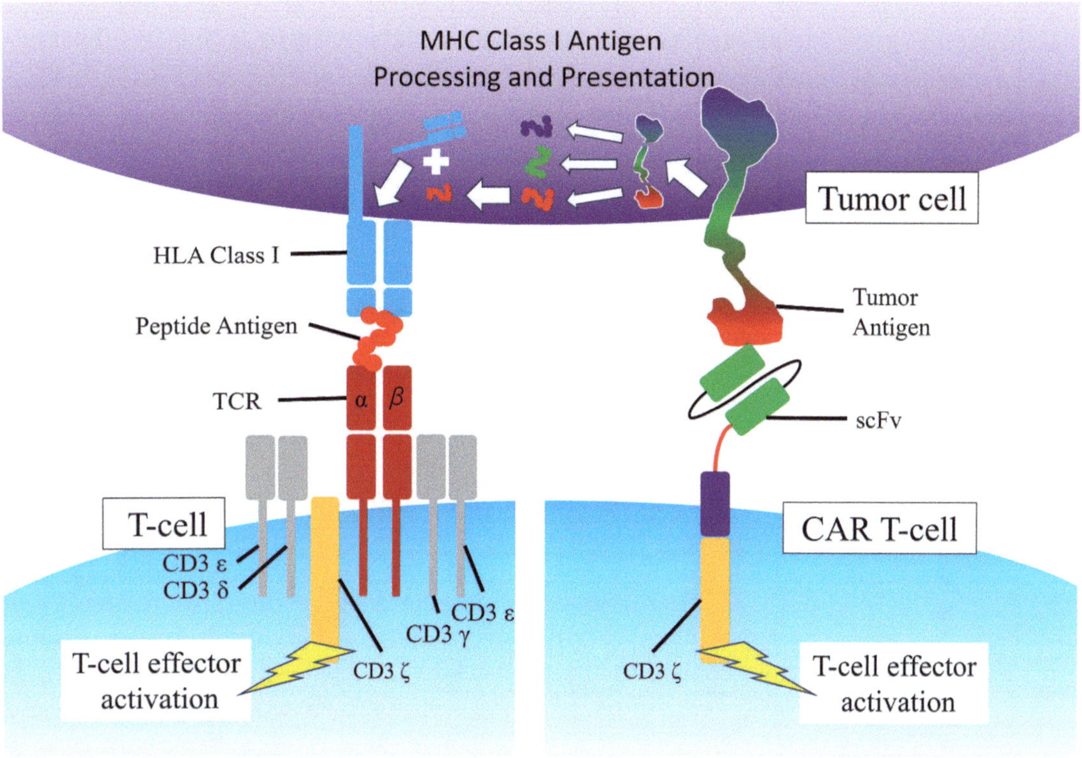

Fig. 2 Comparison of antigen recognition mechanisms between conventional T cells and CAR-T cells. (**a**) Conventional T cells recognize peptide antigens presented by MHC class I molecules through their T cell receptor (TCR) complex after intracellular processing of tumor molecules. (**b**) CAR-T cells directly recognize cell surface tumor antigens in an MHC-independent manner through their chimeric antigen receptor. Both recognition mechanisms lead to T cell effector activation, but CAR-T cells bypass the requirement for MHC presentation, enabling more direct tumor cell recognition

heterogeneity of tumor antigen expressions, and the limited ability of CAR-T cells to reach tumor sites [5]. To address these challenges, attempts have been made to improve the molecular structure of CARs, identify novel target antigens, and make CAR-T cells tolerant to the immunosuppressive environment of tumor tissues. This section describes these technological points in detail, starting with the basic structure and modification of CARs.

2.2 Basic Structure of CARs

CAR consists of three basic structures (Fig. 1). The first structure is the extracellular domain, which is responsible for antigen-recognition region and a hinge region. The second structure is the transmembrane domain, which penetrates the cell membrane. The third structure is the intracellular domain, which is responsible for signal transduction. These three fundamental elements collectively serve as an essential core framework for CAR-T cell function.

The extracellular domain of CAR contains a single-chain variable fragment (scFv), which consists of variable heavy (VH) and variable light (VL) chain regions connected by a flexible linker. The VH and VL regions determine the antigen specificity, the most important factor in determining therapeutic efficacy of CAR-T cells. The ideal target antigen is one that is expressed only on tumor cells and not on normal cells. However, such an antigen is extremely difficult to identify; most tumor-associated antigens (TAAs) are expressed not only on tumor cells but also on normal cells [8]. From this point of view, the affinity of the scFv against the target antigen has a significant impact on the function of CAR-T cells. Too high an affinity leads to an increased risk of off-target toxicity, such as aggression against normal cells [9, 10]. Conversely, insufficient affinity decreases binding to tumor cells and therefore reduces the therapeutic efficacy.

The hinge region, which is part of the extracellular domain, connects the scFv to the transmembrane domain [8, 11, 12]. This domain is important for the formation of immune synapses by providing flexibility in the positioning of the scFv [12, 13]. Their original structures are derived from IgG, CD28, and CD8 [12, 13].

The transmembrane domain anchors the CAR to the T cell membrane. Their original structures are derived from type I transmembrane proteins, such as CD3ζ, CD28, CD4, and CD8α [8, 11, 12]. The choice of these transmembrane domains significantly affects stability, signal transduction, T cell activation, cell survival, and persistence of the CAR.

The intracellular domain transmits activation signals to T cells [8, 11, 12]. The CD3ζ chain is particularly important for this function. The CD3ζ chain contains an immunoreceptor tyrosine activation motif (ITAM) whose tyrosine residues undergo phosphorylation by lymphocyte-specific protein tyrosine kinase (LCK) to initiate downstream signaling pathways.

2.3 Advanced Modifications of CARs

CARs that have only the basic structure described above are classified as first-generation CARs (Fig. 3a). They have limited proliferative capacity and persistence in vivo as well as limited therapeutic efficacy [14]. To address these issues, second-generation CARs have been developed by incorporating co-stimulatory domains such as CD28, 4-1BB, OX40, or inducible co-stimulator (ICOS) to enhance T cell activation, proliferation, differentiation, cytokine production, and persistence (Fig. 3b) [1]. These modifications have resulted in superior antitumor efficacy, and second-generation CARs have become the basis for all FDA-approved CAR-T cell products [8, 15]. The functional characteristics of second-generation CARs depend on the specific type of co-stimulatory domain selected.

Third-generation CARs have also been developed by combining multiple co-stimulatory domains (Fig. 3c). These new CARs were developed for better biological and/or clinical outcomes than the second-generation CARs, but no clear advantages have been

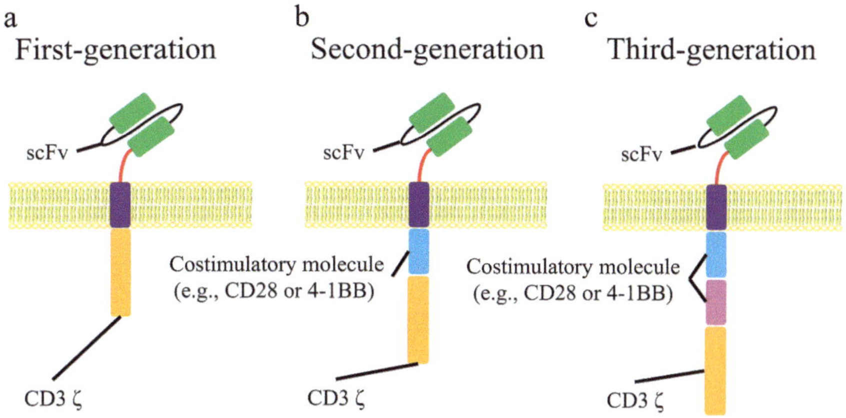

Fig. 3 Evolution of CAR generations and their structural modifications. (**a**) First-generation CARs contain only the CD3ζ signaling domain for T cell activation. (**b**) Second-generation CARs incorporate a single co-stimulatory molecule (e.g., CD28 or 4-1BB) to enhance T cell proliferation and persistence. (**c**) Third-generation CARs feature multiple co-stimulatory domains to further improve T cell function and survival

demonstrated to date [16, 17]. Rather, because the gene size of third-generation CARs is larger than that of second-generation CARs, it has raised concerns about increased manufacturing complexity and cost to overcome lower gene transfer efficiency.

Other advanced approaches are currently being applied to develop "next-generation" CARs (Fig. 4) [8, 11, 12, 18, 19]. These include tandem CARs with multiple scFvs to recognize multiple tumor antigens [20], bivalent CARs that allow the simultaneous expression of two different CARs [21], logic-gated activation systems that allow precise control of CAR activation [18], switchable control mechanisms that allow external regulation [18], and universal CAR platforms that provide controllable antigen targeting (Fig. 4a) [11]. There are two transgenic strategies for these modifications. One is the single vector systems incorporating multiple genes in tandem, which simplifies manufacturing and increases cell uniformity (Fig. 4b) [22–24]. The other is the multi-vector sequential approaches that allow independent control of gene expression levels and timing while avoiding vector payload limitations [25]. The latter multi-vector approach increases manufacturing complexity and cellular stress [26].

3 CAR Gene Delivery Methods

3.1 Overview of CAR Gene Delivery Methods

CAR gene delivery methods can be categorized into viral vectors, nonviral vectors, and mRNA-based approaches. Viral vectors, particularly lentiviral vectors (LVVs) and retroviral vectors (RVVs), are the primary methods of CAR gene delivery in clinical applications. LVVs are particularly well suited to produce CAR-T cells because

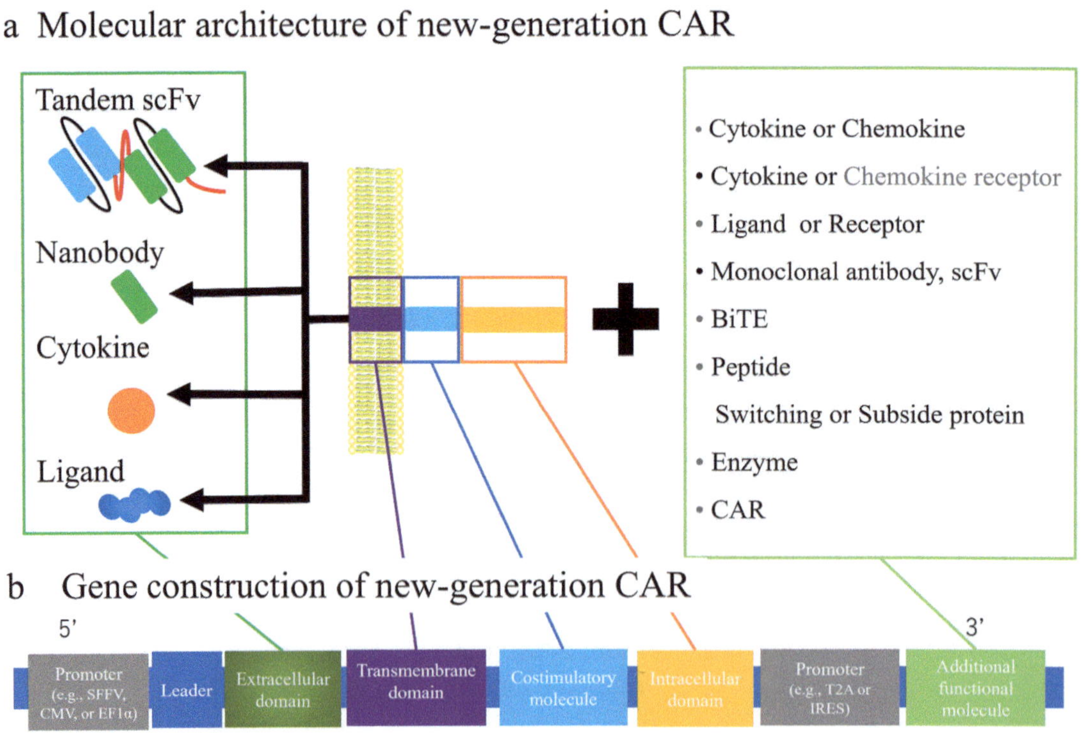

Fig. 4 Molecular architecture and gene construction of next-generation CAR. (**a**) Next-generation CARs incorporate diverse modifications to address challenges in solid tumor therapy. The extracellular domain is modified with tandem scFvs, nanobodies, cytokines, or ligands to recognize multiple tumor antigens. Additional functional molecules co-expressed with the CAR enhance T cell trafficking to tumor sites, improve tumor penetration, overcome the immunosuppressive microenvironment, and maintain T cell cytotoxicity. (b) A single lentiviral vector (LVV) enables uniform CAR-T cell production with simplified manufacturing. The vector contains tandemly arranged genes for CAR and additional functional molecules, each controlled by separate promoters for simultaneous expression

they can efficiently infect both dividing and nondividing cells and also deliver relatively large DNA [27]. All current FDA-approved CAR-T cell products utilize either LVVs or RVVs for gene delivery [12, 28].

Nonviral approaches such as transposon systems and programmable nucleases have safety and cost advantages. However, their lower gene delivery efficiency and shorter duration of CAR gene expression have limited their widespread use [26]. This section describes these methods in detail, focusing on vector systems and their practical applications in CAR-T cell production.

3.2 Vector Systems

The most promising method for CAR gene delivery is the use of LVVs, which allows efficient gene delivery into T cells at both mitotic and non-mitotic stages and stable integration of CAR gene cassettes into the host genome [27]. They are also capable of carrying relatively large genes of 8–10 kb [29]. LVVs are designed as self-inactivating (SIN) vectors by the following mechanism. The LVV genome artificially lacks the U3 region of the 3′ long terminal repeat (LTR). This deletion inactivates the promoter activity of the complementary 5′ LTR after it is incorporated into the host genome. In this way, it prevents the activation of LVV-derived unwanted genes by eliminating their ability to initiate transcription [30, 31]. Approved products based on LVVs include tisagenlecleucel (Kymriah®), lisocabtagene maraleucel (Breyanzi®), and idecabtagene vicleucel (Abecma®), and ciltacabtagene autoleucel (Carvykti®) [32].

RVVs are another promising method for CAR gene delivery and are characterized by their ability to specifically deliver genes only into dividing cells [26]. RVVs have the advantage of simpler structure and easier scale production than LVVs, but they are more immunogenic than LVVs. In addition, RVVs tend to be randomly integrated in the genome, increasing the risk of unintended gene expression or insertional mutagenesis. Despite these limitations, RVV-based products have been approved with clinical success, including brexucabtagene autoleucel (Tecartus®) and axicabtagene ciloleucel (Yescarta®) [28].

Nonviral vectors are an alternative approach to CAR gene delivery that is safer and less inexpensive than the viral vectors [26]. These include transposon systems like Sleeping Beauty (SB) and piggyBac (PB), programmable nucleases such as CRISPR/Cas9, and various forms of DNA vectors including plasmids, minicircles, and nanoplasmids. Nonviral vectors have limited gene delivery efficiency and gene expression duration, primarily due to inefficient integration of CAR genes into the host cell genome [26]. Current research is focusing on addressing these issues.

3.3 Practice of CAR Gene Delivery by LVVs

Clinical-grade CAR-T cell production is performed under strict quality control standards in accordance with cGMP guidelines (Fig. 5) [28]. Patient-derived peripheral blood mononuclear cells (PBMCs) are isolated by apheresis or leukapheresis (Fig. 5a). They contain various immune cells, such as B cells, macrophages, monocytes, NK cells, and T cells. The first manufacturing step is to isolate T cell subsets from these PBMCs by magnetic bead-based selection or selective expansion. Magnetic bead-based selection uses antibody-bound magnetic beads that bind to target T cell populations, and the bead-bound T cells are then separated by magnetic columns. In CAR-T cell production, T cells are selected with anti-CD3 antibodies alone or a combination of anti-CD4 and anti-CD8 antibodies. Meanwhile, LVVs are produced using HEK293T cells as the host (Fig. 5b). LVV particles are harvested from the HEK293 culture medium, purified, and quality tested. The purified LVVs are then transduced into patient-derived activated T cells, and the CAR gene is integrated into the genome. These genetically

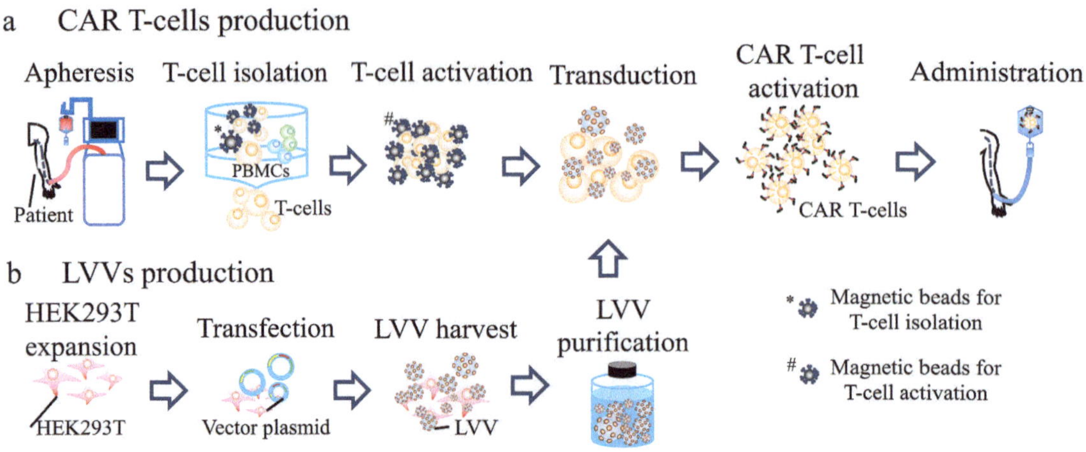

Fig. 5 Manufacturing process of CAR-T cells and LVVs. (**a**) The CAR-T cell production process: Patient T cells are isolated from peripheral blood mononuclear cells (PBMCs) by apheresis, followed by magnetic bead-based T cell isolation. The isolated T cells undergo activation using anti-CD3/CD28 antibody-coated beads, then transduction with LVVs carrying the CAR gene. The CAR-expressing T cells are activated, expanded to therapeutic numbers, and administered back to the patient. (**b**) The LVV production process: HEK293T cells are expanded in culture, transfected with vector plasmids, and used to harvest LVVs. The vectors undergo purification steps before CAR-T cell transduction

modified T cells are activated in culture by exposure to beads coated with monoclonal antibodies targeting CD3 and CD28 and expanded until they reach therapeutic numbers.

High quality, high titer LVVs are required for CAR-T cell production [25]. Modified HEK293T cell lines have been developed for efficient mass production and cost reduction in LVV purification. These specialized cell lines can be grown in serum-free suspension medium and contain pre-integrated viral genes (gag/pol, rev, envelope) to eliminate some gene transfection steps [33, 34]. For LVV purification process, tangential flow filtration and chromatography replaced conventional ultracentrifugation, allowing mass production while increasing concentration and recovery rates [34].

4 CAR-T Cell Therapy for CNS Tumors

4.1 Overview of CAR-T Cell Therapy for CNS Tumors

The traditional view of CNS immune privilege stems from Medawar's 1940s experiments showing prolonged survival of skin allografts in brain tissues [35]. The presence of the blood–brain barrier (BBB) and the absence of lymphatic vessels initially led to the belief that the CNS was immunologically isolated. However, recent discoveries have revealed that the CNS has a "glymphatic" system for waste clearance [36]. Furthermore, the identification of dural lymphatic vessels established a clear immunological pathway between the CNS and peripheral immune systems [37].

The BBB acts as a selective gateway for CNS immune cell traffic, especially during inflammation and infection, while maintaining a restrictive barrier function against immune cell infiltration [38]. This is also a limitation for CAR-T cell therapy for CNS tumors. In fact, intravenous (IV) administration of CAR-T cells often fails to adequately reach CNS tumor tissues. To overcome this issue, more direct delivery approaches have emerged. One is stereotactic intratumoral administration via subcutaneous reservoir-connected catheters, and the other is intracerebroventricular (ICV) administration via intrathecal or intraventricular routes [39]. ICV administration reduces the severity of cytokine release syndrome but requires careful management of inflammation-associated tumor neurotoxicity. These new cell administration techniques, in addition to a better understanding of CNS immunity, have established CAR-T cell therapy as a viable treatment option for CNS tumors, despite the unique challenges posed by the CNS immune environment. This section reviews specific CNS tumor types, their targetable antigens, and the clinical outcomes of CAR-T cell therapy trials.

4.2 Glioblastoma (GBM)

Glioblastoma (GBM) is the most aggressive primary brain tumor, with a worldwide incidence of 3 per 100,000 population and a peak occurrence in the 55–60 years age group [40]. Despite standard treatment including surgery, radiation therapy, and chemotherapy, the prognosis remains dismal, with a median survival of 14–-15 months and a 5-year survival rate of only 6.8% [41].

Several target antigens have been identified for CAR-T cell therapy for GBM. Interleukin-13 receptor alpha 2 (IL13R alpha 2) shows high expression levels in almost all GBM cases but is marginally present in normal CNS tissues [42]. Epidermal growth factor receptor variant III (EGFRvIII), a variant lacking exons 2–7 of wild-type EGFR, appears in 30% of GBM cases. This variant lacks the extracellular ligand-binding domain while preserving its intracellular signaling domain, resulting in sustained activation of proliferative signals in the absence of cytokine stimulation [43]. HER2 appears in 80% of GBM cases and is expressed particularly on CD133-positive tumor stem cells [44]. Disialoganglioside 2 (GD2) is a glycolipid antigen highly expressed in neuroectodermal tumors, including GBM, and accelerates tumor progression [45]. EphA2 appears in 90% of GBM tissues but is rarely found in healthy CNS tissues. B7-H3 appears in more than 50% of GBM cases, but its expression is also limited in normal CNS tissues [46, 47].

Clinical trials targeting these antigens have shown promising results, especially in recurrent GBM. Twelve trials focusing primarily on IL13R alpha 2 and EGFRvIII have been conducted to date [48, 49]. Median survival was 6 months at baseline, which improved to 7.7 months with IL13R alpha 2-targeted therapy

[50, 51], 8 months with EGFRvIII-targeted therapy [51, 52], 11.1 months with HER2-targeted therapy [51], and 10 months with GD2-targeted therapy [53]. Among the recent developments in CAR-T cell therapy, the CARv3-TEAM-E system is a particularly notable advancement, in which a single LVV carries both the CAR gene and the T-cell-engaging antibody molecules (TEAMs) gene [54]. A phase I clinical trial using CAR-T cells generated by this system in recurrent GBM demonstrated a safe but transient antitumor effect. In this system, the antitumor effect is enhanced by a dual targeting mechanism. First, CARv3-TEAM-E T cells directly target tumor cells expressing the EGFRvIII variant. Second, these cells secrete TEAMs, which act as bridges to recruit T cells responsible for tumor cells expressing wild-type EGFR. In this way, the CARv3-TEAM-E-T cells overcome the issue of EGFRvIII antigen loss.

4.3 Diffuse Midline Glioma (DMG)

DMG is an aggressive CNS tumor that arises along the midline structures of the CNS. Originally classified as diffuse intrinsic pontine glioma (DIPG), molecular analysis has revealed its presence in the thalamus, midbrain, cerebellum, and spinal cord. The H3 K27M mutation has been found in this disease, which led to a new name for the disease in the 2021 WHO classification, "H3 K27-altered diffuse midline glioma" [55]. Death from DMG accounts for 25% of pediatric cancer deaths, with a median survival of approximately 1 year [56]. The invasive nature of the disease precludes surgical resection, and the only standard treatment is radiation therapy (54–60 Gy over 6 weeks), with no significant improvement in outcomes for decades [56].

Three antigens are known as target candidates for CAR-T cell therapy for DMG. GD2 is uniquely highly expressed across DMG cases, particularly in tumors harboring the H3K27M mutation [57]. This antigen is considered a promising target because its expression level is variable in GBM while it is constant in DMG. In addition, DMG tissue is poorly infiltrated by immune cells, and the immunosuppressive microenvironment itself, consisting primarily of immunosuppressive cells, is considered as a therapeutic target of DMG, which makes B7-H3 as a new therapeutic candidate in this disease [58]. HER2 is also targeted in DMG, but its therapeutic application is similar to that of GBM; clinical trials usually include both tumor types in the cohort [59, 60].

In parallel, clinical trials have investigated both IV and ICV administration routes of CAR-T cells in DMG. In the GD2-targeted therapy, the combination of IV and ICV improved intratumoral infiltration of CAR-T cells, while simultaneously reducing systemic inflammatory responses and cytokine release syndrome [61]. Most recently, CAR-T cells simultaneously targeting multiple antigens, including B7-H3, HER2, EGFR806, and IL13Rα2, have been investigated to address tumor antigen heterogeneity and potential resistance [62].

4.4 Medulloblastoma

Medulloblastoma is the most common pediatric CNS malignancy and the second most frequent pediatric cancer after leukemia [63]. The current standard of care is a combination of surgery, radiation therapy, and chemotherapy, with long-term survival rates of 70–80% achieved in those without tumor disseminations [64]. The 2021 WHO classification classifies the disease into four molecular subgroups: WNT, SHH, Group 3, and Group 4, each with distinct biological and clinical characteristics [55]. Approximately 30% of patients die from the disease, and survivors suffer long-term treatment-related sequelae [65].

GD2 is highly expressed in approximately 80% of medulloblastoma cases as well as GBM as described earlier [66]. HER2 appears in approximately 40% of medulloblastoma cases and correlates with poor survival [67]. Other targets include EphA2, IL13Rα2, B7-H3, ROBO1, and PRAME, all of which have shown favorable results in preclinical studies [68, 69].

Clinical trials of CAR-T cell therapy for medulloblastoma are limited. A trial of two patients with recurrent medulloblastoma who received GD2-targeeted CAR-T cell therapy demonstrated temporary clinical improvement, including reduction in ataxia and stable disease progression on imaging after two to three treatment cycles of treatment, but both patients eventually experienced disease progression [70].

4.5 CNS Metastasis

CNS metastases account for 30% of adult CNS tumors [71]. Lung cancer, breast cancer, and melanoma are the most common, with 20–40% of those developing CNS metastases during their disease course [72]. Median survival without treatment is only 1–2 months. Current standards of care include surgery, radiation therapy (stereotactic radiosurgery and whole-brain radiation), and chemotherapy, but outcomes remain poor [73].

HER2 is a primary target for CAR-T cell therapy in CNS metastases, particularly breast cancer metastasis. HER2 is overexpressed in approximately 25% of breast cancer cases [74]. Up to 50% of patients with HER2-positive breast cancer develop CNS metastases [75]. The HER2 signaling cascade activates several pathways, including Ras-MAPK and PI3K, whose alteration leads to increased metastasis formation [76].

Historically, clinical trials of CAR-T cell therapy have often excluded cases of CNS metastases due to concerns about neurological complications from cytokine release syndrome. However, several clinical trials of primary CNS tumors have shown that CAR-T cells can pass the BBB, rekindling interest in the treatment of CNS metastases [77]. Recent preclinical studies using HER2-trageted CAR-T cells for CNS metastases have also shown its safety profiles, suggesting the feasibility of this approach [78].

5 Current Issues of CAR-T Cell Therapy for CNS Tumors

Although CAR-T cell therapy has revolutionized cancer treatment, several issues remain to be resolved. These issues span many aspects, from cellular function to safety concerns. This section discusses seven major issues: tumor microenvironment (TME)-mediated inhibition, T cell exhaustion, on-target off-tumor (OTOT) toxicity, antigen heterogeneity, neurotoxicity, secondary malignancy risks, and gene delivery issues of CARs.

5.1 TME-Mediated Inhibition

TME immunologically impairs CAR-T cell function [79]. TME consists of tumor cells, immune cells, stromal cells, and their surrounding extracellular matrix and humoral factors, which collectively affect tumor growth, metastasis, and therapy resistance [80]. These inhibitory effects are manifested through two distinct mechanisms: physical barriers that prevent CAR-T cell infiltration and immunological barriers that suppress their function.

Dense extracellular matrix, poor vasculature, and fibroblasts are physical barriers in TME that prevent CAR-T cell infiltration into CNS tumors. As new strategies to address these barriers, new CAR-T cells have been developed that promote intratumor infiltration by expressing matrix-degrading enzymes and chemokine receptors [2, 81–83]. The intracerebroventricular (ICV), mentioned earlier, is an excellent method to directly bypass the physical barriers of the BBB and TME.

TME in CNS tumor promotes an immunosuppressive state through multiple mechanisms [79]. Myeloid-derived suppressor cells (MDSCs), tumor-associated macrophages (TAMs), and microglia form the core of immunosuppression, promoting tumor growth while suppressing immune responses [84]. Regulatory T cells (Tregs) and PD-L1-expressing tumor cells potently suppress antitumor immunity [85]. Immunosuppressive cytokines, including TGF-beta, also involve [18, 86]. The new-generation CARs incorporate various functions to regulate TME [11, 12, 18, 87].

5.2 T-Cell Exhaustion

T cell exhaustion is often discussed in the context of TME; this is a fundamental issue in CAR-T cell therapy because they are derived from T cells themselves [88, 89]. Continuous antigen exposure in TME can exhaust CAR-T cells and impair their function. Attempts to mitigate CAR-T cells exhaustion include optimization of co-stimulatory domains [90], genome editing of CAR-T cells [91], modulating CAR expression [92], and targeting transcription factors and molecules involved in exhaustion [2, 12, 93].

5.3 OTOT Toxicity

When the target antigens are expressed in normal tissues, CAR-T cells may attack these normal cells and cause serious side effects [8]. This phenomenon, in which CAR-T cells correctly recognize the target antigen but attack normal tissues other than the tumor, is called on-target, off-tumor toxicity (OTOT) [8]. To minimize this issue, it is important to select appropriate target antigens and optimize the CAR design. To increase target specificity, novel designs have been proposed, such as two-step antigen recognition systems and adaptable receptor designs to control CAR-T cell activation after administration [8, 11]. In addition, regulation of CAR-T cell activity by drug-mediated activation, suicide switches, or temporary expression systems has been attempted [8].

5.4 Antigen Heterogeneity of Tumor Cells

Tumor antigen heterogeneity is a major obstacle for CAR-T cell therapy in CNS tumors. The mechanisms underlying this phenomenon are multifaceted. Tumor cells can downregulate antigens, clonally expand antigen-negative cells, and mutate antigens [94]. These adaptive responses allow the tumor cells to evade T cell recognition, ultimately limiting therapeutic efficacy. The clinical impact of this phenomenon has been demonstrated in GBM patients, where targeted therapy against EGFRvIII resulted in the emergence of antigen-negative GBM cells and treatment resistance [52].

One solution to tumor antigen heterogeneity is a multitargeting strategy. For example, the aforementioned tandem CARs incorporate multiple scFvs within a single CAR gene, thereby allowing CAR-T cells to recognize multiple tumor antigens simultaneously. Specifically, multivalent CAR-T cells have been developed that target combinations of antigens, such as HER2, IL13Rα2, and EphA2 [95]. The aforementioned two-step antigen recognition system is also a promising approach to address tumor antigen heterogeneity.

5.5 Neurotoxicity

Neurotoxicity can occur as a side effect of CAR-T cell therapy and manifests by two distinct mechanisms: a systemic inflammatory response and a tumor-associated inflammatory response. The systemic response, known as immune effector cell-associated neurotoxicity syndrome (ICANS), is primarily due to the excessive production of inflammatory cytokines, particularly IL-6 and IL-1, during CAR-T cell activation [96, 97]. These cytokines impair the integrity of the BBB and cause neurological dysfunction.

Tumor inflammation-associated neurotoxicity (TIAN) poses a unique issue for CNS tumors [97]. TIAN is manifested by local inflammation at the tumor site and causes severe edema leading to increased intracranial pressure, tissue herniation, and CSF flow obstruction [60]. This condition can lead to severe consequences, particularly in the brainstem region.

The recent CAR-T cell administration methods described above are also useful in managing these toxicities. ICV administration reduces systemic toxicity compared to IV administration [59]. TIAN can occur due to the higher local densities of CAR-T cells with this method, but these effects usually resolve with repeated cell administration [98]. TIAN occurrence is associated with tumor location to some extent [99].

Anti-IL-6 therapy (tocilizumab) is effective for systemic toxicity when intervened early, and concomitant corticosteroids are useful in managing local inflammatory responses [100]. Such careful monitoring and early intervention of neurological symptoms has become standard practice in managing both ICANS and TIAN, leading to improved safety of CAR-T cell therapy in recent clinical trials.

5.6 Gene Delivery Issues of Complex CARs

Increasing complexity of CAR design has created a bottleneck in the current gene delivery system. While conventional second-generation CARs (3–4 kb) can be efficiently packaged into LVVs or RVVs, newer CAR designs face payload limitations [101]. These advanced CARs often require multiple functional elements, such as additional co-stimulatory domains, regulatory switches, or supplementary genes for enhanced functionality. When combining CAR genes with additional functional elements in a single vector, the total construct size often approaches or exceeds the packaging limits of LVVs (8–10 kb) and RVVs [102].

In contrast, the use of multiple vectors to deliver different genes separately poses its own complications. Each additional transduction step reduces production efficiency, increases cell manipulation, and complicates manufacturing [25]. Multiple rounds of gene transfer result in decreased cell viability and recovery, eventually reducing the final product quality [103]. These multiple steps extend manufacturing timelines, making the overall production process more time-consuming, complex, and costly [26].

To address these limitations, several alternative approaches have been attempted. The use of CRISPR-CAS9 system allows precise gene integration at specific loci and has the potential to improve gene expression stability and regulation [102]. Transposon-based systems offer greater cargo capacity but lower integration efficiency; they require further optimization for clinical applications [26].

5.7 Secondary Malignancy Risk and Long-Term Safety

Viral vectors used in CAR-T cell therapy are at potential risks for secondary malignancies via insertional oncogenesis [104]. Both LVVs and RVVs are randomly integrated into the host genome and can disrupt tumor suppressor genes or activate proto-oncogenes. RVVs tend to insert near promoter regions and regulatory elements, increasing the risk of oncogene activation. LVVs, in contrast, are preferentially incorporated near transcriptionally active genes and transcription initiation sites and also can disrupt tumor suppressor genes [26].

Recent surveillance data from over 27,000 CAR-T cell therapies identified 22 cases of T cell malignancies, most of which occurred within 2 years of treatment [105]. In some cases, incorporation of the CAR gene into lymphoma cells was observed, suggesting vector-mediated oncogenesis.

Self-inactivating (SIN) LVVs have human-derived promoters such as EF-1α and phosphoglycerate kinase, reducing the risks of insertional mutations [25]. In a long-term follow-up study of 724 cases reported, only 1 case of T cell lymphoma has been reported, with no CAR gene insertions observed [106]. This finding indicates the relative safety of LVVs. Nevertheless, new safety strategies are being explored, such as suicide genes for selective CAR-T cell elimination and reversible CAR activation systems using soluble antigen-binding modules [28, 103]. Non-integrating vectors, such as mRNA, only induce CAR expression transiently, reducing long-term risks, but require multiple administrations to maintain antitumor efficacy [103].

Acknowledgments

This work was supported by the Grants-in-Aid for Scientific Research from the Ministry of Education, Culture, Sports, Science and Technology of Japan (MEXT; 22 K09223 to MO, and 21 K09167 to MF).

Funding This study was funded by the Grants-in-Aid for Scientific Research from the Ministry of Education, Culture, Sports, Science and Technology of Japan (Grant number 25K12376).

Conflict of Interest Statement The authors have no conflicts of interest to declare that are relevant to the content of this chapter.

References

1. June CH, Sadelain M (2018) Chimeric antigen receptor therapy. N Engl J Med 379:64–73. https://doi.org/10.1056/NEJMra1706169
2. Albelda SM (2024) CAR T cell therapy for patients with solid tumours: key lessons to learn and unlearn. Nat Rev Clin Oncol 21:47–66. https://doi.org/10.1038/s41571-023-00832-4
3. Omuro A, Brandes AA, Carpentier AF et al (2023) Radiotherapy combined with nivolumab or temozolomide for newly diagnosed glioblastoma with unmethylated MGMT promoter: an international randomized phase III trial. Neuro-Oncology 25:123–134. https://doi.org/10.1093/neuonc/noac099
4. Proulx ST (2021) Cerebrospinal fluid outflow: a review of the historical and contemporary evidence for arachnoid villi, perineural routes, and dural lymphatics. Cell Mol Life Sci 78:2429–2457. https://doi.org/10.1007/s00018-020-03706-5
5. Majzner RG, Mackall CL (2019) Clinical lessons learned from the first leg of the CAR T cell journey. Nat Med 25:1341–1355. https://doi.org/10.1038/s41591-019-0564-6
6. Kuwana Y, Asakura Y, Utsunomiya N et al (1987) Expression of chimeric receptor composed of immunoglobulin-derived V regions and T-cell receptor-derived C regions. Biochem Biophys Res Commun 149:960–968.

https://doi.org/10.1016/0006-291x(87)90502-x

7. Eshhar Z, Waks T, Gross G, Schindler DG (1993) Specific activation and targeting of cytotoxic lymphocytes through chimeric single chains consisting of antibody-binding domains and the gamma or zeta subunits of the immunoglobulin and T-cell receptors. Proc Natl Acad Sci USA 90. https://doi.org/10.1073/pnas.90.2.720
8. Flugel CL, Majzner RG, Krenciute G et al (2023) Overcoming on-target, off-tumour toxicity of CAR T cell therapy for solid tumours. Nat Rev Clin Oncol 20:49–62. https://doi.org/10.1038/s41571-022-00704-3
9. Castellarin M, Sands C, Da T et al (2020) A rational mouse model to detect on-target, off-tumor CAR T cell toxicity. JCI Insight 5. https://doi.org/10.1172/jci.insight.136012
10. Caruso HG, Hurton LV, Najjar A et al (2015) Tuning sensitivity of CAR to EGFR density limits recognition of Normal tissue while maintaining potent antitumor activity. Cancer Res 75:3505–3518. https://doi.org/10.1158/0008-5472.CAN-15-0139
11. Rafiq S, Hackett CS, Brentjens RJ (2020) Engineering strategies to overcome the current roadblocks in CAR T cell therapy. Nat Rev Clin Oncol 17:147–167. https://doi.org/10.1038/s41571-019-0297-y
12. Larson RC, Maus MV (2021) Recent advances and discoveries in the mechanisms and functions of CAR T cells. Nat Rev Cancer 21:145–161. https://doi.org/10.1038/s41568-020-00323-z
13. Dwivedi A, Karulkar A, Ghosh S et al (2019) Lymphocytes in cellular therapy: functional regulation of CAR T cells. Front Immunol 9. https://doi.org/10.3389/fimmu.2018.03180
14. Kershaw MH, Westwood JA, Parker LL et al (2006) A phase I study on adoptive immunotherapy using gene-modified T cells for ovarian cancer. Clin Cancer Res 12:6106–6115. https://doi.org/10.1158/1078-0432.CCR-06-1183
15. Baker DJ, Arany Z, Baur JA et al (2023) CAR T therapy beyond cancer: the evolution of a living drug. Nature 619:707–715. https://doi.org/10.1038/s41586-023-06243-w
16. Milone MC, Fish JD, Carpenito C et al (2009) Chimeric receptors containing CD137 signal transduction domains mediate enhanced survival of T cells and increased antileukemic efficacy in vivo. Mol Ther 17: 1453–1464. https://doi.org/10.1038/mt.2009.83
17. Zhong X-S, Matsushita M, Plotkin J et al (2010) Chimeric antigen receptors combining 4-1BB and CD28 signaling domains augment PI3kinase/AKT/Bcl-XL activation and CD8+ T cell-mediated tumor eradication. Mol Ther 18:413–420. https://doi.org/10.1038/mt.2009.210
18. Hou AJ, Chen LC, Chen YY (2021) Navigating CAR-T cells through the solid-tumour microenvironment. Nat Rev Drug Discov 20:531–550. https://doi.org/10.1038/s41573-021-00189-2
19. Yang J, Chen Y, Jing Y et al (2023) Advancing CAR T cell therapy through the use of multidimensional omics data. Nat Rev Clin Oncol 20:211–228. https://doi.org/10.1038/s41571-023-00729-2
20. Hegde M, Mukherjee M, Grada Z et al (2016) Tandem CAR T cells targeting HER2 and IL13Rα2 mitigate tumor antigen escape. J Clin Invest 126:3036–3052. https://doi.org/10.1172/JCI83416
21. Bagley SJ, Logun M, Fraietta JA et al (2024) Intrathecal bivalent CAR T cells targeting EGFR and IL13Rα2 in recurrent glioblastoma: phase 1 trial interim results. Nat Med 1–10. https://doi.org/10.1038/s41591-024-02893-z
22. Koneru M, Purdon TJ, Spriggs D et al (2015) IL-12 secreting tumor-targeted chimeric antigen receptor T cells eradicate ovarian tumors in vivo. Onco Targets Ther 4:e994446. https://doi.org/10.4161/2162402X.2014.994446
23. Hu B, Ren J, Luo Y et al (2017) Augmentation of antitumor immunity by human and mouse CAR T cells secreting IL-18. Cell Rep 20:3025–3033. https://doi.org/10.1016/j.celrep.2017.09.002
24. Zimmermann K, Kuehle J, Dragon AC et al (2020) Design and characterization of an "all-in-one" Lentiviral vector system combining constitutive anti-GD2 CAR expression and inducible cytokines. Cancers (Basel) 12:375. https://doi.org/10.3390/cancers12020375
25. Irving M, Lanitis E, Migliorini D et al (2021) Choosing the right tool for genetic engineering: clinical lessons from chimeric antigen receptor-T cells. Hum Gene Ther 32:1044–1058. https://doi.org/10.1089/hum.2021.173
26. Metanat Y, Viktor P, Amajd A et al (2024) The paths toward non-viral CAR-T cell manufacturing: a comprehensive review of state-of-the-art methods. Life Sci 348:

122683. https://doi.org/10.1016/j.lfs.2024.122683

27. Naldini L, Blömer U, Gallay P et al (1996) In vivo gene delivery and stable transduction of nondividing cells by a lentiviral vector. Science 272:263–267. https://doi.org/10.1126/science.272.5259.263
28. Labbé RP, Vessillier S, Rafiq QA (2021) Lentiviral vectors for T cell engineering: clinical applications, bioprocessing and future perspectives. Viruses 13:1528. https://doi.org/10.3390/v13081528
29. Powell SK, Rivera-Soto R, Gray SJ (2015) Viral expression cassette elements to enhance transgene target specificity and expression in gene therapy. Discov Med 19:49–57
30. Zufferey R, Dull T, Mandel RJ et al (1998) Self-inactivating lentivirus vector for safe and efficient in vivo gene delivery. J Virol 72:9873–9880. https://doi.org/10.1128/JVI.72.12.9873-9880.1998
31. Dull T, Zufferey R, Kelly M et al (1998) A third-generation lentivirus vector with a conditional packaging system. J Virol 72:8463–8471. https://doi.org/10.1128/JVI.72.11.8463-8471.1998
32. O'Leary MC, Lu X, Huang Y et al (2019) FDA approval summary: Tisagenlecleucel for treatment of patients with relapsed or refractory B-cell precursor acute lymphoblastic leukemia. Clinical Cancer Res: Official J Amer Assoc Cancer Res 25. https://doi.org/10.1158/1078-0432.CCR-18-2035
33. Bauler M, Roberts JK, Wu C-C et al (2020) Production of Lentiviral vectors using suspension cells grown in serum-free media. Mol Ther Methods Clin Dev 17:58–68. https://doi.org/10.1016/j.omtm.2019.11.011
34. Perry C, Rayat ACME (2021) Lentiviral vector bioprocessing. Viruses 13:268. https://doi.org/10.3390/v13020268
35. Medawar PB (1948) Immunity to homologous grafted skin. III. The fate of skin homographs transplanted to the brain, to subcutaneous tissue, and to the anterior chamber of the eye. Br J Exp Pathol 29:58–69
36. Iliff JJ, Wang M, Liao Y et al (2012) A Paravascular pathway facilitates CSF flow through the brain parenchyma and the clearance of interstitial solutes, including amyloid β. Sci Transl Med 4:147ra111. https://doi.org/10.1126/scitranslmed.3003748
37. Betsholtz C, Engelhardt B, Koh GY et al (2024) Advances and controversies in meningeal biology. Nat Neurosci 27:2056–2072. https://doi.org/10.1038/s41593-024-01701-8
38. Engelhardt B, Ransohoff RM (2005) The ins and outs of T-lymphocyte trafficking to the CNS: anatomical sites and molecular mechanisms. Trends Immunol 26:485–495. https://doi.org/10.1016/j.it.2005.07.004
39. Ungar L, Nachum O, Zibly Z et al (2022) Comparison of frame-based versus frameless image-guided intracranial stereotactic brain biopsy: a retrospective analysis of safety and efficacy. World Neurosurg 164:e1–e7. https://doi.org/10.1016/j.wneu.2021.07.063
40. Alexander BM, Cloughesy TF (2017) Adult Glioblastoma. J Clin Oncol 35:2402–2409. https://doi.org/10.1200/JCO.2017.73.0119
41. Wen PY, Weller M, Lee EQ et al (2020) Glioblastoma in adults: a Society for Neuro-Oncology (SNO) and European Society of Neuro-Oncology (EANO) consensus review on current management and future directions. Neuro-Oncology 22:1073–1113. https://doi.org/10.1093/neuonc/noaa106
42. Debinski W, Gibo DM, Hulet SW et al (1999) Receptor for interleukin 13 is a marker and therapeutic target for human high-grade gliomas. Clin Cancer Res 5:985–990
43. Padfield E, Ellis HP, Kurian KM (2015) Current therapeutic advances targeting EGFR and EGFRvIII in glioblastoma. Front Oncol 5:5. https://doi.org/10.3389/fonc.2015.00005
44. Ahmed N, Salsman VS, Kew Y et al (2010) HER2-specific T cells target primary glioblastoma stem cells and induce regression of autologous experimental tumors. Clin Cancer Res 16:474–485. https://doi.org/10.1158/1078-0432.CCR-09-1322
45. Iwasawa T, Zhang P, Ohkawa Y et al (2018) Enhancement of malignant properties of human glioma cells by ganglioside GD3/GD2. Int J Oncol 52:1255–1266. https://doi.org/10.3892/ijo.2018.4266
46. Wykosky J, Gibo DM, Stanton C, Debinski W (2005) Eph A2 as a novel molecular marker and target in glioblastoma multiforme. Mol Cancer Res 3:541–551. https://doi.org/10.1158/1541-7786.MCR-05-0056
47. Nehama D, Di Ianni N, Musio S et al (2019) B7-H3-redirected chimeric antigen receptor T cells target glioblastoma and neurospheres. EBioMedicine 47:33–43. https://doi.org/10.1016/j.ebiom.2019.08.030
48. Montoya M, Gallus M, Phyu S et al (2024) A roadmap of CAR-T-cell therapy in glioblastoma: challenges and future perspectives.

Cells 13:726. https://doi.org/10.3390/cells13090726

49. Agosti E, Garaba A, Antonietti S et al (2024) CAR-T cells therapy in glioblastoma: a systematic review on molecular targets and treatment strategies. Int J Mol Sci 25:7174. https://doi.org/10.3390/ijms25137174
50. Brown CE, Hibbard JC, Alizadeh D et al (2024) Locoregional delivery of IL-13Rα2-targeting CAR-T cells in recurrent high-grade glioma: a phase 1 trial. Nat Med 30: 1001–1012. https://doi.org/10.1038/s41591-024-02875-1
51. Bagley SJ, Desai AS, Linette GP et al (2018) CAR T-cell therapy for glioblastoma: recent clinical advances and future challenges. Neuro-Oncology 20:1429–1438. https://doi.org/10.1093/neuonc/noy032
52. O'Rourke DM, Nasrallah MP, Desai A et al (2017) A single dose of peripherally infused EGFRvIII-directed CAR T cells mediates antigen loss and induces adaptive resistance in patients with recurrent glioblastoma. Sci Transl Med 9:eaaa 0984. https://doi.org/10.1126/scitranslmed.aaa0984
53. Liu Z, Zhou J, Yang X et al (2023) Safety and antitumor activity of GD2-specific 4SCAR-T cells in patients with glioblastoma. Mol Cancer 22:3. https://doi.org/10.1186/s12943-022-01711-9
54. Choi BD, Gerstner ER, Frigault MJ et al (2024) Intraventricular CARv3-TEAM-E T cells in recurrent glioblastoma. N Engl J Med 390:1290–1298. https://doi.org/10.1056/NEJMoa2314390
55. Louis DN, Perry A, Wesseling P et al (2021) The 2021 WHO classification of tumors of the central nervous system: a summary. Neuro-Oncology 23:1231–1251. https://doi.org/10.1093/neuonc/noab106
56. Hoffman LM, Veldhuijzen van Zanten SEM, Colditz N et al (2018) Clinical, radiologic, pathologic, and molecular characteristics of Long-term survivors of diffuse intrinsic pontine glioma (DIPG): a collaborative report from the international and European Society for Pediatric Oncology DIPG registries. J Clin Oncol 36:1963–1972. https://doi.org/10.1200/JCO.2017.75.9308
57. Mount CW, Majzner RG, Sundaresh S et al (2018) Potent antitumor efficacy of anti-GD2 CAR T-cells in H3K27M+ diffuse midline gliomas. Nat Med 24:572–579. https://doi.org/10.1038/s41591-018-0006-x
58. Haydar D, Houke H, Chiang J et al (2020) Cell-surface antigen profiling of pediatric brain tumors: B7-H3 is consistently expressed and can be targeted via local or systemic CAR T-cell delivery. Neuro-Oncology 23:999–1011. https://doi.org/10.1093/neuonc/noaa278
59. Vitanza NA, Johnson AJ, Wilson AL et al (2021) Locoregional infusion of HER2-specific CAR T cells in children and young adults with recurrent or refractory CNS tumors: an interim analysis. Nat Med 27: 1544–1552. https://doi.org/10.1038/s41591-021-01404-8
60. Thomas BC, Staudt DE, Douglas AM et al (2023) CAR T cell therapies for diffuse midline glioma. Trends Cancer 9:791–804. https://doi.org/10.1016/j.trecan.2023.07.007
61. Monje M, Mahdi J, Majzner R et al (2024) Intravenous and intracranial GD2-CAR T cells for H3K27M+ diffuse midline gliomas. Nature 1–8. https://doi.org/10.1038/s41586-024-08171-9
62. van den Bent M, Saratsis AM, Geurts M, Franceschi E (2024) H3 K27M-altered glioma and diffuse intrinsic pontine glioma: semi-systematic review of treatment landscape and future directions. Neuro-Oncology 26: S110–S124. https://doi.org/10.1093/neuonc/noad220
63. Ostrom QT, Patil N, Cioffi G et al (2020) CBTRUS statistical report: primary brain and other central nervous system tumors diagnosed in the United States in 2013–2017. Neuro-Oncology 22:iv1–iv96. https://doi.org/10.1093/neuonc/noaa200
64. Ellison DW, Clifford SC, Gajjar A, Gilbertson RJ (2003) What's new in neuro-oncology? Recent advances in medulloblastoma. Eur J Paediatr Neurol 7:53–66. https://doi.org/10.1016/s1090-3798(03)00014-x
65. Hill RM, Richardson S, Schwalbe EC et al (2020) Time, pattern, and outcome of medulloblastoma relapse and their association with tumour biology at diagnosis and therapy: a multicentre cohort study. Lancet Child Adolesc Health 4:865–874. https://doi.org/10.1016/S2352-4642(20)30246-7
66. Longee DC, Wikstrand CJ, Mansson J-E et al (1991) Disialoganglioside GD2 in human neuroectodermal tumor cell lines and gliomas. Acta Neuropathol 82:45–54. https://doi.org/10.1007/BF00310922
67. Nellan A, Rota C, Majzner R et al (2018) Durable regression of Medulloblastoma after regional and intravenous delivery of anti-HER2 chimeric antigen receptor T cells. J Immunother Cancer 6:30. https://doi.org/10.1186/s40425-018-0340-z

68. Guzman G, Pellot K, Reed MR, Rodriguez A (2023) CAR T-cells to treat brain tumors. Brain Res Bull 196:76–98. https://doi.org/10.1016/j.brainresbull.2023.02.014
69. Chokshi CR, Shaikh MV, Brakel B et al (2024) Targeting axonal guidance dependencies in glioblastoma with ROBO1 CAR T cells. Nat Med 1–11. https://doi.org/10.1038/s41591-024-03138-9
70. Lin FY, Stuckert A, Tat C et al (2024) Phase I trial of GD2.CART cells augmented with constitutive Interleukin-7 receptor for treatment of high-grade pediatric CNS tumors. J Clin Oncol 42:2769–2779. https://doi.org/10.1200/JCO.23.02019
71. Scoccianti S, Ricardi U (2012) Treatment of brain metastases: review of phase III randomized controlled trials. Radiother Oncol 102:168–179. https://doi.org/10.1016/j.radonc.2011.08.041
72. Kamp MA, Slotty PJ, Cornelius JF et al (2018) The impact of cerebral metastases growth pattern on neurosurgical treatment. Neurosurg Rev 41:77–86. https://doi.org/10.1007/s10143-016-0760-5
73. Steindl A, Brunner TJ, Heimbach K et al (2022) Changing characteristics, treatment approaches and survival of patients with brain metastasis: data from six thousand and thirty-one individuals over an observation period of 30 years. Eur J Cancer 162:170–181. https://doi.org/10.1016/j.ejca.2021.12.005
74. Eroglu Z, Tagawa T, Somlo G (2014) Human epidermal growth factor receptor family-targeted therapies in the treatment of HER2-overexpressing breast cancer. Oncologist 19:135–150. https://doi.org/10.1634/theoncologist.2013-0283
75. Leone JP, Lin NU (2019) Systemic therapy of central nervous system metastases of breast cancer. Curr Oncol Rep 21:49. https://doi.org/10.1007/s11912-019-0802-6
76. Bryan S, Witzel I, Borgmann K, Oliveira-Ferrer L (2021) Molecular mechanisms associated with brain metastases in HER2-positive and triple negative breast cancers. Cancers (Basel) 13:4137. https://doi.org/10.3390/cancers13164137
77. Huang Z, Dewanjee S, Chakraborty P et al (2023) CAR T cells: engineered immune cells to treat brain cancers and beyond. Mol Cancer 22:22. https://doi.org/10.1186/s12943-022-01712-8
78. Priceman SJ, Tilakawardane D, Jeang B et al (2018) Regional delivery of chimeric antigen receptor-engineered T cells effectively targets HER2+ breast cancer metastasis to the brain. Clin Cancer Res 24:95–105. https://doi.org/10.1158/1078-0432.CCR-17-2041
79. Quail DF, Joyce JA (2017) The microenvironmental landscape of brain tumors. Cancer Cell 31:326–341. https://doi.org/10.1016/j.ccell.2017.02.009
80. Binnewies M, Roberts EW, Kersten K et al (2018) Understanding the tumor immune microenvironment (TIME) for effective therapy. Nat Med 24:541–550. https://doi.org/10.1038/s41591-018-0014-x
81. Caruana I, Savoldo B, Hoyos V et al (2015) Heparanase promotes tumor infiltration and antitumor activity of CAR-redirected T lymphocytes. Nat Med 21:524–529. https://doi.org/10.1038/nm.3833
82. Zhao R, Cui Y, Zheng Y et al (2021) Human hyaluronidase PH20 potentiates the antitumor activities of Mesothelin-specific CAR-T cells against gastric cancer. Front Immunol 12:660488. https://doi.org/10.3389/fimmu.2021.660488
83. Zhao Y, Dong Y, Yang S et al (2022) Bioorthogonal equipping CAR-T cells with hyaluronidase and checkpoint blocking antibody for enhanced solid tumor immunotherapy. ACS Cent Sci 8:603–614. https://doi.org/10.1021/acscentsci.2c00163
84. Karimi E, Yu MW, Maritan SM et al (2023) Single-cell spatial immune landscapes of primary and metastatic brain tumours. Nature 1–9. https://doi.org/10.1038/s41586-022-05680-3
85. Filippone A, Lanza M, Mannino D et al (2022) PD1/PD-L1 immune checkpoint as a potential target for preventing brain tumor progression. Cancer Immunol Immunother 71:2067–2075. https://doi.org/10.1007/s00262-021-03130-z
86. Chen W-W, Chu TSM, Xu L et al (2022) Immune related biomarkers for cancer metastasis to the brain. Exp Hematol Oncol 11:105. https://doi.org/10.1186/s40164-022-00349-z
87. Chalise L, Kato A, Ohno M et al (2022) Efficacy of cancer-specific anti-podoplanin CAR-T cells and oncolytic herpes virus G47Δ combination therapy against glioblastoma. Mol Ther Oncolytics 26:265–274. https://doi.org/10.1016/j.omto.2022.07.006
88. Dg T, Aj O, B von S, et al (2021) Understanding T cell phenotype for the design of effective chimeric antigen receptor T cell therapies. J Immunother Cancer 9:. doi:https://doi.org/10.1136/jitc-2021-002555

89. Gumber D, Wang LD (2022) Improving CAR-T immunotherapy: overcoming the challenges of T cell exhaustion. EBioMedicine 77:103941. https://doi.org/10.1016/j.ebiom.2022.103941
90. Long AH, Haso WM, Shern JF et al (2015) 4-1BB costimulation ameliorates T cell exhaustion induced by tonic signaling of chimeric antigen receptors. Nat Med 21:581–590. https://doi.org/10.1038/nm.3838
91. Agarwal S, Aznar MA, Rech AJ et al (2023) Deletion of the inhibitory co-receptor CTLA4 enhances and invigorates chimeric antigen receptor T cells. Immunity S1074761323004077. https://doi.org/10.1016/j.immuni.2023.09.001
92. Weber EW, Parker KR, Sotillo E et al (2021) Transient rest restores functionality in exhausted CAR-T cells through epigenetic remodeling. Science 372:eaba1786. https://doi.org/10.1126/science.aba1786
93. Shah NN, Fry TJ (2019) Mechanisms of resistance to CAR T cell therapy. Nat Rev Clin Oncol 16:372–385. https://doi.org/10.1038/s41571-019-0184-6
94. Majzner RG, Mackall CL (2018) Tumor antigen escape from CAR T-cell therapy. Cancer Discov 8. https://doi.org/10.1158/2159-8290.CD-18-0442
95. Goutnik M, Iakovidis A, Still MEH et al (2024) Advancements in chimeric antigen receptor-expressing T-cell therapy for glioblastoma multiforme: literature review and future directions. Neurooncol Adv 6:vdae025. https://doi.org/10.1093/noajnl/vdae025
96. Norelli M, Camisa B, Barbiera G et al (2018) Monocyte-derived IL-1 and IL-6 are differentially required for cytokine-release syndrome and neurotoxicity due to CAR T cells. Nat Med 24:739–748. https://doi.org/10.1038/s41591-018-0036-4
97. Taraseviciute A, Tkachev V, Ponce R et al (2018) Chimeric antigen receptor T cell-mediated neurotoxicity in nonhuman primates. Cancer Discov 8:750–763. https://doi.org/10.1158/2159-8290.CD-17-1368
98. Monje M, Mahdi J, Majzner R et al (2024) Sequential intravenous and intracerebroventricular GD2-CAR T-cell therapy for H3K27M-mutated diffuse midline gliomas. medRxiv 2024.06.25.24309146. https://doi.org/10.1101/2024.06.25.24309146
99. Majzner RG, Ramakrishna S, Yeom KW et al (2022) GD2-CAR T cell therapy for H3K27M-mutated diffuse midline gliomas. Nature 603:934–941. https://doi.org/10.1038/s41586-022-04489-4
100. Pawlowski KD, Duffy JT, Gottschalk S, Balyasnikova IV (2023) Cytokine modification of adoptive chimeric antigen receptor immunotherapy for glioblastoma. Cancers (Basel) 15:5852. https://doi.org/10.3390/cancers15245852
101. Canté-Barrett K, Mendes RD, Smits WK et al (2016) Lentiviral gene transfer into human and murine hematopoietic stem cells: size matters. BMC Res Notes 9:312. https://doi.org/10.1186/s13104-016-2118-z
102. Dong W, Kantor B (2021) Lentiviral vectors for delivery of gene-editing systems based on CRISPR/Cas: current state and perspectives. Viruses 13:1288. https://doi.org/10.3390/v13071288
103. Michels A, Ho N, Buchholz CJ (2022) Precision medicine: in vivo CAR therapy as a showcase for receptor-targeted vector platforms. Mol Ther 30:2401–2415. https://doi.org/10.1016/j.ymthe.2022.05.018
104. Poorebrahim M, Sadeghi S, Fakhr E et al (2019) Production of CAR T-cells by GMP-grade lentiviral vectors: latest advances and future prospects. Crit Rev Clin Lab Sci 56:393–419. https://doi.org/10.1080/10408363.2019.1633512
105. Verdun N, Marks P (2024) Secondary cancers after chimeric antigen receptor T-cell therapy. N Engl J Med 390. https://doi.org/10.1056/NEJMp2400209
106. Hamilton MP, Takeshi S, Troy N et al (2024) Risk of second tumors and T-cell lymphoma after CAR T-cell therapy. N Engl J Med 390:2047–2060. https://doi.org/10.1056/NEJMoa2401361

Index

A

B

C

D

E

F

G

H

I

L

Kazuto Kobayashi et al. (eds.), *Lentiviral Vectors: Applications for Basic and Clinical Neuroscience*, Methods in Molecular Biology, vol. 2974, https://doi.org/10.1007/978-1-0716-4807-0,

Index

MIX
Papier aus verantwortungsvollen Quellen
Paper from responsible sources
FSC® C105338

If you have any concerns about our products,
you can contact us on
ProductSafety@springernature.com

In case Publisher is established outside the EU,
the EU authorized representative is:
Springer Nature Customer Service Center GmbH
Europaplatz 3, 69115 Heidelberg, Germany

Printed by Libri Plureos GmbH
in Hamburg, Germany